U. T. Bornscheuer, R. J. Kazlauskas

Hydrolases in Organic Synthesis

WILEY-VCH

U. T. Bornscheuer, R. J. Kazlauskas

Hydrolases in Organic Synthesis

Regio- and Stereoselective
Biotransformations

**WILEY-VCH**

Weinheim · New York · Chichester · Brisbane · Singapore · Toronto

Dr. Uwe T. Bornscheuer
Institute for Technical Biochemistry
University of Stuttgart
Allmandring 31
D-70569 Stuttgart
Germany
email: itbubo@po.uni-stuttgart.de

Prof. Romas J. Kazlauskas
Department of Chemistry
McGill University
801 Sherbrooke St. West
Montreal, Québec H3A 2K6
Canada
email: cxrk@musica.mcgill.ca

Cover: The cover picture shows the open structure of lipase from *Rhizomucor miehei* with a triglyceride bound to the active site. The structure is surrounded by four important industrial intermediates obtained by hydrolase-catalyzed synthesis.

Library of Congress Card No.: applied for

British Library Cataloguing-in-Publication Data:
A catalogue record for this book
is available from the British Library

Die Deutsche Bibliothek – CIP-Einheitsaufnahme
Bornscheuer, Uwe Theo:
Hydrolases in organic synthesis : regio- and stereoselective biotransformations / U. T. Bornscheuer ;
R. J. Kazlauskas. – Weinheim ; New York ; Chichester ; Brisbane ; Singapore ; Toronto :
Wiley-VCH, 1999
ISBN 3-527-30104-6

© WILEY-VCH Verlag GmbH, D-69469 Weinheim (Federal Republic of Germany), 1999

Printed on acid-free and chlorine-free paper

Printing: Strauss Offsetdruck GmbH, D-69509 Mörlenbach
Bookbindung: Großbuchbinderei J. Schäffer, D-67269 Grünstadt
Printed in the Federal Republic of Germany

To

Tanja and Aleli

and also to

Annika, Lottie and Anna

who can't read yet

About the Authors

Uwe Bornscheuer (born 1964) studied chemistry at the University of Hannover, Germany, where he graduated with his Diploma in 1990. After receiving his Ph. D. in Chemistry in 1993 at the Institute of Technical Chemistry (with Prof. K. Schügerl and T. Scheper) at the same University, he spent a postdoctoral year at the University of Nagoya, Japan, with Prof. T. Yamane. He then joined the group of Prof. R. D. Schmid at the Institute of Technical Biochemistry, University of Stuttgart, Germany, where he finished his Habilitation in late 1998. He is married and has one daughter.

Romas Kazlauskas (born 1956) studied at Cleveland State University, Cleveland, OH, USA, and received his Ph. D. in Chemistry in 1982 from MIT, Cambridge, MA, USA. He then spent a postdoctoral year with Prof. George Whitesides at Harvard University, Cambridge, MA, USA. After working at General Electric Co. (Schenectady, NY, USA) in 1988 he joined the Chemistry Department at McGill University, Montreal, Canada, where he is currently an Associate Professor. He is married and has two daughters.

Preface

Each traveller to a city seeks something different. One wants to see that special painting in the museum, another wants to drink the local beer, a third wants to meet a soulmate.

Each organic chemist also seeks something different from the field of biocatalysis. One wants high enantioselectivity, another wants reaction under mild conditions, a third wants to scale up to an industrial scale. We hope this book can be a guide to organic chemists exploring the field of biocatalysis. Enzyme-catalyzed reactions, especially hydrolase-catalyzed reactions, have already solved hundreds of synthetic problems usually because of their high stereoselectivity.

The organization is aimed at the chemist – by reaction type and by different functional groups. This information should help organic chemists identify the best hydrolase for their synthetic problem. In addition, we suggest how to choose an appropriate solvent, acyl donor, immobilization technique and other practical details. We hope that learning how others solved synthetic problems will generate ideas that solve the next generation of problems.

Although this book has more than 1 700 references, we might have missed important hydrolase-catalyzed reactions. The choices on what to include usually reflect our own research interests, but were sometimes arbitrary or even inadvertent. We will post corrections and additions on a web site: *http://pasteur.chem.mcgill.ca/hydrolases.html*.

Montreal/Stuttgart, February 1999   Romas J. Kazlauskas, Uwe T. Bornscheuer

Acknowledgments

This book evolved from our review started during RJK's sabbatical year in Prof. Rolf D. Schmid's laboratories at the University of Stuttgart. RJK thanks Prof. Schmid and his group for their warm hospitality and support. UTB also thanks Prof. Schmid for the opportunity to work in his laboratories and his generous support over the years. We thank the staff at Wiley-VCH, especially Ms. Karin Dembowsky for all her help. We are also grateful to Erik Henke, Frank Zocher and Markus Enzelberger for reading the manuscript for final corrections.

Contents

1   Introduction

Hydrolases are the group of enzymes that catalyze bond cleavage by reaction with water. The natural function of most hydrolases is digestive – to break down nutrients into smaller units for digestion. For example, proteases hydrolyze proteins to smaller peptides and then to amino acids and lipases hydrolyze lipids (triglycerides) to glycerol and fatty acids (Fig. 1). Because of the need to break down a wide range of nutrients, hydrolases usually have a broad substrate specificity.

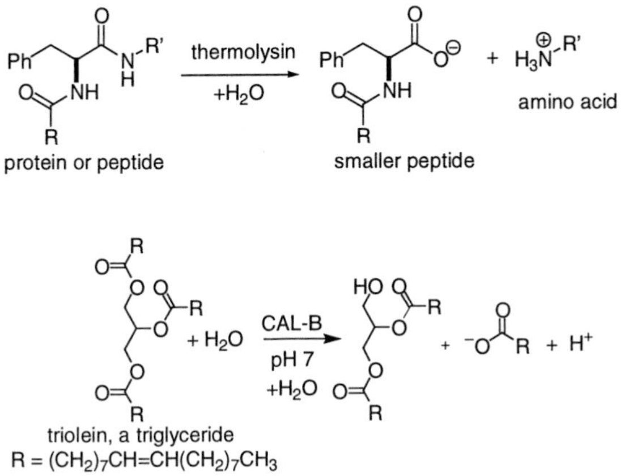

Fig. 1.   The natural role of most hydrolases is digestive – to break down nutrients into smaller units. Thermolysin, a protease secreted by thermophilic bacteria, catalyzes the hydrolysis of proteins to peptides and then further to amino acids. Lipase from *Candida antarctica* (CAL-B) catalyzes the stepwise hydrolysis of triglycerides (e.g., triolein) to fatty acids and glycerol. The reaction shows only the first step from a triglyceride to a diglyceride.

Several characteristics make hydrolases useful to the organic chemist. First, because of their broad substrate specificity, hydrolases often accept as substrates various synthetic intermediates. Second, hydrolases often show high stereoselectivity, even toward unnatural substrates. Third, besides hydrolysis, hydrolases also catalyze several related reactions – condensations (reversal of hydrolysis) and alcoholysis (a cleavage using an alcohol in place of water). Two examples from industry are shown in Fig. 2. Thermolysin catalyzes the condensation of two amino acid derivatives to make an aspartame derivative (Isowa et al., 1979). The reaction proceeds in the condensation direction because the product precipitates from solution. The high enantioselectivity permits using racemic starting materials and the high regioselectivity of thermolysin eliminates the need to protect the β-carboxyl group of the aspartic acid derivative. The second example is an alcoholysis reaction (Morgan et al., 1997a). The ester, vinyl acetate, is cleaved not by water, but by the substrate alcohol. The liberated vinyl alcohol (not shown) tautomerizes to acetaldehyde. These alcoholysis reactions are also called transesterification reactions.

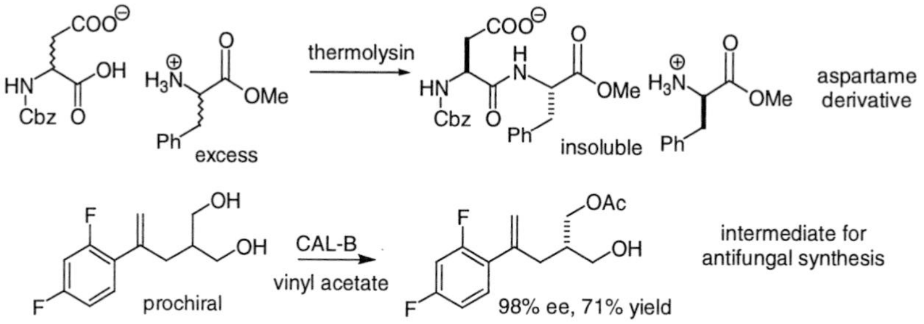

Fig. 2.   Several unnatural, synthetically useful reactions catalyzed by hydrolases. Thermolysin catalyzes the regio- and enantioselective coupling of *N*-benzyloxycarbonyl-L-aspartate with L-phenylalanine methyl ester. Precipitation of the product drives this reaction in the condensation direction instead of the normal hydrolysis direction. This condensation is a key step in the manufacture of aspartame, a low-calorie sweetener. Because of the high enantioselectivity of thermolysin, racemic substrates may be used. Because of the high regioselectivity of thermolysin for the α-carboxyl group, the β-carboxyl group in the aspartic acid derivative needs no protection. Lipase from *Candida antarctica* (CAL-B) catalyzes the enantioselective acetylation of a prochiral diol yielding an intermediate for the synthesis of antifungal agents. This example is an alcoholysis where the ester, vinyl acetate, is cleaved not by water, but by the substrate alcohol. This reaction is run in an organic solvent to avoid the competing hydrolysis.

Several other features make hydrolases convenient to use as synthetic reagents. Many hydrolases (approximately several hundred) are commercially available. They do not require cofactors and they tolerate the addition of water-miscible solvents (e.g., DMSO, DMF). Lipases, esterases and some proteases are also stable and active in neat organic solvents.

Enzymes are often classified according to the reaction catalyzed using an Enzyme Commission (EC) number. According to this classification, hydrolases form group 3 and are further classified according to the type of bond hydrolyzed. For example enzymes in the group 3.1 hydrolyze ester bonds (Tab. 1). Further classification into subcategories yields a four number EC number. For example, lipases have the number EC 3.1.1.3. Classification of the more useful enzymes for organic synthesis is given in Tab. 1. The most useful hydrolases are italicized. A convenient web site to look up numbers and classification is at *http://129.195.254.61/sprot/enzyme.html*. One disadvantage of this classification is that all enzymes catalyzing the same reaction have the same number, even though they may have very different structures, properties and other characteristics. For example, all lipases have the same number even though there are more than one hundred different lipases.

Tab. 1.   Selected Hydrolases Useful in Organic Synthesis.

EC Number	Type of bond hydrolyzed	Examples
3.1	**Ester**	
3.1.1	in carboxylic acid esters	*triacylglycerol lipase, acetylcholine esterase,* phospholipase A_2, gluconolactonase, lipoprotein lipase
3.1.3–4	in phosphoric acid mono- or diesters	phospholipase C, *phospholipase D*
3.2	**Glycosidic**	
3.2.1	in *O*-glycosides	α-amylase, oligo-1,6-glucosidase, lysozyme, neuraminidase, *α-glucosidase, β-galactosidase,* α-mannosidase, *N*-acetyl-β-glucosaminidase, sucrose α-glucosidase, nucleosidases
3.3	**Ether**	
3.3.2	in epoxides	*epoxide hydrolase*
3.4	**Peptide**	
3.4.11	aminopeptidase	leucine aminopeptidase
3.4.16, 21	serine proteinase	*subtilisin, chymotrypsin, thermitase*
3.4.18, 22	cysteine proteinase	*papain*
3.4.17, 24	metalloproteinase	*thermolysin*
3.5	**Other amides**	
3.5.1	in linear amides	*penicillin amidase* (penicillin G acylase)
3.5.2	in cyclic amides	*hydantoinase*
3.5.5	in nitriles	*nitrilase*[a]
3.8	**Halide bonds**	
3.8.1	carbon-halide bonds	haloalkane dehalogenase

[a]Nitrile hydratase (EC 4.2.1.84), which catalyzes addition of water to a nitrile yielding an amide, is not a hydrolase, but a lyase.

This book describes the application of lipases and proteases in organic syntheses, but also surveys esterases, epoxide hydrolases, nitrile hydrolyzing enzymes and glycosidases. The emphasis is on examples that are synthetically useful, especially those that exploit the regio- and stereoselectivity of hydrolases.

2 Availability and Structure of Lipases, Esterases, and Proteases

2.1 Lipases and Esterases

2.1.1 Introduction

Both lipases (EC 3.1.1.3) and esterases (EC 3.1.1.1) catalyze the hydrolysis of esters, but lipases preferentially catalyze hydrolysis of water-insoluble esters such as triglycerides. For example, lipases catalyze the hydrolysis of triolein to diolein (Eq. 1).

$$\text{triolein} \xrightarrow[\text{pH 7}]{\text{lipase}} \text{1,2- or 2,3-diolein} + \text{oleic acid} \tag{1}$$

triolein   1,2- or 2,3-diolein   oleic acid
a triglyceride   a diglyceride   a fatty acid

$$R = (CH_2)_7CH=CH(CH_2)_7CH_3$$

In addition, lipases also catalyze the hydrolysis of a broad range of natural and unnatural esters, while retaining high enantio- or regioselectivity. This combination of broad substrate range and high selectivity makes lipases an ideal catalyst for organic synthesis. Chemists use lipase-catalyzed biotransformations to prepare enantiomerically-pure pharmaceuticals and synthetic intermediates (Sect. 5), to protect and deprotect synthetic intermediates (Sect. 6.1), to modify natural lipids (Sect. 6.2) as well as for more specialized uses. A survey of these reactions is the main focus of this book.

Besides high selectivity and broad substrate range, another major advantage of lipases for synthetic reactions is that they act efficiently on water-insoluble substrates. Lipases need this ability because the natural substrates of lipases – triglycerides – are insoluble in water. Lipases bind to the water-organic interface and catalyze hydrolysis at this interface. This binding not only places the lipase close to the substrate, but also increases the catalytic power of the lipase, a phenomenon called interfacial activation. Most lipases are poor catalysts in the absence of an interface such as an organic droplet or a micelle. A conformational change in the lipase probably causes the interfacial activation (see Sect. 2.1.7). In contrast, efficient reactions with proteases often require chemical modification of the substrate to increase water solubility.

Cheese manufacturers use lipase-catalyzed hydrolysis of milk fat to enhance flavors, accelerate cheese ripening and to manufacture cheese-like products (for reviews containing sections on cheese making and detergents see Berry and Paterson, 1990;

Cheetham, 1993, 1997; Vulfson, 1994; Haas and Joerger, 1995). Traditional cheese-making adds extracts containing lipases to the raw cheeses to impart characteristic fla-vors. For example, extracts of the pregastric gland of a calf imparts a buttery and slightly peppery flavor, while a kid extract imparts a sharp flavor and a lamb extract imparts a strong 'dirty sock' flavor. In addition, microbes responsible for cheese ripening secrete lipases. For example, lipase from *Penicillium roquefortii* liberates short and medium chain fatty acids which add flavor both directly and by serving as precursors for δ-lac-tones and methylketones. Modern cheese-makers can substitute commercial lipases (e.g., lipases from *Aspergillus niger* or *Rhizomucor miehei*) for the pregastric gland extracts and for the microbes. Also, addition of lipases to cow's milk can mimic the flavor of goat's or sheep's milk. Addition of lipases to cheese followed by incubation at high tem-perature yields a concentrated cheese flavor that can be used to flavor sauces and other prepared foods.

Some detergents include microbial lipases (e.g., lipase from *Humicola lanuginosa*) to aid removal of fat stains, but the advantage accumulates only after multiple washing. The wash cycle is too short for significant hydrolysis, but the lipase remains on the fat in the subsequent drying where it hydrolyzes the fats. The next wash cycle removes these fats. Lipases may also prevent redeposition of fats on textiles. Recently, Novo introduced a new lipase preparation (LipoPrime[TM]) optimized by protein engineering proven to be stable during the washing process in the presence of protease, high ionic strength, bleach, chlorinated water and within a broad range of water hardness. Moreover, the enzyme is active in the temperature range of 10–50 °C, which meets the washing conditions in many countries.

Using lipases for biotransformations is a smaller market, so biotransformations often use lipases that were originally developed for other uses. Two exceptions are lipase from *Rhizomucor miehei*, which Novo developed specifically for lipid modification and lipase from *Candida antarctica* B, which Novo produces for applications in organic syntheses.

A number of books on lipases, or with large sections on lipases and esterases, and ex-tensive reviews are available (Borgstrom and Brockman, 1984; Boland et al., 1991; Al-berghina et al., 1991; Collins et al., 1992; Poppe and Novak, 1992; Roberts, 1992–1996; Sheldon, 1993; Woolley and Petersen, 1994; Jaeger et al., 1994; Wong and Whitesides, 1994; Drauz and Waldmann, 1995; Faber, 1997; Gandhi, 1997; Theil, 1997; Kazlauskas and Bornscheuer, 1998; Jaeger and Reetz, 1998; Schmid and Verger, 1998). More spe-cialized reviews will be cited in the appropriate sections.

2.1.2   Occurrence and Availability of Lipases

Lipases occur in plants, animals and microorganism where the biological role of lipases is probably digestive. Most biotransformations use commercial lipases, about 70 of which are available. Tab. 2 lists the most popular of these. Pancreatic cholesterol esterase is included with the lipases because its sequence and biochemical properties are identical to bile-salt stimulated lipase (Nilsson et al., 1990; Hui and Kissel, 1990).

Lipases are usually named according to the (micro)organism that produces the lipase. The classification, and thus name, of a microorganism can change as researchers learn more about it. Likewise, the name of the lipase sometimes changes which can be frustrating to organic chemists accustomed to molecules whose name rarely changes. For example, Amano researchers first classified the microorganism that produces 'Amano P' (ATCC 21808) as *Pseudomonas fluorescens*, but have since reclassified it as *P. cepacia*. For this reason pre-1990 papers on this lipase refer to it as *P. fluorescens* lipase. Confusingly, some researchers continue to refer to this lipase as *P. fluorescens* lipase and Fluka sells SAM-II under the name of lipase from *Pseudomonas fluorescens*. Unfortunately, ATCC 21808 has been again renamed to *Burkholderia cepacia*. For this book, we will continue to use *Pseudomonas cepacia*. Furthermore, a lipase isolated from a strain initially designated as *Pseudomonas* sp. KWI-56 (Iizumi et al., 1991) was reclassified as *Burkholderia cepacia*. The enzyme is now produced by Boehringer Mannheim and sold as Chirazyme L-1, however the enzyme is probably identical to the Amano lipase PS.

A recent reclassification of the *Rhizopus* fungi renamed *R. niveus*, *R. delemar* and *R. javanicus* all as *Rhizopus oryzae* (review: Haas and Joerger, 1995). Consistent with this reclassification, the lipases isolated from *R. delemar*, *R. javanicus* and *R. niveus* have identical amino acid sequences and lipase from *R. oryzae* (ROL) differs only by two conservative substitutions (His134 is Asn and Ile234 is Leu in ROL). In spite of these similarities, Amano sells three different lipases from this group, and they show slightly different selectivities, perhaps due to cleaving the prolipase at different positions (Uyttenbröck et al., 1993; Beer et al., 1998). The prolipase contains extra amino acid residues to guide folding and secretion of the lipase. After folding, proteases cleave the extra amino acid residues to give the mature lipase. This cleavage may not always occur at the same amino acid residue. Furthermore, the prolipases ProROL and PreProROL had considerably higher thermostability and it could be shown that the natural leader sequence of ROL is able to inhibit the folding supporting properties of the prosequence, resulting in a retardation of folding (Beer et al., 1998). Expression of ROL in *E. coli* proceeds with formation of inclusion bodies, which have to be refolded in order to get active enzyme. This could be circumvented by expression of the ROL gene in the yeasts *Saccharomyces cerevisiae* (Takahashi et al., 1998) or *Pichia pastoris* (Minning et al., 1998), which also facilitated secretion of active enzyme into the culture supernatant.

Tab. 2.   Selected Examples of Commercially-Available Lipases.

Abbreviation	Origin of Lipase	Other Names	Commercial Source and Name
	Mammalian Lipases		
PPL	porcine pancreas		Amano, Boehringer Mannheim (Chirazyme® L-7), Fluka, Sigma, Genzyme, Sigma
CE (BSSL)	pancreatic cholesterol esterase		
	Fungal Lipases		
CRL	Candida rugosa[a]	Candida cylindracea	Altus Biologics (ChiroCLEC-CR), Amano (lipase AY), Meito Sangyo (lipase MY, lipase OF-360), Boehringer Mannheim (Chirazyme® L-3)
GCL	Geotrichum candidum		
HLL	Humicola lanuginosa	Thermomyces lanuginosa	Boehringer Mannheim (Chirazyme® L-8), Novo Nordisk (SP 524, Lipolase®)
PcamL	Penicillium camembertii	Penicillium cyclopium	Amano (lipase G)
RJL	Rhizomucor javanicus	Mucor javanicus	Amano (lipase M)
RML	Rhizomucor miehei	Mucor miehei	Boehringer Mannheim (Chirazyme® L-9), Amano (MAP), Novo Nordisk (SP 523, Lipozyme®), Fluka
ROL	Rhizopus oryzae	R. javanicus, R. delemar, R. niveus[a]	Amano (lipase F), Amano (lipase D), Amano (lipase N), Fluka, Sigma, Seikagaku Kogyo Co. (Japan)
CAL-A	Candida antarctica A		Boehringer Mannheim (Chirazyme® L-5), Novo Nordisk (SP 526)[b]
CAL-B	Candida antarctica B		Boehringer Mannheim (Chirazyme® L-2), Novo Nordisk (SP 525 or SP435)[b] Sigma
ANL	Aspergillus niger		Amano (lipase A, AP), Röhm, Novo Nordisk (Palatase®)
CLL	Candida lipolytica		Amano (lipase L)
ProqL	Penicillium roquefortii		Amano (lipase R)

Tab. 2.   Selected Examples of Commercially-Available Lipases (continued).

Abbreviation	Origin of Lipase	Other Names	Commercial Source and Name
	Bacterial Lipases[c]		
PCL	Pseudomonas cepacia	Burkholderia cepacia[d]	Altus Biologics (ChiroCLEC-PC), Amano (P, P-30, PS, LPL-80, LPL-200S), Boehringer Mannheim (Chirazyme® L-1), Fluka (SAM-II), Sigma, Fluka (sol-gel immobilized)
PCL-AH	Pseudomonas cepacia		Amano (lipase AH)
PFL	Pseudomonas fluorescens		Amano (lipase AK), Amano (lipase YS), Biocatalysts Ltd.
PfragiL	Pseudomonas fragi		lipase B, Wako Pure Chemical (Osaka)
CVL[e]	Chromobacterium viscosum	Pseudomonas glumae	Sigma, Genzyme, Asahi Chemical, Biocatalysts Ltd.
	Pseudomonas sp.		Amano (K-10)
BTL2	Bacillus thermocatenulatus		Boehringer Mannheim (Chirazyme® L-1)
	Alcaligenes sp.		Meito Sangyo (lipase QL)

[a] The amino acid sequences of lipases R. delemar, R. javanicus and R. niveus are identical. The sequence of lipase from R. oryzae differ by only two conservative substitutions. [b] SP 525 is a powder containing 40 wt % protein, while SP 435 is the same enzyme immobilized on macroporous polypropylene (1 w/w % protein). For a while Novo Nordisk supplied a lipase SP 382 from Candida sp. This lipase was a mixture of lipases A and B from Candida antarctica. [c] Boehringer Mannheim sells two lipases from Pseudomonas species – Chirazyme® L-4 and Chirazyme® L-6. It is not clear from the product literature which of the Pseudomonas lipases they correspond to. [d] Lipase from microorganism ATCC21808. Early reports classified this microorganism as Pseudomonas fluorescens, later as Pseudomonas cepacia, most recently as Burkholderia cepacia. Neither the microorganism nor the lipase has changed by the change in name. [e] The amino acid sequence and biochemical properties of lipase from Pseudomonas glumae and lipase from Chromobacterium viscosum are identical (Taipa et al., 1995; Lang et al., 1996).

Note that even when the microorganism classification is settled, the same species may produce different lipases. Amano sells lipase AH from *Pseudomonas cepacia* which differs from lipase P in the amino acid sequence in 16 of 320 residues. These two lipases had opposite selectivity for a dihydropyridine substrate (Sect. 4.2.2) (Hirose et al., 1995).

For the purposes of this book, we will simplify the lipase names. The properties of all commercial preparations of lipase from *Candida rugosa* seem similar, for this reason we will refer to all of them as CRL. Amano P, purified forms of this lipase, and SAM-II (Fluka) all come from microorganism ATCC 21808. We will refer to all of these as PCL, even if the authors did not. The amino acid sequence and biochemical properties of lipase from *Pseudomonas glumae* and lipase from *Chromobacterium viscosum* are identical (Taipa et al., 1995; Lang et al., 1996), and we will refer to both of these as CVL. We will refer to all the *Rhizopus* lipases as ROL. For the other lipases, we will use the abbreviations shown in Tab. 2 or the full name.

2.1.3 Classification of Lipases

Naming lipases according to their microbial source sometimes obscures structural similarities. A better classification uses protein sequence alignments (Tab. 3), which is also consistent with the 3-D structures of lipases (see Sects. 2.1.6 and 2.1.8). The mammalian (pancreatic) lipases form one group, the fungal lipases form two – the *Candida rugosa* and the *Rhizomucor* families – and the bacterial lipases also form two – the *Pseudomonas* and the *Staphylococcus* families. The *Candida rugosa* family includes CRL, GCL and, even though it is a mammalian lipase, pancreatic cholesterol esterase. These lipases are large (60–65 kDa). Note that *Candida antarctica* lipase B does not belong to this family, even though it comes from a *Candida* yeast. The *Rhizomucor* family includes lipases from a wide range of fungi: the *Rhizopus* lipases, the *Rhizomucor* lipases, *Penicillium camembertii* lipase, HLL, CAL-B. These lipases are all small (30–35 kDa). The *Pseudomonas* lipases are also small and include all the *Pseudomonas* lipases and CVL. The *Staphylococcus* lipases are medium-sized (40–45 kDa), but none are commercially-available. One lipase in this group, a thermostable lipase from *Bacillus thermocatenulatus* (BTL2), is available from Boehringer Mannheim. A number of lipases remain unclassified. For some, e.g., ANL, the amino acid sequence is not known, for others, e.g., CAL-A, the sequence is known (Hoegh et al., 1995), but it shows little similarity to the other lipases.

The most useful lipases for organic synthesis are: porcine pancreatic lipase (PPL), lipase from *Pseudomonas cepacia* (Amano lipase PS, PCL), lipase from *Candida rugosa* (CRL), and lipase B from *Candida antarctica* (CAL-B). For lipid modification, lipase from *Rhizomucor miehei* (RML) is the most important. For this reason, we emphasize these five lipases in this book. Note that the synthetically-useful lipases include examples from all the classifications in Tab. 3 except the *Staphyloccocus* family. Two examples – RML and CAL-B – come from the *Rhizomucor* family.

Tab. 3   Classification of Commercial Lipases According to Similarities in Protein Sequence[a]

Classification	Characteristics	Examples
Mammalian (pancreatic) lipases	50 kDa	PPL
Fungal lipases		
Candida rugosa family	60–65 kDa	CRL, GCL, CE
Rhizomucor family	30–35 kDa	CAL-B, RML, ROL, HLL, PcamL
Unclassified		ANL, CAL-A, CLL
Bacterial lipases		
Pseudomonas family	30–35 kDa	PCL, PFL, CVL
Staphylococcus family	40–45 kDa	BTL2

[a] Classification according to Cygler et al. (1993) and Svendsen (1994) with some additions.

2.1.4   General Features of PPL, PCL, CRL, CAL-B, and RML

Researchers use crude, rather than purified lipases, in most biocatalytic applications for two reasons. First, crude enzymes are less expensive. Microbes secrete lipases into the growth medium. To isolate the lipase, manufacturers simply remove the cells and concentrate. Crude preparations often contain other proteins, but they usually contain only one hydrolase. The second reason researchers often use crude preparations is that they often work better than purified enzymes. The crude preparations contain sugars and other inert carriers which increase the surface area and stabilize the lipases, especially for reactions in organic solvents. A bound calcium ion stabilizes the 3-D structure of PCL, for this reason, crude preparation of PCL often contain added calcium salts. Because of the ill-defined nature of the product, most commercial lipases remain proprietary products. Suppliers sometimes create different preparations of the same lipase intended for different applications. In addition, lipases from different suppliers may be identical due to cross-licensing agreements or may be different due to separate patents on different strains of the same species. Two values for wt % protein in crude lipases are listed below. The higher value is the Lowry assay on the crude sample. This assay overestimates protein content due to interferences in the Lowry assay by sugars and other additives. The lower value refers to the Lowry assay after precipitation of the proteins with trichloroacetic acid (Weber et al., 1995b). This assay will underestimate the protein content if the proteins do not precipitate completely.

Crystallographers have solved the X-ray crystal structures of all five lipases (see below).

2.1.4.1   PPL

Porcine pancreatic lipase has a molecular weight of 50 kDa. PPL from Sigma contains 8–20 wt % protein (Weber et al., 1995b). Of all the commonly-used lipases for synthesis, PPL is the least pure. Microbial lipases, even when they are not recombinant lipases, are purer because microbes secrete the lipase into the medium. Removal of the cells and precipitation of the lipase yields much purer lipase. In contrast, PPL must be isolated from pancreas or bile which contains numerous hydrolases. SDS gel electrophoresis of crude PPL from Sigma shows four or five major proteins. Cholesterol esterase, trypsin, and chymotrypsin are likely contaminating hydrolases. Several groups reported increased enantioselectivity upon purification of PPL (e.g., Ramos-Tombo et al., 1986; Cotterill et al., 1991; Quartey et al., 1996, see also Bornemann et al., 1992).

2.1.4.2   CRL

Commercial samples contain 2–11 wt % protein (Weber et al., 1995b), the rest is sugars and inert carriers. Gel electrophoresis shows a single protein with molecular weight of 63 kDa when stained with Coomassie blue, but more sensitive staining reveals small amounts of other proteins. Molecular biologists have cloned five different isozymes of CRL from the *Candida rugosa* yeast (Lotti et al., 1993), which all have similar molecular weights. However, heterologous expression of these clones failed because *Candida rugosa* uses an unusual codon for serine which leads to incorrect translation into leucine in other microorganisms. Recently, this was overcome by designing a synthetical gene (*lip1*, 1647 bp) encoding the most prominent isozyme. This enabled functional expression in the yeasts *Saccharomyces cerevisae* and *Pichia pastoris*. Furthermore, expression in the methylotrophic yeast *P. pastoris* allows secretion of active (150 U/ml) and highly pure CRL into the cultivation medium facilitating downstream processing (Brocca et al., 1998). However, commercial samples of CRL are still non-recombinant enzymes and may contain more than one isozyme. Protein chemists have isolated several different lipases from commercial samples (Rúa et al., 1993; Chang et al., 1994). Differences in glycosylation of these lipases may contribute to these differences. In addition, some purification procedures appear to change the conformation of the lipase (Wu et al., 1990; Colton et al., 1995). One group also reported a small amount of contaminating protease (Lalonde et al., 1995). In spite of this complexity, commercial CRL is a useful and reproducible biocatalyst. Altus Biologics, Inc. sells cross-linked crystals of purified CRL as CLEC-CR (Lalonde et al., 1995). A detailed study using purified isozymes of CRL in kinetic resolutions of secondary alcohols showed that they differ in their reactivity and enantioselectivity, but exhibited the same enantiopreferences (Lundell et al., 1998).

2.1.4.3   RML

RML has a molecular weight of 33 kDa (Huge-Jensen et al., 1987) and commercial material contains 25–57 wt % protein (Weber et al., 1995b). RML is a recombinant lipase produced in *Aspergillus* fungus (Huge-Jensen et al., 1989).

2.1.4.4   CAL-B

CAL-B has a molecular weight of 33 kDa and commercial material contains 16–51 wt %
protein (Weber et al., 1995b). CAL-B is a recombinant protein produced in *Aspergillus*
fungus (Hoegh et al., 1995). CAL-B shows little or no interfacial activation and hydrolyzes long chain triglycerides only slowly. For this reason, it may be better classified as
an esterase. It shows very high activity and high enantioselectivity toward a wide range
of alcohols. Its enantioselectivity is usually low toward carboxylic acids. The application
of CAL-B in organic synthesis has been reviewed recently (Anderson et al., 1998).

2.1.4.5   PCL

PCL is 320 amino acids long with a molecular weight of 33 kDa. Amano lipase P or PS
is the industrial grade which contains 1–25 wt % protein as well as diatomaceous earth,
dextran, and $CaCl_2$. LPL-80 and LPL-200S are diagnostic grades that contain glycine.
LPL-200S contains no detectable amounts of any other proteins. SAM-II from Fluka
differs from lipase P or PS only in the purification method. Four groups have cloned and
expressed PCL starting from different *Pseudomonas* strains, but the amino acid sequences of all four are very similar (for original references see Iizumi et al., 1991;
Jorgensen et al., 1991; Nakanishi et al., 1991; Hom et al., 1991; for reviews see Gilbert,
1993; Svendsen et al., 1995). Expression of active lipase required stoichiometric
amounts of an additional protein which guides the proper folding of the prolipase
(Hobson et al., 1993; Quyen et al., 1999). Commercial PCL is probably not a recombinant protein. PCL shows interfacial activation with an increase in activity of ~25 in the
presence of an interface (Curtis and Kazlauskas, unpublished data). A single step purification yields crystalline PCL, but this pure material is no longer active in organic solvents (Bornscheuer et al., 1994a). Cross-linking of the crystals gives CLEC-PC (Altus
Biologics, Inc), which are active in organic solvents (see Sect. 4.4.2.1). Xie (1991) reviewed the application of this lipase in organic synthesis.

2.1.5   General Features of Esterases

Esterases (Carboxylester hydrolases, EC 3.1.1.1) catalyze like lipases the hydrolysis of
carboxylic acid esters and can be isolated from the same sources. Esterases and lipases
show many similarities with respect to their biochemical and structural properties. All
esterases, from which the structures are known have the characteristic α/β-hydrolase fold
(Sect. 2.1.6, Fig. 3) and a similar catalytic triad.
   The physiological role of most esterase is still unknown. The only exceptions are ace-
tyl- and butyryl choline esterases, both hydrolyze *in vivo* these neutrotransmitters. Some
acetyl- and cinnamic acid esterases are involved in metabolic pathways giving access to
carbon sources through degradation of hemicelluloses (Dalrymple et al., 1996). Esterases

also catalyze the detoxification of biocides. For instance, an insectizide resistence was related to an amplification of esterase genes (Blackman et al., 1995) and an esterase from *Bacillus subtilis* is capable of cleaving the phytotoxin Brefeldin A (Wie et al., 1996). Also, an esterase was described, which converts heroin into morphin with high specificity (Rathbone et al., 1997). Esterases might also be involved in the formation of ω-hydroxy acids from lactones (Griffin and Trudgill, 1976; Onakunle et al., 1997; Khalameyzer et al., 1999), which are produced *in vivo* in an enzymatic Baeyer-Villiger oxidation (Taschner and Black, 1988; Kelly et al., 1998; Roberts and Wan, 1998). This might enable growth on carbon sources such as cyclic alkanes or cyclic alkanones or is required for the production of flavor lactones.

In contrast to lipases, only a few esterases have practical use in organic synthesis. The most widely used mammalian esterase is isolated from pig liver (Sect. 9.1), examples for the use of other mammalian esterases appear to a much lesser extent and are summarized in Sect. 9.2 (acetylcholine esterase, AChE) and Sect. 1.1. The use of microbial esterases is reviewed in Sect. 9.4.

2.1.6   Lipases and Esterases are α/β Hydrolases

Although lipases differ significantly in their amino acid sequences, all 11 lipases whose structures have been solved show similar 3-D structures (Tab. 4) (for reviews see Cygler et al., 1992; Dodson et al., 1992; Cambillau and Tilbeurgh, 1993; Derewenda et al., 1994b; Derewenda, 1994; Ransac et al., 1996). This fold, called the α/β-hydrolase fold (Ollis et al., 1992), consists of a core of eight mostly-parallel β-sheets, which are surrounded on both sides by α-helices. The connectivity of the sheets and helices is the same in all α/β-hydrolases (Fig. 3). A similar structural pattern was found for an esterase from *Pseudomonas fluorescens* (pdb codes: 1aur (PMSF-inhibited enzyme), 1auo (non-inhibited enzyme) Kim et al., 1997a).

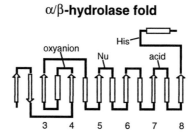

α/β-hydrolase fold

Fig. 3   Schematic diagram of the α/β-hydrolase fold. Oxyanion: residues that stabilize the oxyanion, Nu: nucleophilic residue; for lipases, esterases, and proteases this is a serine, α-helices are shown as rectangles, β-sheets as arrows.

Lipases are serine esterases. The catalytic machinery consists of a triad – Ser, His, and Asp(Glu) – and several oxyanion-stabilizing residues. These residues occur in the same order in all lipase amino acid sequences and orient in the same three-dimensional way in all the structures as shown schematically in Fig. 3. The 3-D orientation of the catalytic machinery is approximately the mirror image of that in the subtilisin and chymotrypsin families of proteases.

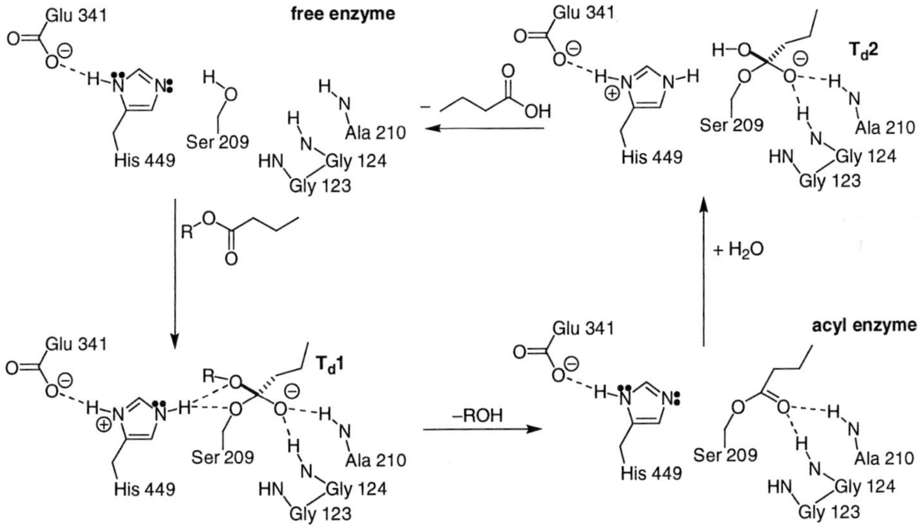

Fig. 4.   Hydrolysis of a butyric acid ester catalyzed by lipase or esterase involves an acyl enzyme intermediate and two different tetrahedral intermediates. Formation of the acyl enzyme involves the first tetrahedral intermediate, T_d1. Alcohol is released in this step, thus, this step determines the selectivity of lipases toward alcohols. Release of the acyl enzyme involves the second tetrahedral intermediate, T_d2. When deacylation limits the rate, this step determines the selectivity of the lipase toward acids. The amino acid numbering corresponds to the active site of lipase from *Candida rugosa*, CRL.

Tab. 4. X-Ray Crystal Structures of Lipases.

Lipase	Comments	pdb Code[a]	Reference
Mammalian pancreatic lipases			
HumanPL	with colipase & phospholipid	1lpa	Tilbeurgh et al. (1993)
HumanPL	with colipase & phosphonate	1lpb	Egloff et al. (1995a, b)
HumanPL	closed form	none	Winkler et al. (1990)
HorsePL	closed form	1hpl	Bourne et al. (1994)
PigPL	with colipase & surfactant	1eth	Hermoso et al. (1996)
CE (BSSL)[b]	open form with bile salts	none	Wang et al. (1997a)
Candida rugosa family			
CRL	closed form	1trh	Grochulski et al. (1993)
CRL	open form	1crl	Grochulski et al. (1994)
CRL[c]	linoleate complex	none	Ghosh et al. (1995)
CRL	sulfonate complexes	1lpn, 1lpo, 1lpp	Grochulski et al. (1994)
CRL	phosphonate complexes	1lpm, 1lps	Cygler et al. (1994); Grochulski et al. (1994)
GCL	closed form	1thg	Schrag et al. (1991)
Rhizomucor family			
CAL-B	open form	1tca, 1tcb, 1tcc	Uppenberg et al. (1994)
CAL-B	phosphonate complex	1lbs	Uppenberg et al. (1995)
CAL-B	Tween 80 complex	1lbt	Uppenberg et al. (1995)
RML	closed form	3tgl	Brady et al. (1990)
RML	phosphonate complex	4tgl, 5tgl	Brzozowski et al. (1991); Derewenda et al. (1992)
PcamL		1tia	Derewenda et al. (1994a)
HLL	disordered lid	1tib	Derewenda et al. (1994a, c)
ROL	closed and partially open forms	1tic, 1lgy	Derewenda et al. (1994a, c); Kohno et al. (1996)
ROL	homology model	none	Beer et al. (1996)
Pseudomonas family			
CVL (PGL)	closed form	1tah, 1cvl	Lang et al. (1996); Noble et al. (1993, 1994)
PCL	open form	1oil, 2lip, 3lip	Kim et al. (1997b); Schrag et al. (1997)
PAL	homology model	none	Misset et al. (1994)

[a] Accession code for the Brookhaven protein data bank. [b] Human pancreatic cholesterol esterase is identical to human bile salt stimulated lipase in milk. The structure is for the bovine pancreatic cholesterol esterase. A homology model for pancreatic cholesterol esterase from salmon has also been reported (Gjellesvik et al., 1994). [c] Isozyme of CRL.

The catalytic mechanism for lipase-catalyzed hydrolysis is similar to that for serine proteases (Dodson and Wlodawer, 1998). First, the ester binds to the lipase and the catalytic serine attacks the carbonyl forming a tetrahedral intermediate (Fig. 4). Collapse of this tetrahedral intermediate releases the alcohol and leaves an acyl enzyme intermediate. In a hydrolysis reaction, water attacks this acyl enzyme to form a second tetrahedral intermediate. Collapse of this intermediate releases the acid. Alternatively, another nucleophile such as an alcohol can attack the acyl enzyme thereby yielding a new ester (a transesterification reaction). In most cases, it appears that formation of the acyl enzyme is fast; thus, deacylation is the rate-determining step.

2.1.7   Lid or Flap in Interfacial Activation of Lipases

The x-ray structures of lipases usually show the 'closed' conformation where a lid or flap (a helical segment) blocks the active site. However, x-ray structures of lipases containing bound transition state analogs or bound lipids show the 'open' conformation where the lid is opened to permit access to the active site. For this reason researchers believe a lipid-induced change in the lid orientation causes interfacial activation. Lipases show poor activity toward soluble substrates in aqueous solution because the lid is closed. Upon binding to a hydrophobic interface such as a lipid droplet, the lid opens and the catalytic activity of the lipase increases. In addition, the opening of the lid places one of the oxyanion-stabilizing residues into the catalytic orientation. Cutinase and acetylcholine esterase, which show no interfacial activation, lack a lid and contain a pre-formed oxyanion hole (Martinez et al., 1992, 1994). However, the interfacial activation mechanism may be more complex. A number of lipases (for example, lipase from *Pseudomonas aeruginosa*, CVL, and CAL-B) do not show interfacial activation even though they contain a (small) lid. Lipase from *Staphylococcus hyicus* shows interfacial activation with some substrates, but not with others (for reviews see Ransac et al., 1996; Verger, 1997).

2.1.8   Substrate Binding Site in Lipases and Esterases

X-ray crystal structures of transition state analogs bound to the active site of lipases have identified distinct binding sites for the alcohol and acid portion of esters. The alcohol binding site is similar in all lipases. It is a crevice containing two regions – a large hydrophobic pocket which is open to the solvent and a small pocket that faces the floor of the crevice. As discussed in Sect. 5.1.1, the shape of this pocket sets the stereoselectivity of lipases toward secondary alcohols. The alcohol binding site corresponds to the S1' site in proteases (see Sect. 2.2.2.1).

The binding site for the acid portion of the ester varies considerably among the lipases. In CRL the acyl chain binds in a tunnel long enough to accommodate at least an eighteen carbon chain (Fig. 3). In RML and CAL-B, this region is only a short trough on the surface. In all three structures the α-carbon of an acyl chain binds just below the large hydrophobic region of the alcohol binding site. Substituents at the α-carbon would extend into the hydrophobic pocket. This acyl binding region, formed by the tunnel or trough and the hydrophobic pocket, corresponds to the S_1 site in proteases (Fig. 5).

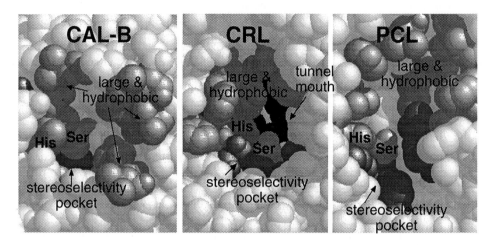

Fig. 5.   Proposed substrate binding site in three synthetically-useful lipases. The catalytic Ser lies at the bottom of a crevice with the catalytic His on the left. Although the details of this crevice differ for each lipase, each crevice contains a large hydrophobic pocket (light gray) and smaller pocket (medium gray), labeled "stereoselectivity pocket". This crevice is the alcohol binding site and the two pockets resemble the empirical rule discussed in Sect. 5.1.1.1. The regions that bind the acyl chain of the ester differ significantly among the three structures. For CRL, the acyl chain binds in a tunnel, the mouth of which is shown in dark gray. In CAL-B and PCL, the acyl chain binds in the large hydrophobic pocket of the crevice. Pictures were drawn with RasMac v2.6 using the Brookhaven protein data bank files cited in Tab. 4.

Pleiss et al. (1998) analyzed and compared shape and physico-chemical properties of the scissile fatty acid binding site of six lipases and two serine esterases. The lipases were subdivided into three groups, (1) those with a funnel-like binding site (CAL-B, PCL, mammalian pancreatic lipase and cutinase), (2) lipases with a hydrophobic, crevice-like binding site located near the protein surface (RML, ROL), and (3) lipase with a tunnel-like binding site (CRL). The 2-dimensional model (Fig. 6) also allowed identification of residues, which mediate chain length specificity and thus may guide protein engineering to alter this specificity.

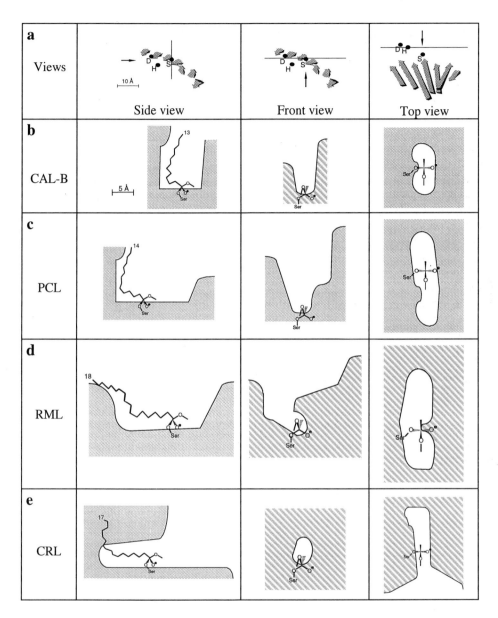

Fig. 6.   Shape of the binding site of four lipases (CAL-B , PCL, RML, CRL). **a** Orientation of the cross-sections which are planes perpendicular to the paper plane and indicated by a straight line (D, Asp; H, His; S, Ser). The direction of the view is indicated by an arrow. **b-e** Shape of the binding sites in side, front and top view; the number indicates the length of the longest fatty acid which completely binds inside the binding pocket (Pleiss et al., 1998).

2.2 Proteases and Amidases

2.2.1 Occurrence and Availability of Proteases and Amidases

By far the most important commercial proteases are the subtilisins. Most laundry detergents contain subtilisins to help remove protein-based stains. Subtilisins are produced on a multi ton scale and are thus very inexpensive. Because of their commercial importance, industrial researchers devoted a lot of effort in protein engineering of subtilisins, especially to improve stability and activity at high pH and at high temperatures. The first recombinant subtilisin, added to detergents starting in 1988, was a subtilisin BPN' variant developed by Genencor with Procter & Gamble. Current subtilisins for laundry applications tolerate high pH, high temperatures, surfactants, oxidants and organic cosolvents.

Proteases and amidases are grouped into four families based on their catalytic mechanism: serine, cysteine, aspartic, or metallo proteases. Subtilisin, subtilisin relatives, chymotrypsin, and penicillin amidase are serine proteases. Papain is a cysteine protease, while acylase, thermolysin and aminopeptidase are metallo proteases. Researchers rarely use the aspartic proteases in organic synthesis.

Tab. 5.   Some Commercially-Available Proteases and Amidases[a]

Enzyme	Biological Source	Synonyms
Subtilisins		
Subtilisin BL	*Bacillus lentus*	Savinase
Subtilisin BPN'	*Bacillus amyloliquefaciens*	subtilisin Novo
Subtilisin Carlsberg	*Bacillus licheniformis*	Alcalase, subtilisin A, Optimase
Proteases structurally related to subtilisins		
Thermitase	*Thermoactinomyces vulgaris*	
Proteinase K	*Tritirachium album* Limber	
Other proteases		
Chymotrypsin	bovine pancreas	
Thermolysin	*Bacillus thermoproteolyticus*	
Papain	papaya	
Amidases		
Amino acid acylase	porcine kidney, *Aspergillus melleus*	acylase 1
Penicillin amidase	*Escherichia coli*	penicillin acylase
Leucine aminopeptidase		

[a] Most proteases are available in research quantities from common suppliers such as Sigma, Fluka, Amresco, Boehringer Mannheim and others. Industrial quantities are available from Genencor, Novo Nordisk, and Biocatalysts. Thermitase is available from Life Technologies as PreTaq Thermophilic protease.

Unlike lipases, proteases and amidases act only on soluble substrates. Many substrates of interest to synthetic organic chemists dissolve only slightly in water, so researchers often add cosolvents, such as DMF, DMSO or acetone, to help dissolve the substrate. Both subtilisins and chymotrypsin tolerate low levels of organic cosolvents (usually < 10 vol %). Some solvents, e.g., dioxane, sharply reduce catalytic activities even at low levels (Bonneau et al., 1993).

The two main applications of proteases and amidases in organic synthesis are: enantioselective hydrolysis of natural and unnatural α-amino acid esters and other carboxylic acid esters (Sects. 8.1 and 8.2) and synthesis of di- and oligopeptides by coupling of *N*-protected amino acids and peptides esters (Sect. 4.2.5). To a lesser extent organic chemists also use proteases for enantioselective hydrolyis of esters of secondary alcohols and for regioselective reactions of sugars (Sect. 8.1.1).

2.2.2   General Features of Subtilisin, Chymotrypsin, and Other Proteases and Amidases

2.2.2.1   Substrate Binding Nomenclature in Proteases and Amidases

Proteases usually contain a channel on their surface which binds the polypeptide substrate. Schechter and Berger (1967) suggested numbering the different regions within this channel according to the amino acids residues that bind there and the distance of these amino acids from the amide link to be cleaved (Fig. 7). Thus, the acyl group of the amino acid undergoing cleavage binds at the S_1 site and this amino acid is the P_1 residue. The amino group to be released belongs to the P_1' residue, which binds at the S_1' site.

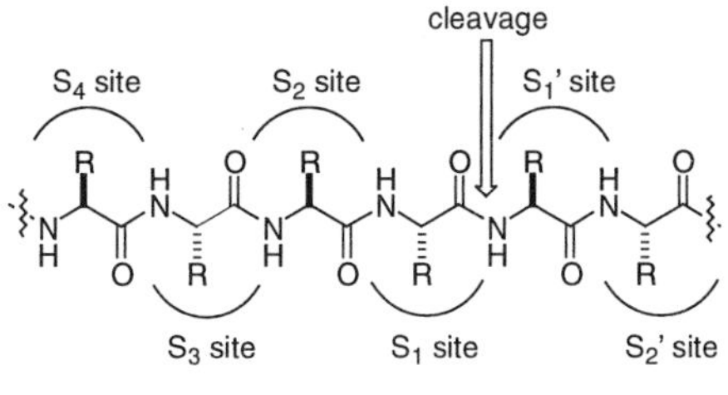

Fig. 7.   Naming of the binding site of proteases according to Schechter and Berger (1967). The acyl part of the amide link to be cleaved lies in the S_1, S_2, S_3, etc. binding sites, while the amino part of the amide link to be cleaved lies in the S_1', S_2', etc binding sites. The substrate residues are called P_1, P_2, P_3, etc, and P_1', P_2', etc. according to their location relative to the amide link being cleaved.

2.2.2.2   Subtilisin and Related Proteases

Subtilisins are a family of bacterial serine proteases secreted by various *Bacillus* species (Tab. 5). Siezen et al. (1991) reviewed the structures and sequence alignments of amino acid sequences of subtilisins (see also: Siezen and Leunissen, 1997). The mature subtilisin contains approximately 270 amino acids. Subtilisins are endopeptidases with a broad specificity and contain a structural calcium ion. Like chymotrypsins, they favor a large uncharged residue at the P_1 position, for example, phenylalanine. An approximate order of preference is Tyr, Phe > Leu, Met, Lys > His, Ala, Gln, Ser >> Glu, Gly (Estell et al., 1986; Wells et al., 1987). Subtilisins show little preference for amino acids at the P_2 and P_3 positions, but favor hydrophobic residues at the P_4 position (Perona and Craik, 1995). The most important commercial subtilisins are subtilisin BL (subtilisin from *Bacillus lentus*), subtilisin Carlsberg (subtilisin from *B. licheniformis*) and subtilisin BPN' (subtilisin from *B. amyloliquefaciens*). In spite of differences in amino acid sequence (84 of 275 amino acids differ between subtilisins Carlsberg and BPN'), the structures and substrate specificities are very similar.

Subtilisins are alkaline serine proteases, that is, proteases that show maximum activity at alkaline pH. There is no optimum pH. The reaction rate increases in the pH range 6–9 and then remains constant. Above pH 11, both subtilisin Carlsberg and subtilisin BPN' denature, but subtilisin BL remains stable to at least pH 12. This difference likely stems from the fact that *Bacillus lentus* is an alkalophilic bacteria that can grow at higher pH than other *Bacillus* species.

While the amino acid sequences of subtilisin BPN' and subtilisin Carlsberg are 75 % identical (for the 194 core residues (Siezen et al., 1991)), the amino acid sequences of thermitase (secreted by a *Thermoactinomyces* bacteria) and proteinase K (secreted by a *Tritirachium* fungus) show respectively only 52 % and 44 % sequence identity to subtilisin BPN'.

2.2.2.3   Chymotrypsin

The most widely studied form is α-chymotrypsin (241 amino acids, 25 kDa) from bovine pancreas (α-CT) (review by Jones and Beck, 1976). The pancreas secretes the catalytically-inactive chymotrypsinogen A. Proteases, including trypsin, remove two peptides yielding active α-CT, which consists of three polypeptide chains linked by five disulfide bonds. α-CT is an endopeptidase favoring peptide links with Phe, Tyr, or Trp as the acyl group.

Although crystalline α-CT is stable indefinitely, α-CT digests itself in solution near neutral pH (pH 6–9), so solutions of α-CT should not be stored. The pH optimum for amide hydrolysis is pH 7.8 and most researchers carry out reactions near this pH. For more reactive substrates like esters, reactions also proceed at pH 5, where α-CT is more stable.

Like other proteases, chymotrypsin only acts on dissolved substrates. To dissolve nonpolar organic substrates in aqueous solutions, researchers often add organic cosolvents. Organic cosolvents decrease the rate of α-CT-catalyzed reactions usually by increasing the K_m for the substrate. Some aromatic compounds also inhibit α-CT, presumably by binding to the hydrophobic binding pocket.

Above a concentration of ~0.25 mg/mL, the catalytic efficiency of α-CT decreases because it associates into inactive dimers, similar to that found in the crystal structure. High ionic strength, such as 0.1 M NaCl typically used in synthetic reactions, decreases this association thereby increasing the rate of reaction. For reactions in organic solvent using suspended lyophilized α-CT, Khmelnitsky et al. (1994) dramatically increased the rate of reactions by lyophilizing a salty solution of α-CT. Presumably the salts minimized the association of chymotrypsin in the lyophilized form.

2.2.2.4 Thermolysin

Thermolysin is a thermostable metalloprotease secreted by *Bacillus thermoproteolyticus*. Thermolysin contains a zinc ion at the active site and and four calcium ions that stabilize the structure. Like subtilisin and chymotrypsin, thermolysin also favors hydrophobic amino acids at the P_1 residue.

2.2.2.5 Penicillin G Acylase

Penicillin G acylase (PGA, penicillin amidase, for a review see Baldaro et al., 1992) catalyzes hydrolysis of the phenylacetyl group in penicillin G (benzylpenicillin) to give 6-aminopenicillanic acid (6-APA) (Eq. 2).

penicillin G   6-APA   (2)

Penicillin G acylase also cleaves the side chain in penicillin V, where the phenylacetyl group is replaced by a phenoxyacetyl group. The commercially available enzyme are derived from *E. coli* strains. Both penicillin G and V are readily available from fermentation, so penicillin manufacturers carry out a PGA-catalyzed hydrolysis to make 6-APA on a scale of approximately 5 000 metric tons per year (Matsumoto, 1992). They use 6-APA to prepare semisynthetic penicillin such as ampicillin, where a D-phenylglycine is linked to the free amino group of 6-APA, or amoxicillin, where a D-4-hydroxyphenyl glycine is linked.

PGA favors hydrolysis of phenylacetyl esters and amides, but accepts acyl groups that are structurally similar to phenylacetyl, such as 4-pyridylacetyl and phenoxyacetyl. On the other hand, PGA accepts a wide range of structures as the leaving group (alcohol part of an ester or amine part of an amide) (Sect. 8.2.2.3).

2.2.2.6   Amino Acid Acylases

Acylases catalyze the hydrolysis of *N*-acetyl-L-amino acids to the L-amino acid. Derivatives of D-amino acids do not react. Two amino acid acylases are available commercially – from porcine kidney and from *Aspergillus melleus*. Both are metallo proteases containing a zinc ion in the active site. On a lab scale, Chenault et al. (1989) recommended using the porcine kidney enzyme because it accepts a slightly wider range of substrates. On an industrial scale, researchers usually use the *Aspergillus* acylase because it is more stable. Researchers have found several other interesting acylases in microorganisms. Acylases from a *Pseudomonas* strain or a *Comamonas* strain accept a wider range of amino acids than the two commercial acylases, including derivatives of cyclic amino acids such as proline and *N*-alkyl amino acids (Kikuchi et al., 1983; Groeger et al., 1990, 1992).

2.2.3   Structures of Proteases and Amidases

Crystallographers have solved the structures of most of the proteases important for organic synthesis (Tab. 6). In most cases, several structures are available, including structures of mutants and of proteases with bound inhibitors. One notable absence from this list are amino acid acylases.

Tab. 6.   X-Ray Crystal Structures of Selected Proteases and Amidase.

Proteinase	Number of structures	pdb code (example)	Reference
Subtilisins and relatives			
Subtilisin BL	3	1jea	Bott et al. (1996)
Subtilisin BPN'	23	1sbt	Wright et al. (1969)
			Drenth et al. (1972)
Subtilisin Carlsberg	19	1sec	McPhalen et al. (1985)
Thermitase	5	2tec	Gros et al. (1989)
Proteinase K	6	2prk	Betzel et al. (1992)
Other proteinases			
α-Chymotrypsin	9	2cha	Matthews et al. (1967)
Penicillin amidase	10	1pnk	Duggleby et al. (1995)
Thermolysin	20	4tmn	review: Matthews (1988)

2.2.3.1   Serine Proteases – Subtilisin and Chymotrypsin

Although the catalytic residues in both subtilisins and chymotrypsins have a similar three-dimensional arrangement, their protein folds are not related. Chymotrypsin has a β/β fold – two antiparallel β-barrel domains. On the other hand, subtilisin has an α/β fold – a core of parallel β-sheets surrounded by four α-helices. The similar three dimensional arrangement of catalytic residues in these two proteases is an example of convergent evolution where completely different loop regions, attached to different framework structures, form similar active sites. Note that this fold is not the same as the α/β-fold of lipases (see Sect. 2.1.6) (Branden and Tooze, 1991).

Subtilisin belongs to the family of subtilases, a superfamily of subtilisin-like serine proteases. The main members of this family are the subtilisins, thermitase and proteinase K. Many subtilisin mutants and variations are available. For examples, see the recent sequence alignment of subtilases (Siezen and Leunissen, 1997).

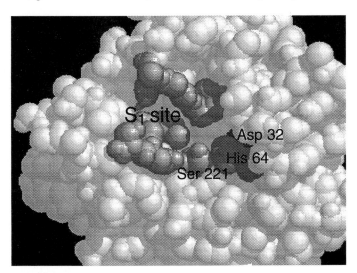

Fig. 8.   The crystal structure of subtilisin Carlsberg shown in a space-filling representation. The labels and coloring show the amino acid residues of the catalytic triad and the residues forming the S_1 binding site. This binding site is a shallow groove lined with nonpolar amino acid residues. In contrast, the S_1 binding site of chymotrypsin (Fig. 161, Sect. 8.2.1.2) is a well defined hydrophobic pocket. For this reason subtilisin accepts a wider range of substrates than chymotrypsin. Coordinates are from Brookhaven protein data bank file 1sbc (Neidhart and Petsko, 1988) and the figure was created using RasMac v 2.6 (Sayle and Milner-White, 1995).

The substrate preference in subtilisin is dominated by the S_1 and S_4 sites, which both favor aromatic or large nonpolar residues such as phenylalanine (see Sect. 2.2.2.1) for explanation of the active site nomenclature.) A typical substrate for colorimetric assay of subtilisin activity (see Sect. 4.4.1.3), has a Phe residue at P_1 and an Ala residue at P_4 (Graham et al., 1993). For organic synthesis applications, the substrates are usually to small to reach the S_4 site, so the S_1 site is the most important one. Since most synthetic organic intermediates are hydrophobic, it is not surprising that the most useful proteases are those that favor hydrophobic substrates. Fig. 8 shows a structure of subtilisin Carlsberg with the S_1 site highlighted. This site is a shallow groove lined with nonpolar amino acids.

Subtilisin's substrate specificity stems mainly from the acyl binding site, not the amide binding site. For this reason, subtilisin shows higher stereoselectivity toward acids and a lower, broader selectivity toward the amine (or alcohol in the case of an ester).Like subtilisin, chymotrypsin also favors hydrophobic residues at the P_1 position. However, the S_1 site in chymotrypsin is a well defined hydrophobic pocket (Blow, 1976).

Penicillin amidase is also a serine protease, but its active site structure is unusual – instead of triad, it contains only an N-terminal serine residue (Duggleby et al., 1995). Researchers suggest that the amino group of the serine may act as the base during catalysis.

2.3   How to Distinguish Between Lipase, Esterase, and Protease

The far most reliable feature to distinguish these enzymes is their substrate specificity. Proteases usually cleave peptide or amide bonds. Lipases preferentially hydrolyze triglycerides composed of long chain fatty acids and esterases usually only accept water-soluble esters or short-chain fatty acid triglycerides, such as tributyrin. For lipases and esterases, a similar observation can be made in the hydrolysis of p-nitrophenyl ester: lipases hydrolyze p-nitrophenyl palmitate, esterases do not, but both hydrolyze p-nitrophenyl acetate. This substrate specificity might be expanded to water-soluble (lipases and esterases) vs. water-insoluble compounds (lipases), but this is not valid for all enzymes. Pig liver esterase mainly accepts methyl esters of carboxylic acids and acetates of alcohols. Proteases usually do not accept any of these typical lipase or esterase substrates. In contrast, some lipases are capable to convert amines or amides (Sect. 5.1.3.4).

Another distinction can be made on the basis of the protein structure. Most lipases possess a lid covering the active site, esterase do not have a lid (Sect. 2.1.7). Exceptions are lipases from *Candida antarctica* B (CAL-B) and Cutinase from *Fusarium solani pisi*, which have only a small or no lid. A further criteria is the interfacial activation phenomenon observed for most lipases. Measurement of enzyme activity at different substrate concentrations should reveal, whether the enzyme is a lipase or an esterase. However, no interfacial activation was found for CAL-B. Esterases also obey normal Michaelis-Menten kinetics. Proteases and some esterases are inhibited by phenylmethylsulfonyl fluoride (PMSF), but not lipases.

2.4   Screening and Directed Evolution

Traditionally, enzymes are isolated from sample collections of microorganisms by enrichment through culturing, isolation and growth. Isolated cultures permit extended and reproducible growth that in turn allows phenotypic and genotypic characterization (Satoshi, 1992). This strategy has been a productive source of commercially valuable enzymes. However, only a small fraction of the worlds microbial diversity has been described in the literature so far – some 5 000 microbial species. In contrast current estimates of the number of microbial species range from 1 to 100 million (Short, 1997).

Recent molecular approaches to genes bypass culturing and focus on direct isolation of genomes from environmental samples. This strategy avoids problems in developing specific culture techniques (e.g., for microorganisms from hot springs or volcanoes) and preserves sample diversity. The advantage of these techniques might be demonstrated by the fact that Diversa (San Diego, USA) discovered 171 genes encoding tentative esterases out of only 6 % of an environmental library prepared from a DNA sample from pH 10 soil. They estimated that this sample alone contains the genes of up to several thousand different microbial species and subspecies (Short, 1997). Thus, these technologies provide access to a huge number of new enzymes, which might possess new properties and substrate specificities (Dalbøge and Lange, 1998).

Another source of new enzymesis mutagenesis of known enzymes. Evolvution adapted these enzyme for their natural substrates and environments, but not for organic synthesis applications. For instance, natural enzymes may be unstable at high termperatures or in the presences of organic solvent or their stereoselectivity toward unnatural substrates may be too low. Over the last decade, researchers learned how to improve proteins by site-directed mutagenesis in combination with computer modeling. However, this rational approach requires knowing the three-dimensional structure of the enzyme as well as the detailed mechanism. Successful improvements typically require several rounds of computer modeling and mutagenesis and are thus slow. In addition, computer modeling often cannot predict how to change many important properties such as pH, thermo- and solvent stability

Recently, researchers have turned to directed (or molecular) evolution as a way to improve existing enzymes. Directed evolution is the combination of random mutagenesis with an efficient assay to identify improved enzymes. In contrast to classical random mutagenesis by e.g., x-ray or chemical mutagens, directed evolution restricts mutations to the gene encoding the protein of interest or even smaller parts. Directed evolution requires an effective mutation strategy, the functional expression of the protein in a suitable – usually – microbial host and a fast and reliable assay for the identification of desired mutants out of a pool of 10^4 to more than 10^6 variants (Kuchner and Arnold, 1997).

The strategies developed so far can be divided into two major approaches: asexual and sexual evolution. In an asexual evolution, random mutagenesis is targeted preferentially towards the gene encoding *one* parent protein. An enzyme library is generated and then screened for improved properties. The best enzymes identified in a first generation are then subjected to further mutation cycles until a desired biocatalyst is found (Fig. 9, left).

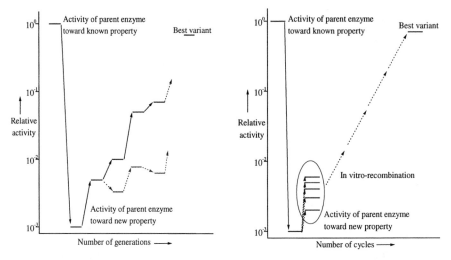

Fig. 9:   Strategies for the improvement of a wild-type enzyme by the production of sequential generations from one parent gene (asexual evolution, *left*) or by *in vitro* recombination through, e.g., DNA-shuffling in   sexual evolution (*right*). During asexual evolution a step-by-step improvement (*lined arrows*) as well as a worsening (*dotted arrows*) can occur in higher generations. For sexual evolution, a pool of homologous variants from several parent genes (*circle*) is useful before *in vitro* recombination should be performed (after Arnold and Moore (1997), modified).

However, this implies a step-by-step improvement of variants from the first generation by introducing positive and deleting negative mutations, which might not always be the case. The most frequently used method for this asexual evolution is an error-prone polymerase chain reaction (PCR). By using non-optimal reaction conditions during PCR, the error rate for e.g., *Taq* (*Thermus aquaticus*)-polymerase can be increased from 0.001–0.02 % under standard conditions to more than 1 % (Cadwell and Joyce, 1992). Error-prone PCR is rather easy to perform, mutations can be limited to the nucleotide sequence encoding the protein of interest and the mutation rate can be adjusted by altering the PCR conditions. Disadvantages are the non-statistic exchange of nucleotides and an often observed low ligation efficiency after PCR. For the latter problem, modified PCR protocols have been described (Spee et al., 1993; Vartanian et al., 1996).

As an alternative, mutator strains with defects in their DNA repair mechanisms can be used. Insertion of a plasmid bearing the gene of the target protein into such a mutator strain followed by cultivation leads to the introduction of mutations during replication. Only one mutator strain is commercially available as *Escherichia coli* variant *Epicurian coli* XL1-Red (Stratagene, La Jolla, USA) (Greener et al., 1996). The use of a mutator strain is even simpler than error-prone PCR, but the mutation rate is difficult to adjust. Due to a mutation of the entire plasmid, also promotor and/or plasmid defects as well as improvements might occur.

In contrast, sexual evolution starts from a *pool* of homologous parent genes. These can originate from asexual evolution, from related natural sequences or from enzyme variants generated by rational protein-design. The common method for this approach is DNA or gene shuffling (Stemmer, 1994a, b), which is based on a partial DNAse I digestion of the genes followed by a recombination and amplification of fragments by PCR. Again, accumulation of positive and elimination of negative mutations is desired (Fig. 9, right). Main disadvantages are the control of DNAseI digestion to generate fragments of suitable size and problems in the ligation efficiency after PCR. A further, recently described (Zhao et al., 1998) strategy for *in vitro* mutagenesis and recombination is based on a modified PCR protocol only. This staggered extension process (StEP) method allows the production of full-length genes, which carry different sequence information. Beside the use of homologous parent genes, especially short reaction times and the use of different templates during PCR are the basis for this StEP-method.

The number of enzyme variants, which can be generated by directed evolution, grows exponentially with the size of the enzymes and the number of amino acids exchanged simultaneously. Even for a small protein of 200 amino acids more than 9 billion theoretically possible variants can be generated simply by introducing three substitutions at the same time.

An identification of desired biocatalysts cannot be performed by using common tedious and time-consuming methods like HPLC or gas chromatography. The first law of directed evolution 'You only get what you screen for' as proposed by F. Arnold (Arnold and Moore, 1997), clearly emphasizes that an efficient assay system is of upmost importance. Identification can be based on an altered antibiotics resistance (Crameri et al., 1996) or growth on media lacking components essential for growth (MacBeath et al., 1998; Bornscheuer et al., 1999). Problems related to organic synthesis usually require assay systems, which provide direct information about the mutant properties such as improved stereoselectivity (see below) or altered substrate selectivity. Also, fluorescense correlation spectroscopy (Eigen and Rigler, 1994), phage display techniques (Schwienhorst, 1998) or time-resolved IR-thermographic detection (Reetz et al., 1998) might allow the rapid identification of desired enzyme variants within a library.

Some other problems that can make directed evolution difficult include: (1) microorganisms produce the enzymes intracellularly, thus requiring the researcher to lyse the cells before assaying for improved properties, (2) uncertainties about a correct processing and folding of all mutants and (3) uncertainties about the stability of all mutants under assay conditions.

Directed evolution of enzymes is a very new area and researchers have improved only a few lipases, proteases or esterases. These will be discussed in more detail below. Additional examples for the application of directed evolution are summarized in a number of recent reviews (Joyce, 1992; Stemmer, 1995; Kuchner and Arnold, 1997; Schwienhorst, 1998; Harayama, 1998; Arnold, 1998; Bornscheuer, 1998; Reetz and Jaeger, 1999).

Subtilisin is a useful catalyst for organic synthesis, particularly in the presence of organic solvents (see Sect. 8.1.1.2). However, wild-type subtilisin E is unstable when researchers add a cosolvent like dimethylformamide (DMF) to dissolve the substrate. Directed evolution of subtilisin E using error-prone PCR and a *Bacillus subtilis-Escherichia coli* shuttle vector led to a variant which was 471 times more active in 60 % DMF than the wild-type protease in the hydrolysis of succinyl-Ala-Ala-Pro-Phe-*p*-nitroanilide. Even in the absence of DMF, the mutant was ca. 15-fold more active (You and Arnold,

1994). In another example, the half-life of Subtilisin E at 65 °C was increased 50 times through recombination of two subtilisin E variants using the StEP-method (Zhao et al., 1998). Nakano et al. (1999) increased the stability of lipase from *Pseudomonas* sp. KWI-56 in DMSO by subjecting the gene to error-prone PCR. The best variant containing four mutations was ~40 % more stable in 80 % DMSO compared to the wild-type enzyme.

However, also a combination of site-directed mutagenesis and chemical modification was described to influence selectivity and activity of subtilisin by alteration of the S_2 and S_1' binding sites of subtilisin (Berglund et al., 1997). First, site-directed mutagenesis introduced a cysteine residue; next, alkylation of the thiol group with alkylthiomethane sulfonates ($R–S–SO_2Me$) introduced different alkyl groups. The authors reported changes in selectivity and – in several cases – higher activity than for wild type subtilisin. This approach allows researchers to generate 'mutants' faster than site-directed mutagenesis and also allows researchers to introduce functional groups not present in the 20 proteinogenic amino acids.

For organic synthesis, the alteration of stereoselectivity and substrate specificity is most important.

Reetz et al. (1997) increased the enantioselectivity of a lipase from *Pseudomonas aeruginosa* (PAL) towards 2-methyl decanoate from E ~ 1, to E ~ 10 (Fig. 10). They generated mutants using error-prone PCR and estimated the enantioselectivity by comparing the rates of release of *p*-nitrophenol for each enantiomer.

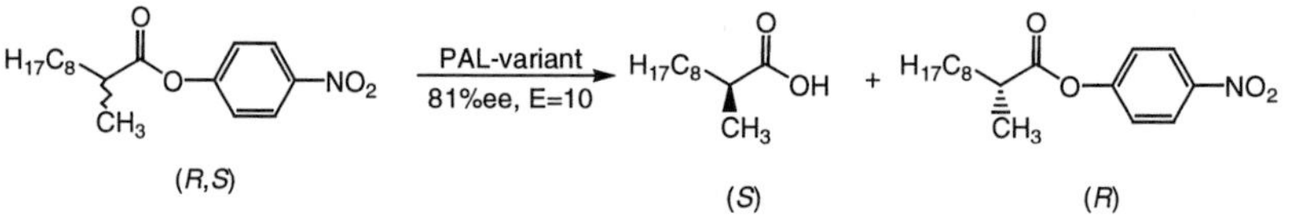

Fig. 10:   Enantioselectivity of a lipase from *Pseudomonas aeroginosa* PAO1 towards 2-methyldecanoate improved from 2 %ee (E ~ 1) to 81 %ee (E ~ 10) by error-prone PCR.

Directed evolution of an esterase from *Pseudomonas fluorescens* (PFE) expanded its substrate range. Neither wild-type PFE, nor 20 other hydrolases catalyzed the hydrolysis of the 3-hydroxy ester in Fig 11, a precursor to epothilones. Mutation using special strain of *E. coli* that lacks DNA repair mechanisms (*Epicurian coli* XL1-Red) followed by screening using both an indicator assay and a growth assay produced a double mutant of PFE. This mutant did catalyze hydrolysis with moderate stereoselectivity (25 %ee for remaining 3-hydroxy ester, E ~ 5, Fig. 11) (Bornscheuer et al., 1998a, 1999).

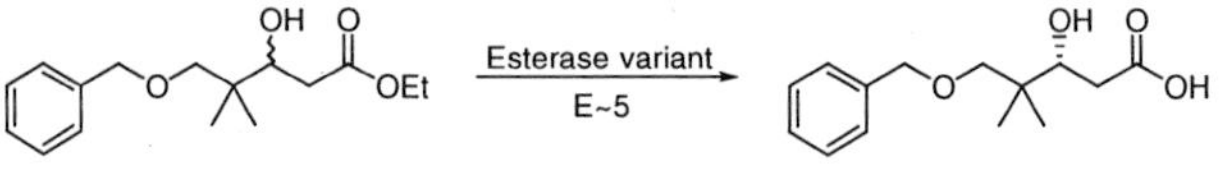

Fig. 11.   Expanding the substrate range of an esterase from *Pseudomonas fluorescens* by using a mutator strain.

A combination of error-prone PCR and DNA-shuffling led to the generation of a more stable and active variant of an esterase from *Bacillus subtilis* (Moore and Arnold, 1996).

This enzyme hydrolyzes the *p*-nitrobenzyl ester of Loracarbef, a cephalosporin antibiotic, with 150 times higher activity compared to the wild-type in 15 % DMF (Fig. 12) (Arnold and Moore, 1997). In a recent paper, the thermostability of this esterase could also be increased by ~14 °C by directed evolution (Giver et al., 1998).

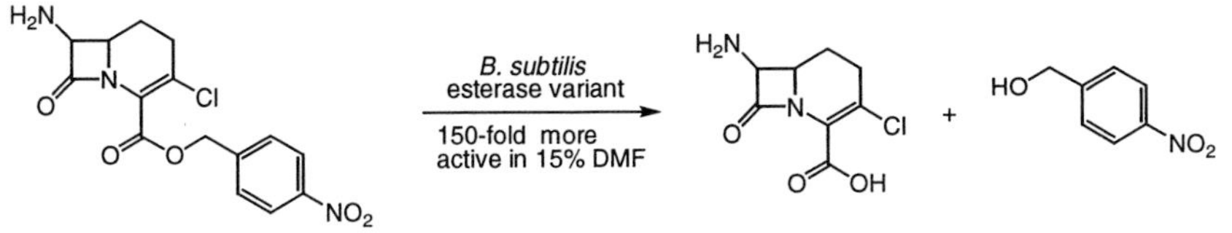

Fig. 12:   A combination of error-prone PCR and DNA-shuffling led to a variant of an esterase from *Bacillus subtilis*, which exhibits 150-fold higher activity in 15 % DMF compared to the wild-type in the cleavage of the *p*-nitrobenzyl ester of Loracarbef. For practical reasons, the corresponding *p*-nitrophenyl ester was used in the assays.

Thus, directed evolution can be also used to solve problems related to the technical application of biocatalysts, such as sufficient stability under process conditions with respect to pH profile, temperature activity and stability and solvent tolerance. These deficits are extremely difficult to overcome by site-directed mutagenesis and directed evolution can contribute to the fast generation and identification of improved enzymes.

Despite the methods developed so far and the successful examples given above, several problems have to be solved to allow a broader application of this method. This includes the further optimization of methods for the generation of mutants and enzyme libraries and the development of highly efficient assay systems.

3 Designing Enantioselective Reactions

3.1 Quantitative Analysis

3.1.1 Kinetic Resolutions

In a kinetic resolution, the enantiomeric purity of the product and starting material varies as the reaction proceeds (reviewed by Kagan and Fiaud, 1988). Thus, comparing enantiomeric purities for two kinetic resolutions is meaningful only at the same extent of conversion. To more conveniently compare kinetic resolutions, Charles Sih's group developed equations to calculate their inherent enantioselectivity (Chen et al., 1982, 1987; reviewed by Sih and Wu, 1989). This enantioselectivity, called the enantiomeric ratio, E, measures the ability of the enzyme to distinguish between enantiomers. A non-selective reaction has an E of 1, while resolutions with E's above 20 are useful for synthesis. To calculate E, one measures two of the three variables: enantiomeric purity of the starting material (ee_s), enantiomeric purity of the product (ee_p), and extent of conversion (c) and uses one of the three equations below (Eq. 3). Often enantiomeric purities are more accurately measured than conversion; in these cases, the third equation is more accurate.

$$E = \frac{\ln[1 - c(1 + ee_p)]}{\ln[1 - c(1 - ee_p)]}; \quad E = \frac{\ln[(1 - c)(1 - ee_s)]}{\ln[(1 - c)(1 + ee_s)]}; \quad E = \frac{\ln\left[\dfrac{1 - ee_s}{1 + (ee_s/ee_p)}\right]}{\ln\left[\dfrac{1 + ee_s}{1 + (ee_s/ee_p)}\right]} \qquad (3)$$

High E values (≥ 100) are less accurately measured than low or moderate E values because the enantiomeric ratio is a logarithmic function of the enantiomeric purity. When $E \geq 100$, small changes in the measured enantiomeric purities give large changes in the enantiomeric ratio. Thus, the survey below avoids reporting E values above 100. In practice, we found that even E values near 50 were sometimes difficult to measure more precisely than ± 10. A simple program to calculate enantiomeric ratio using the above equations is freely available at *http://www-orgc.tu-graz.ac.at* (Kroutil et al., 1997a). In spite of the fact that these equations include assumptions such as an irreversible reaction, one substrate and product, and no product inhibition, they are reliable in the vast majority of cases, especially for screening studies. Recently, faster spectrophotometric methods for measuring the enantiomeric ratio (Janes and Kazlauskas, 1997b; Janes et al., 1998) using samples of pure enantiomers were developed. One method, called Quick E, is restricted to *p*-nitrophenyl derivatives of chiral carboxylic acids, but a more recent method can be used for any ester and identification of active and enantioselective hydrolases is based on a pH change using *p*-nitrophenol as pH indicator.

For careful optimization of reactions, three situations require a more careful approach. First, when the biocatalyst is a mixture of enzymes, for example, isozymes, which all act on the substrate, then the calculated E value reflects a weighted average of all the enzymes (Chen et al., 1982). When these enzymes differ significantly in their affinity for the substrate, then different enzymes will dominate the activity at different substrate concentrations. Thus, the apparent enantioselectivity may vary as the reaction depletes the substrate or when the reaction is carried out with different initial substrate concentrations. When enzymes differ in their stability, apparent enantioselectivities for long vs. short reaction times may differ. To measure the true E-value, one must purify the enzymes and measure E separately.

Second, when product inhibits the reaction the apparent enantioselectivity can change (Rakels et al., 1994a; van Tol et al., 1995a, b). For example, addition of 4 v/v % ethanol to a carboxylesterase NP-catalyzed hydrolysis of ethyl 2-chloropropionate increased the enantioselectivity from 4.7 to 5.4 (see also Sect. 9.4.1). Rakels et al. (1994a) attributed this change not to changes in the inherent selectivity of the enzyme, but to selective inhibition of one of the enantiomers by ethanol. In another example van Tol et al. (1995a, b), could not recover enantiomerically pure starting material in the PPL-catalyzed hydrolysis of glycidol butyrate even at high conversion. The enantiomeric purity of the remaining glycidol butyrate reached 95 %ee at 70 % conversion, but did not increase further even at 90 % conversion. In other words, the apparent enantioselectivity dropped from 20 at 31 % conversion to 2.7 at 95 % conversion. Van Tol et al. (1995a, b) attributed this plateau to product inhibition promoting the reverse reaction for the product enantiomer. To include product inhibition in the quantitative analysis, reseachers used more complex equations which take into account the mechanism of lipase-catalyzed reactions (ping-pong bi-bi). Until now few researchers included product inhibition in their analysis, but a readily available computer program (Anthonsen et al., 1995; *http://bendik.mnfak.unit.no*) simplifies this task.

Third, when the reaction is reversible, such as transesterification, one must include the equilibrium constant for the reaction (Chen et al., 1987). One can first measure the equilibrium constant in a separate experiment and then determine E from measurements of ee_s and ee_p. Anthonsen et al. (1995) developed a simpler approach where they determine both K and E by fitting a series of ee_s and ee_p measurements.

3.1.2   Recycling and Sequential Kinetic Resolutions

To enhance the enantiomeric purity, the enriched material can be isolated and resolved again. This double resolution is called recycling. Chen et al. (1982) derived an equation to predict the optimum degree of conversion in recycling reactions and many researchers have used this strategy (for an example see Johnson et al., 1995). Brown et al. (1993) and Kanerva and Vänttinen (1997) reported several examples and a computer program for calculations is available for instance at *http://www-orgc.tu-graz.ac.at* (Kroutil et al., 1997b). Guo (1993) reported plots to predict the maximum chemical yield in various situations. To minimize the work in recycling reactions, several groups used *in situ* recycling where the two resolutions are carried out stepwise, but without isolation of the intermediate products (Chen and Liu, 1991; Sugai et al., 1996; Majeric and Sunjic, 1996). Some authors called these reactions sequential kinetic resolutions, but we favor *in situ* recycling and reserve the term sequential kinetic resolution only for those reactions where both steps occur at the same time, such as the acylation of diols.

Like recycling reactions, sequential kinetic resolutions enhance the enantiomeric purity of the products (Kazlauskas, 1989; Guo et al., 1990; Caron and Kazlauskas, 1991). For example, hydrolysis of *trans*-1,2-diacetoxycyclohexane proceeds stepwise – first hydrolysis to the monoacetate, then to the diol (Fig. 13) (Caron and Kazlauskas, 1991). Both reactions favor the same enantiomer, thus, the two resolutions reinforce each other. Maximum reinforcement occurs when both reactions occur at comparable rates with an overall enantioselectivity of approximately $(E_1 \times E_2)/2$ (Caron and Kazlauskas, 1991). In addition, sequential kinetic resolutions yield both the starting material and product in high enantiomeric purity at the same extent of conversion because the 'mistakes' remain in the intermediate product (monoacetate in the example in Fig. 13). In contrast, single step kinetic resolutions yield high enantiomeric purity for the product at < 50 % conversion, but high enantiomeric purity for the starting material requires > 50 % conversion.

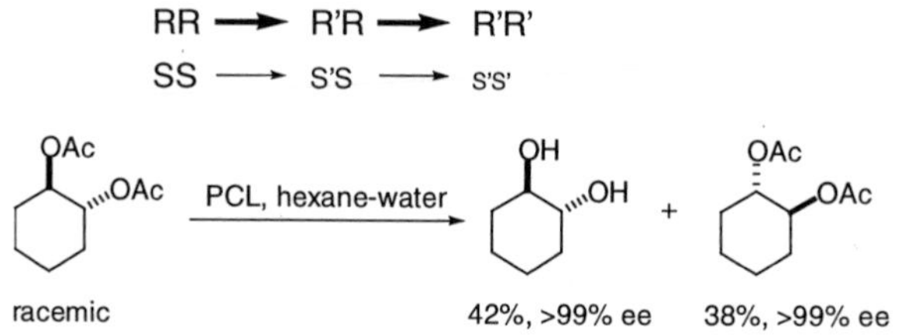

Fig. 13.   Sequential kinetic resolution enhances the enantiomeric purity of the product through two enantioselective steps.

C_2-symmetric diols are especially well suited to sequential kinetic resolution because both steps are likely to have the same enantiopreference (Fig. 14).

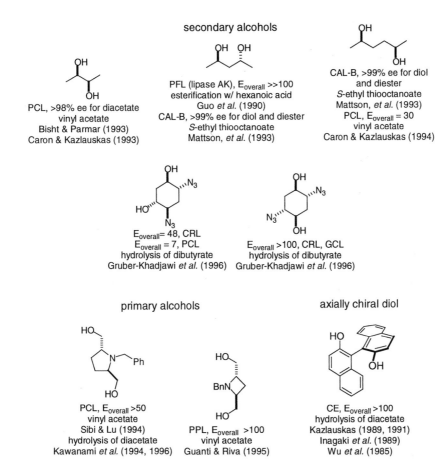

Fig. 14. Examples of C_2-symmetric diols resolved by sequential kinetic resolution include secondary and primary alcohols as well as diols with axial chirality.

Unsymmetrical diols can also undergo a sequential kinetic resolution (Fig. 15).

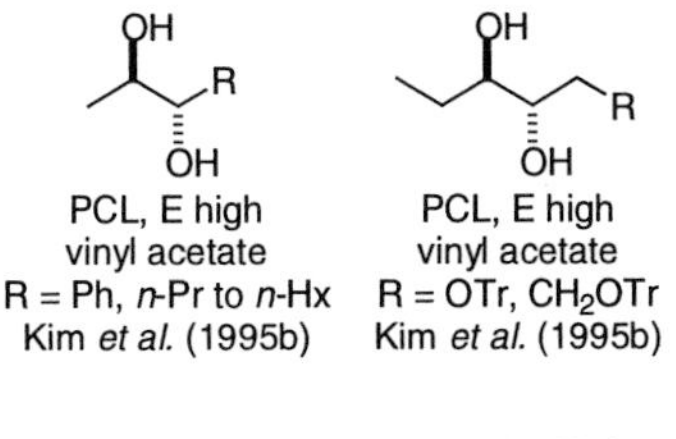

Fig. 15. Sequential kinetic resolution of non-C_2 symmetric diols.

Only one dicarboxylic acids was resolved by lipase-catalyzed sequential kinetic resolution and this was a special case. Node et al. (1995) hydrolyzed a racemic C_2-symmetric tetraester. The non-conjugated ester groups reacted selectively followed by spontaneous decarboxylation. Interestingly, CRL and RJL favored opposite enantiomers. Although Node et al. (1995) suggested possible racemization of the starting tetraester, which would allow a dynamic kinetic resolution (Sect. 5.4), they did not report yields over 50 %. The lack of carboxylic acid examples may be due to more efficient resolution of alcohols by lipases, or to the slow hydrolysis of monoesters containing a charged carboxylate group by lipases (Fig. 16).

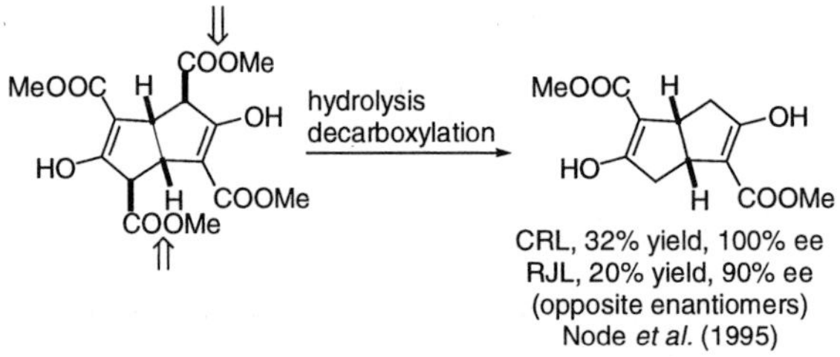

CRL, 32% yield, 100% ee
RJL, 20% yield, 90% ee
(opposite enantiomers)
Node *et al.* (1995)

Fig. 16.   Sequential kinetic resolution of a chiral diacid.

For substrates with a single functional group, researchers demonstrated a sequential kinetic resolution by *in situ* hydrolysis of an ester and reesterification to a new ester (Macfarlane et al., 1990). However, reversibility of these reactions limited the enhancement of enantioselectivity (Straathof et al., 1995). In these cases, an *in situ* recycling reaction (see above) is probably a better way to enhance the enantiomeric purity.

Enantioselective reactions can also separate diastereomers. For example, Wallace et al. (1992) used the (*R*)-enantioselectivity of PCL to separate a mixture of *meso* and racemic diols. The (*R,R*)-diol reacted to the diacetate, the (*R,S*)-diol to the monoacetate, and the (*S,S*)-diol did not react (Fig. 17).

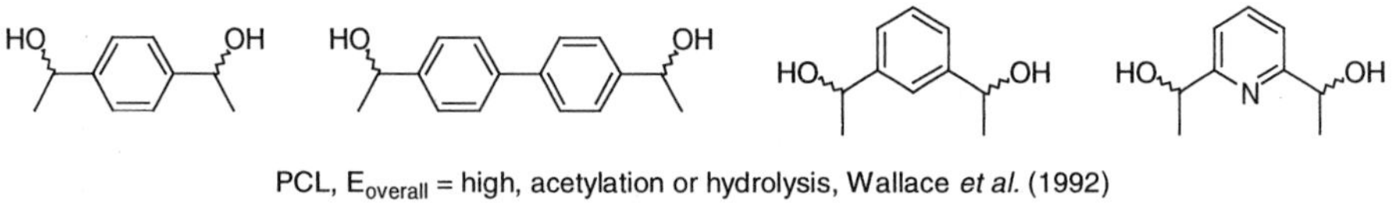

PCL, $E_{overall}$ = high, acetylation or hydrolysis, Wallace *et al.* (1992)

Fig. 17.   Enantioselective reactions separated diastereomers as well as enantiomers.

3.1.3   Asymmetric Syntheses

Lipase-catalyzed asymmetric syntheses start with *meso* compounds or prochiral compounds and yield chiral products in up to 100 % yield. In an asymmetric synthesis the enantiomeric purity of the product remains constant as the reaction proceeds and is given by ee = (E-1)/(E+1), where E is the enantiomeric ratio. For example, an enantioselectivity of 50 yields product with 96 %ee. Rearrangement of this equation gives

$E = (1 + ee)/(1 - ee)$, useful to calculate the enantioselectivity from the enantiomeric purity of the product.

   In practice, however, many lipase-catalyzed asymmetric syntheses undergo a subsequent reaction, a kinetic resolution (Fig. 18). For example, hydrolysis of a *meso* diester first gives the chiral monoester, but this monoester also reacts giving the *meso* diol. Although this overhydrolysis lowers the yield of the monoester, it usually favors the minor enantiomer and thus increases the enantiomeric purity of the monoester by kinetic resolution. For the PPL-catalyzed hydrolysis of 1,5-diacetoxy-*cis*-2,4-dimethylpentane, the enantiomeric ratio for the diacetate to monoacetate hydrolysis was 16 yielding an enantiomeric purity of 88 %ee (Wang et al., 1984). The subsequent kinetic resolution with an enantiomeric ratio of 5 increased the enantiomeric purity to 97 %ee, but lowered the yield of monoacetate to ~70 %. Quantitative analysis of the enantioselectivity in asymmetric syntheses is more difficult than for kinetic resolutions because three variables must be measured: the enantioselectivity of each step and the relative rate of each step. Wang et al. (1984) developed the necessary equations, but most researchers only report the enantiomeric purity and yield of the product. For this reason, we will also report only the enantiomeric purity and yield for asymmetric syntheses in this book.

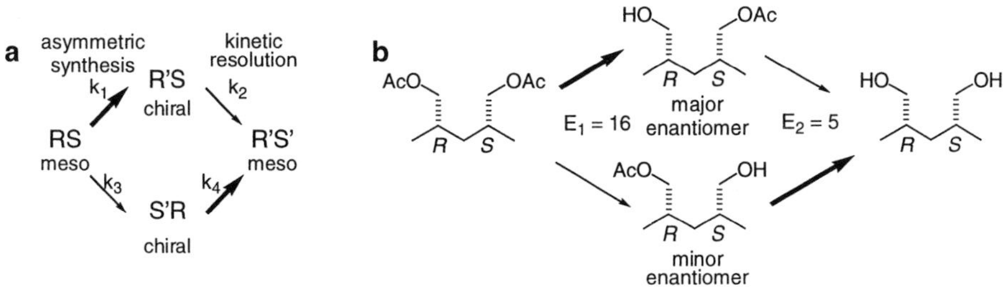

Fig. 18:   Asymmetric synthesis are usually coupled to kinetic resolutions. **a** Schematic diagram; **b** PPL-catalyzed hydrolysis of 1,5-diacetoxy-*cis*-2,4-dimethylpentane.

   Selected examples of lipase-catalyzed asymmetric syntheses are shown in Fig. 19; more are included in the survey of enantioselectivity in Sect. 5.1 and 5.2 and in an excellent review (Schoffers et al., 1996). The lipase-catalyzed asymmetric syntheses include a wide range of primary and secondary alcohols, as well as carboxylic acids. One example of the advantage of the combined asymmetric synthesis and kinetic resolution is the PCL-catalyzed acetylation of *cis*-2-cyclohexen-1,4-diol (Harris et al., 1991), a *meso*-secondary alcohol. Although the enantioselectivity for first acetylation (asymmetric synthesis) is only 4 and the enantioselectivity for the second acetylation (kinetic resolution) is only 10, the monoacetate was isolated in moderate yield (51 %) and high enantiomeric purity (95 %ee). Many of the primary alcohol examples are 2-substituted 1,3-propanediols, which are versatile synthetic starting materials. The dihydropyridine example below is a chiral acid, but the acetyloxymethyl group places the stereocenter in the alcohol part of the ester; thus, this prochiral compound can be classified as a chiral alcohol.

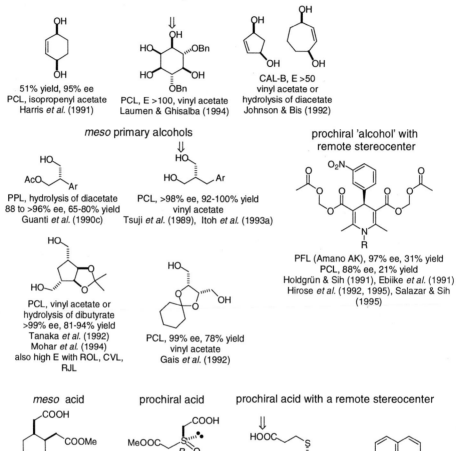

Fig. 19.   Examples of lipase catalyzed asymmetric syntheses.

4 Choosing Reaction Media: Water and Organic Solvents

4.1 Hydrolysis in Water

The simplest hydrolase-catalyzed reaction is the hydrolysis of substrates in water or biphasic mixtures of water and an organic solvent. Although proteases require a soluble substrate, lipases and esterases do not. A second phase is even desirable because it activates most lipases by 10 to 100-fold, probably due to lid opening as discussed above in Sect. 2.1.7. Liquid substrate can also act as the organic phase.

A general experimental procedure for a lipase-catalyzed reaction is as follows. Add 100 mg of liquid ester such as acetate or butyrate (or 1 mL of a solution of ester in a water-immiscible solvent such as toluene or ethyl ether) to 5 mL of 50 mM sodium phosphate buffer at pH 7. Monitor reaction by pH stat until reaction reaches ~30 % conversion. If a pH stat is not available, monitor by TLC or GC and use 100 mM buffer. Work up reaction, measure enantiomeric purity of both starting material and product, and calculate E using Eq. 3 in Sect. 3.1. *Preparative Biotransformations* (Roberts, 1992–1996) contains many detailed and tested procedures.

4.2 Transesterifications and Condensations in Organic Solvents

Researchers reported lipase-catalyzed esterifications in organic solvents containing approximately 10 % water more than 50 years ago (Sym, 1936; Sperry and Brand, 1941), but most of this work was forgotten. In 1976, Unilever researchers patented a process for cocoa butter equivalent using a lipase-catalyzed transesterification of lipids in hydrocarbon solvents (Coleman and Macrae, 1977; see Sect. 6.2.1.1). Klibanov's group discovered many other examples of enzyme-catalyzed reactions in organic solvents and further demonstrated that enzymes require only traces of water (Cambou and Klibanov, 1984; Zaks and Klibanov, 1984, 1985; for reviews see Klibanov, 1989, 1990; Koskinen and Klibanov, 1996). Klibanov's work convinced others that enzyme-catalyzed reactions are not only possible, but also sometimes more convenient, in organic solvents. Today, researchers report slightly more reactions in organic solvents than in water.

One advantage of reactions in organic solvents is the ability to do an esterification reaction instead of hydrolysis. Although lipases favor the same prochiral group in both cases, the two reactions yield opposite enantiomers. For example, acetylation of 2-benzyl glycerol with PCL yields the (S)-monoacetate, while hydrolysis of the diacetate with PPL yields the (R)-monoacetate. The lipases react at the *pro-R* position in both cases (Fig. 20).

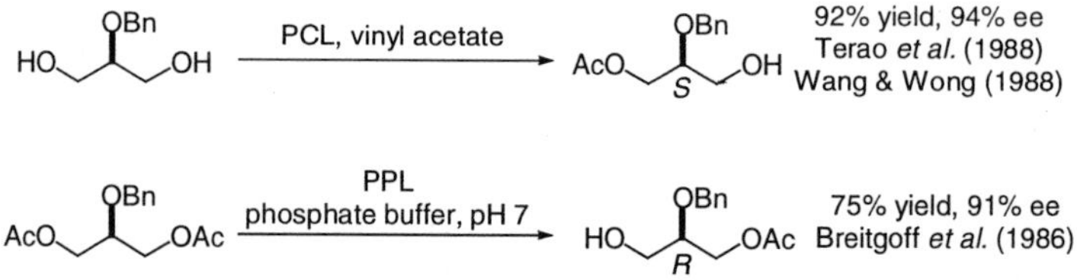

Fig. 20.   Although lipases favor the same prochiral group in both cases, acylation of the *meso* alcohol and hydrolysis of the *meso* ester yield opposite enantiomers.

Another advantage of organic solvents is the potential to change the selectivity of the enzyme in different solvents, sometimes called solvent (or medium) engineering. For example, the regioselectivity of the transesterification of a 2-octyl-1,4-dihydroxybenzene butyric acid ester reversed from favoring the 4-position in cyclohexane to the 1-position in acetonitrile (Rubio et al., 1991) (Fig. 21).

Rubio et al. (1991) rationalized the reversal in selectivity according to differences in substrate solvation. Cyclohexane solvates the octyl group well, thus, the ester at the less hindered 4-position reacts. Acetonitrile solvates the octyl group poorly, thus the substrate binds in a manner that places the octyl group within the hydrophobic lipase active site. For this reason, the ester at the 1-position now reacts more rapidly.

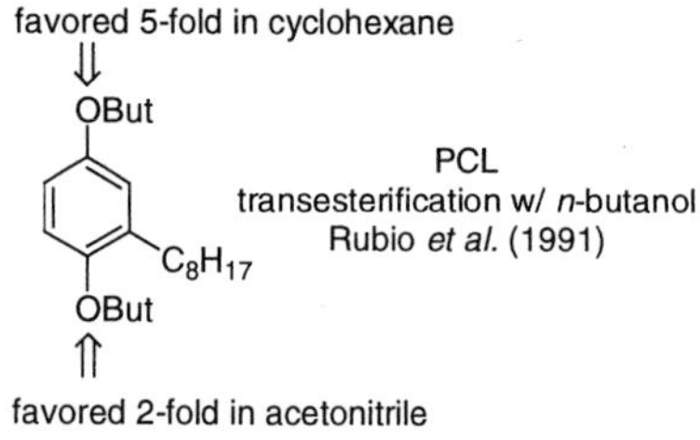

Fig. 21.   Regioselectivity changes in different solvents.

Consistent with this explanation, Halling (1990) found that although the observed reaction rate can vary in different solvents, the true specificity constants of the enzyme vary only slightly after correcting for the activity of the substrate in different solvents. Note that substrate solvation changes do not explain changes in *enantio*selectivity, since both enantiomers are solvated equally in achiral solvents. However, solvation of enzyme-enantiomer complexes may differ and these differences may account for changes in enantioselectivity (see Sect. 4.2.2).

Another reason to avoid water is to prevent decomposition of water-sensitive compounds such as organometallics or to simplify the workup of hydrophilic compounds which are difficult to recover from water. Some organic chemists favor reactions in organic solvents simply because they are less familiar with reactions in water.

4.2.1   Increasing the Catalytic Activity in Organic Solvents

Enzymes suspended in organic solvents are less active than enzymes dissolved in water. For crystalline subtilisin, Schmitke et al. (1996) and Klibanov (1997) attributed this drop in activity to approximately equal contributions from (1) changes in the pH optimum, (2) changes in substrate solvation, and (3) low thermodynamic activity of water. Noncrystalline subtilisin is even less active, probably due to denaturation during lyophilization. Noone has done such a careful study for lipases, but additional possibilities for lipases are (1) diffusional limitations (ability of the substrate to reach the active site) and (2) lid orientation. To minimize diffusional limitations, researchers disperse the lipase on supports with high surface area or modify the lipase so that it dissolves in organic solvents. To ensure an open orientation of the lid, researchers add lipids or surfactants to the lipase. Salts hydrates or other techniques can optimize the water activity (Sect. 4.2.4) and organic phase buffers (Blackwood et al., 1994) or solid $KHCO_3$ (Berger et al., 1990) can control the pH. Addition of water in a manner that slows agglomeration of the enzyme particles increased the rate of reaction by approximately a factor of 10 for CRL (Tsai and Dordick, 1996). Heating the reaction mixture also accelerates the reaction. Several groups reported that heating the reaction mixture with microwaves is more effective than simple heating (Carrillo-Munoz et al., 1996; Parker et al., 1996; Gelo-Pujic et al., 1996).

4.2.1.1   Choosing the Best Organic Solvent for High Activity

Finding the best organic solvent for a enzyme-catalyzed resolution is still a trial and error process, but nonpolar solvents are usually better than polar solvents. Laane (Laane et al., 1987; Laane, 1987) divided solvents into three groups according to their log$P$ value – the logarithm of their partition coefficient between n-octanol and water. Lipases, as well as other biocatalysts, showed low activity in solvents with log$P$ values less than 2, which includes polar solvents like methanol, acetone, pyridine and diethyl ether. Biocatalytic activity was difficult to predict for solvents of moderate polarity ($2 < $ log$P < 4$) such as n-octanol, toluene and n-hexane. Biocatalysts are usually active in nonpolar solvents (log$P > 4$) such as n-decane and diphenyl ether. Good solvents for lipase-catalyzed reactions include n-hexane, vinyl acetate, and toluene. Other researchers correlated activity with solvent parameters such as the dielectric constant or the dipole moment (Fitzpatrick and Klibanov, 1991; Bornscheuer et al., 1993), but log$P$ gave the best correlation.

Researchers believe that enzymes must retain an essential shell of water to remain active in organic solvents. Nonpolar solvents do not affect this shell of water, while polar solvents strip this essential shell thereby inactivating the biocatalyst. Adding water to the polar solvent helps to retain activity (see Sect. 4.2.4 on optimizing the amount of water in an organic solvent). Clumping or aggregation of the enzyme upon addition of water can lower the activity.

4.2.2 Increasing the Enantioselectivity in Organic Solvents

Lowering the temperatures sometimes increases enantioselectivity. For example, Sakai et al. (1997) increased the enantioselectivity of a PCL-catalyzed acylation in diethyl ether from $E = 17$ at 30 °C to $E = 84$–99 at –40 °C. On the other hand, Yasufuku and Ueji (1995, 1996, 1997) increased the temperature to increase enantioselectivity. The enantio-selectivity of a CRL-catalyzed reaction of 2-phenoxypropionic acid increased from $E = 5$ at 10 °C to $E = 33$ at 57 °C.

Changing from water to organic solvent often changes lipase enantioselectivity, as does changing from one organic solvent to another. Many researchers used this 'medium engi-neering' to optimize reactions in organic solvents. For example, Mori et al. (1987) re-ported that PPL showed no enantioselectivity in the hydrolysis of seudenol acetate, but Johnston et al. (1991) reported moderate enantioselectivity ($E = 17$) in the acetylation of seudenol with trifluoroethyl acetate in ethyl ether. The enantioselectivity of the CAL-B-catalyzed acetylation of seudenol with vinyl acetate varied from 8–32 depending on the solvent and the water content. The highest enantioselectivity was in dry benzene (Orrenius et al., 1995a). Occasionally, the enantioselectivity even reverses upon changing the solvent. For example, CRL esterified the (R)-enantiomer of a chiral acid, 2-phenoxy-propionic acid, with butanol in carbon tetrachloride ($E = 16$), but the (S)-enantiomer in acetone ($E = 1.6$) (Ueji et al., 1992). In another example, PCL acetylated the (R)-enanti-omer of a secondary alcohol, methyl 3-hydroxyoctanoate, with vinyl acetate in methylene chloride ($E = 5$), but the (S)-enantiomer in hexane ($E = 16$) (Bornscheuer et al., 1993). In the most dramatic example, Amano lipase AH hydrolyzed the *pro-R* ester of a dihydro-pyridine derivative in cyclohexane ($E\sim20$), but the *pro-S* ester in diisopropyl ether ($E > 100$) (Hirose et al., 1992).

Finding the molecular basis of the enantioselectivity changes is difficult because ener-gies involved are small. Rarely does the enantioselectivity change by more than a factor of ten, almost never by a factor of one hundred. A factor of ten corresponds to a $\Delta\Delta G^{\#}$ of 1.4 kcal/mol, a relatively weak interaction. Nevertheless, these enantioselectivity changes are often enough to tip the balance from a useless reaction to useful one. For this reason, many researchers have tried to understand why enantioselectivity changes in different solvents. They proposed at least four different explanations, but none can predict changes in enantioselectivity reliably (reviewed by Carrea et al., 1995).

First, Klibanov proposed that differences in the solvation of the enzyme-substrate tran-sition state complex determine the selectivity (Rubio et al., 1991; Ke et al., 1996). For example, they explained the variation in enantioselectivity for the lipase-catalyzed hy-drolysis of a prochiral diester (Fig. 22) (Terradas et al., 1993).

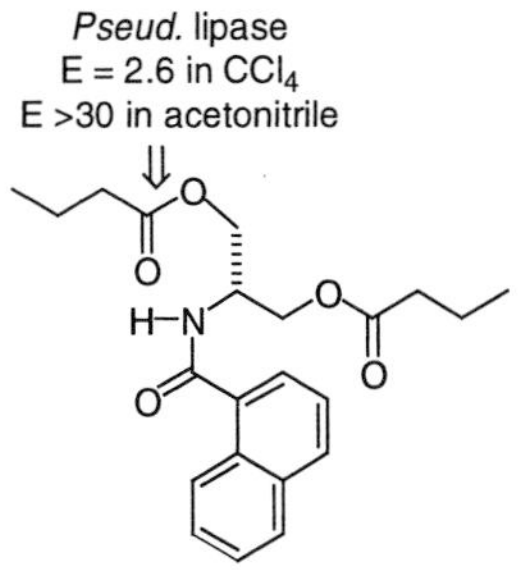

Fig. 22.   Enantioselectivity changes in different solvents.

They suggested that the substrate, especially the hydrophobic naphthyl group, binds tightly to the active site in polar solvents giving high enantioselectivity, but binds loosely, or not at all in nonpolar solvents giving low enantioselectvity. To estimate the differences in solvation the authors used the solvent $\log P$ and found a good correlation with enantioselectivity. Other researchers reported similar correlations (recent examples are given by Hof and Kellog, 1996a; Ema et al., 1996), but Parida and Dordick (1991) found only partial correlation in a CRL-catalyzed esterification of 2-hydroxy acids and, in this case, the enantioselectivity increased in less polar solvents. Carrea et al. (1995) found no correlation of enantioselectivity and $\log P$ in PCL-catalyzed acylation of several secondary alcohols. The lack of correlation suggests that $\log P$ maybe a poor measure of enzyme-substrate solvation in these cases. Recently, Ke et al. (1996) used a more sophisticated approach to estimate solvation of the enzyme-substrate complex. They estimated the portion of the substrate bound to the enzyme and calculated the activity coefficient of this fragment. This approach predicts the enantioselectivity of crystalline proteases in organic solvents, but not for amorphous proteases in organic solvents.

Second, researchers suggested that solvent changes the active site by binding in it or near it (Nakamura et al., 1991; Hirose et al., 1992; Secundo et al., 1992). X-ray structures of protease crystals soaked in solvents indeed showed solvent molecules (hexane or acetonitrile) bound to the active site (Fitzpatrick et al., 1993; Yennawar et al., 1995). Ottolina et al. (1994) found that the activities of lipases (PCL, CVL, PPL, CRL, RML) increased by as much as a factor of eight in (R)-carvone as compared to (S)-carvone. The enantiomeric solvents presumably form different complexes with the lipase. However, the enantioselectivity toward secondary alcohols did not differ in the two solvents. Arroyo and Sinisterra (1995) reported small changes in enantioselectivity of CAL-B toward a carboxylic acid (ketoprofen) in (R)- vs. (S)-carvone.

Third, van Tol et al. (1995a, b) suggested that the changes in enantioselectivity are due to a combination of errors in measuring enantioselectivity and changes in solvation of the substrate. The endpoint method for measuring enantioselectivity can give erroneous results when product inhibits the reaction (Sect. 3.1). After correcting for inhibition and the thermodynamic activity of the substrates, van Tol et al. (1995a, b) found that the enantioselectivity of a PPL-catalyzed acylation of glycidol with vinyl butyrate did not change in hexane, diisopropyl ether, tetrachloromethane, and 2-butanone (E = 5.5). Without these corrections, the apparent enantioselectivity varied between 20 and 2.7 in a single reaction.

Fourth, researchers correlated the increased enantioselectivity of subtilisin in different solvents with the increased flexibility of the active site on the nanosecond time scale (Broos et al., 1995b). Noone has yet made such a correlation for lipases.

The enantioselectivity of serine proteases such as subtilisin and chymotrypsin toward amino acid esters drops from 10^3–10^4 in water to less than 10 in organic solvents (Sakurai et al., 1988). Sometimes the enantioselectivity even inverts, but remains low (Broos et al., 1995a). This lowered enantioselectivity permits coupling of D-amino acids.

4.2.3   Acyl Donor for Acylation Reactions

The ideal acyl donor would be inexpensive, acylate quickly and irreversibly in the presence of lipase, and be completely unreactive in the absence of lipase. No acyl donor fulfills all three criteria. For transformations of inexpensive chemicals (for example, modified lipids, Sect. 6.2), cost is most important, so researchers use acids and simple esters (e.g., methyl, glyceryl). Acylations with these donors are often slow and reversible with an equilibrium constant near one. To drive reactions to completion, researchers removed the water or alcohol by evaporation (Björkling et al., 1989), azeotropic distillation (Bloomer et al., 1992), microwave heating (Carrillo-Munoz et al., 1996) or chemical drying agents such as molecular sieves or inorganic salts (Kvittingen et al., 1992). In other cases, crystallization of the product drives the reaction (McNeill et al., 1991; Cao et al., 1996, 1997).

For resolution reactions of fine chemicals, researchers use activated acyl donors (Fig. 23). Lipase-catalyzed acylations with these donors are one to two orders of magnitude faster than with acids or simple esters. In addition, activated acyl donors shift the equilibrium constant in favor of acylation.

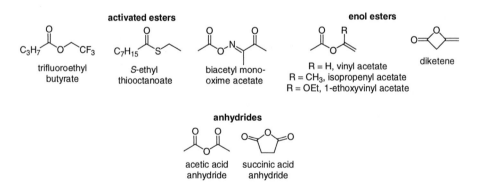

Fig. 23.   Examples of activated acyl donors for irreversible acylation of alcohols.

In the case of enol esters and acid anhydrides, acylation is practically irreversible. An irreversible reaction is important for kinetic resolution of alcohols because the reverse reaction degrades the enantiomeric purity of the remaining starting material thereby lowering the efficiency of the resolution (see Sect. 3.1).

Researchers first used activated esters where the alcohol is a better leaving group. For example, Stokes and Oehlschlager (1987) acylated sulcatol with trifluoroethyl laurate and recovered the unreacted alcohol in 97 %ee (Eq. 4).

However, for another secondary alcohol, de Amici et al. (1989) could not recover the unreacted starting material in high enantiomeric purity even though the transesterification was enantioselective. They attributed this difficulty to the reversibility even for this activated ester. The less expensive trichloroethyl esters are less convenient because the product trichloroethanol is difficult to remove (bp. 151 °C). The thioester S-ethyl thiooctanoate drives the reaction both because the thiol is a good leaving group and because the ethanethiol is easily removed by evaporation (Öhrner et al., 1992; Frykman et al., 1993), but working with volatile thiols requires extra care. The oxime esters react faster than simple esters and even enol esters (Ghogare and Kumar, 1989, 1990), but the non-volatile oxime may complicate separations. Several other leaving groups (not shown) are less useful: Cyanomethyl esters release the toxic formaldehyde cyanohydrin while 2-chloroethyl esters do not activate the ester enough.

The most useful activated acyl donors are enol esters, such as vinyl acetate or isopropenyl acetate. The product alcohol tautomerizes to a carbonyl compound, thereby driving the reaction and eliminating potential product inhibition. The first reports of lipase-catalyzed acylations with enol esters appeared in 1986–1987 (Sweers and Wong, 1986; Degueil-Castaing et al., 1987) and the first enantioselective acylations appeared in 1988 (Wang et al., 1988; Laumen et al., 1988; Terao et al., 1988; Wang and Wong, 1988). Hoechst AG patented the resolution of alcohols using vinyl esters in 1988. Since that time researchers have resolved hundreds of alcohols using this method. For example, Berkowitz et al. (1992) efficiently resolved glycals for the synthesis of artificial oligosaccharides (Eq. 5).

Most lipases except CRL and GCL tolerate the liberated acetaldehyde. Acetaldehyde slowly inactivates these two probably by formation of a Schiff base with Lys residues (Weber et al., 1995a). Acetone from isopropenyl acetate is less reactive and may not inactivate CRL or GCL, but this has not been investigated. Another alternative is 1-ethoxyvinyl acetate which liberates ethyl acetate (Kita et al., 1996; Schudok and Kretzschmar, 1997).

Although the acylation of alcohols by enol esters, such as vinyl acetate, is indeed irreversible, another equilibrium can cause reversibility and lower the enantiomeric purity of the remaining alcohol (Lundh et al., 1995). Since the reaction mixture contains small amounts of water, the lipase can catalyze hydrolysis of the product acetate ester to the alcohol plus acetic acid. Hydrolysis of the faster-reacting acetate lowers the enantiomeric purity of the remaining alcohol. To minimize this hydrolysis, Lundh et al. (1995) recommend dry conditions and an excess of vinyl acetate. In addition, stopping the reaction before 50 % conversion, separating the ester and alcohol and subjecting the alcohol to a second esterification will also minimize hydrolysis.

Acylation with diketene, a cyclic enol ester, is fast and has the advantage that it produces no by-products (Balkenhohl et al., 1993b; Jeromin and Welsch, 1995; Suginaka et al., 1996). However, the reported enantioselectivity was slightly lower than that for vinyl acetate, possibly due to nonenzymic acylation. For example, the acylation of α-phenylethanol with diketene showed an enantioselectivity of 12 to 80 (Eq. 6), while vinyl acetate showed an enantioselectivity > 100 with the same enzyme (Laumen et al., 1988; Nishio et al., 1989).

$$\text{(6)}$$

Acid anhydrides also irreversibly acylate alcohols (Bianchi et al., 1988c), but the release of carboxylic acid may decrease the enantioselectivity of the reaction. For example, CRL-catalyzed acetylation of a bicyclic secondary alcohol with acetic anhydride was moderately enantioselective (E = 20), but in the presence of solid potassium bicarbonate the enantioselectivity increased dramatically to E = 240 (Eq. 7) (Berger et al., 1990). In polar solvents, uncatalyzed acylation by acid anhydrides can lower the overall selectivity.

$$\text{(7)}$$

In some cases, acid anhydrides inactivate lipases probably by depleting essential water. Xu et al. (1995a, b) minimized this inactivation in a CRL-catalyzed resolution of menthol by controlling the rate of anhydride addition. Fast addition promoted hydrolysis of anhydride and decreased the available amount of water, while slow addition promoted reaction between free acid and alcohol which released water. Controlled addition of anhydride kept the water constant at 2–4 mM and the reactor was stable for over two months.

Terao et al. (1989) used succinic acid anhydride to both acylate an alcohol and to simplify the separation of unreacted alcohol and ester. Simple extraction separated the neutral alcohol from the charged succinate half ester.

For resolution of amines, researchers avoid chemical acylation by either using less reactive acyl donors (e.g., ethyl methoxyacetate (Balkenhohl et al., 1997)) or minimizing the contact time between acyl donor and amine (Gutman et al., 1992). An additional problem sometimes encountered with amines is that the resulting amide can be difficult to cleave. To solve this problem, Wong's group developed two acyl donors that yield readily cleaved carbamates or amides (Fig. 24) (Takayama et al., 1996; Orsat et al., 1996).

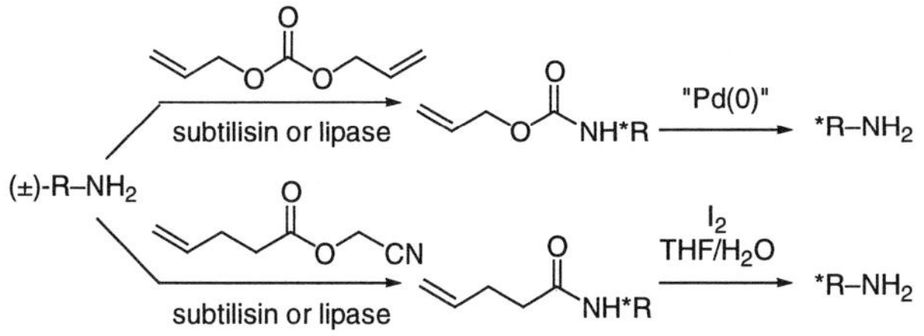

Fig. 24.   Special acyl donors that yield carbamates or amides that can be readily cleaved to the free amine.

Kanerva and Sundholm (1993b) compared the enantioselectivity of PCL-catalyzed acylation with butyric acid anhydride, vinyl butyrate, and trifluoroethyl butyrate. All showed similar enantioselectivity, but for one substrate the rate of reaction with trifluoroethyl butyrate was very slow. On the other hand, the enantioselectivity of a CAL-B-catalyzed acylation of several secondary alcohols was highest with 2-chloroethyl butyrate (Hoff et al., 1996). Vinyl butyrate, butyric acid anhydride, and 2,2,2-trichloroethyl butyrate showed 5 to 35 times lower enantioselectivity. In most cases, the key to high enantioselectivity is to avoid chemical acylation. Sometimes acyl donors with longer chains (butyrates and above) show higher enantioselectivity than those with shorter chains like acetate (for examples see Stokes and Oehlschlager, 1987; Sonnet, 1987; Holmberg et al., 1989a; Yamazaki and Hosono, 1990; Guo et al., 1990; Ema et al., 1996).

4.2.4   Water Content and Water Activity

The amount of water in the reaction mixture strongly influences the reaction rate, and to a lesser extent the enantioselectivity, of enzyme-catalyzed reactions in organic solvent. Polar solvents require typically 1–3 % added water for optimal activity, while nonpolar solvents require only 0.05–1 % added water. Enzymes require this minimum amount of water to maintain their structure and flexibility (Rupley et al., 1983; Affleck et al., 1992; Broos et al., 1995b).

Initially, researchers optimized the amount of water for a lipase-catalyzed reaction in organic solvent by measuring the total water content with Karl-Fischer titration. More recently, researchers found that the thermodynamic water activity (a_w) is a better measure of the amount of water, especially when comparing different reaction conditions (Halling, 1990, 1994, 1996; Bell et al., 1995). For example, the optimum reaction rate for RML in solvents ranging from 3-pentanone to hexane occurred at the same thermodynamic water activity, $a_w = 0.55$, but at widely differing total water content (Valivety et al., 1992a). Similarly, optimal activity of RML immobilized on different supports occurred at $a_w = 0.55$, but at different total water content (Oladepo et al., 1994, 1995). Polar solvents require more added water than nonpolar solvents because polar solvents competed more effectively with the lipase for the available water. Bell et al. (1995) suggested that water activity is like temperature, while water content is like heat content. Two systems may have the same water activity or temperature, but at the same time differ in the water content or heat content.

Of the available methods for controlling the thermodynamic water activity (Tab. 7), the simplest are equilibration of the reaction components with salt solutions of known a_w or pairs of salt hydrate. Equilibration occurs, albeit slower, even without direct contact between the reaction components and the salt solutions or salt hydrates. The pairs of salt hydrates (e.g., $CuSO_4*5\ H_2O$ / $CuSO_4*3\ H_2O$ gives $a_w = 0.32$ at 25 °C; $Na_4P_2O_7*10\ H_2O$ /$Na_4P_2O_7$ gives $a_w = 0.56$ at 35 °C) act as water buffers taking up or releasing water to the reaction components. Halling (1992) lists water activity values for 48 salt hydrate pairs.

Tab. 7.   Methods to Control Water Activity in Organic Media.

Method	Reference
Equilibrate with saturated salt solutions	Goderis et al. (1987); Bloomer et al. (1991); Adlercreutz (1991); Valivety et al. (1992a); Rosell et al. (1996)
Equilibrate with salt hydrates	Kvittingen et al. (1992); Halling (1992); Kim and Choi (1995)
Equilibrate with saturated salt solutions via silicone tubing	Wehtje et al. (1993, 1997)
Equilibrate with wet silica gel	Halling (1994)
Measure a_w with sensor and control drying of headspace	Khan et al. (1990)

It is more difficult to maintain constant water activity in reactions that consume or produce water (e.g., esterification between an acid and alcohol) because water must be removed or added during the reaction. Salt solutions or pairs of salt hydrates can be added directly to the reaction to buffer the water activity. Alternatively, preequilibrated silica particles of known a_w may be used (Halling, 1994). However, these approaches are not practical on a large scale due to cost, low water activity-buffering capacity, and difficulties in recovering the catalyst. One improvement is to add a silicon tube containing a saturated salt solution (e.g., $a_w = 0.75$ for water saturated with NaCl). The circulating salt solution can both take up and release water through the silicon (Wehtje et al., 1993, 1997). On an industrial scale, the best route maybe an a_w sensor (several are commercially available, Halling (1994)) combined with either drying by recirculation of the headspace gases through a drying column or water addition (Khan et al., 1990). These humidity sensors can also monitor the water activity continuously during reactions (Goldberg et al., 1988, 1990; Khan et al., 1990; Bornscheuer et al., 1993; Lamare and Legoy, 1995).

Lipases differ in the amount of water needed to maximize the rate of esterification between decanoic acid and dodecanol in hexane (Valivety et al., 1992b). RML and ROL were most active at low a_w (optimum: $0.32 < a_w < 0.55$), HLL, CRL, and PCL required a_w close to one. Sequence comparison of HLL and ROL suggested that changes in charged residues in the 'hinge and lid' region may be significant in low a_w tolerance. Different reactions may also have different optima. The overall rate of esterification between glycerol and oleic acid using RML did not change significantly with changes in water activity, but the synthesis of diolein from monoolein was fastest at $a_w = 0.5$ and triolein synthesis was fastest at low values of a_w (Dudal and Lortie, 1995). It may also be useful to change a_w as the reaction proceeds. For example, the initial phase of a PCL-catalyzed esterification of decanoic acid and dodecanol proceeds faster at high a_w, but at later stages a lower a_w is useful to obtain higher yields (Svensson et al., 1994).

Water activity also influenced the enantioselectivity of lipase-catalyzed reaction, but not in a consistent manner. For the resolution of 2-methyl alkanoic acids with CRL (Eq. 8), the enantioselectivity was higher at higher water activity (Högberg et al., 1993; Berglund et al., 1994). However, for a resolution of seudenol increasing water activity had different effects in different solvents (Orrenius et al., 1995a). In most cases, the enantioselectivity was higher at low water activity. For example, the enantioselectivity in hexane was $E = 20$ at $a_w < 0.11$, but dropped to less than $E = 10$ at $a_w > 0.75$. However, for dichloromethane and t-amyl alcohol the enantioselectivity was independent of water activity and for vinyl acetate and 3-pentanone the enantioselectivity was higher at high water activity. This variation reflects the fact that the effects of different solvents on enantioselectivity are still not well understood (see Sect. 4.2.2).

4.2.5 Synthesis of Amide Bonds Using Proteases and Amidases

Proteases and amidases catalyze both formation and hydrolysis of amide links. Although their natural role is hydrolysis, researchers also use proteases and amidases to form amide links. They use two different strategies – thermodynamic control or kinetic control (Fig. 25).

In thermodynamically-controlled syntheses, researchers change reaction conditions to shift the equilibrium toward synthesis instead of hydrolysis. Hydrolysis of peptides is favored by ~2.2 kcal/mol and is driven mainly by the favorable solvation of the carboxylate and ammonium ions. One common way to shift the equilibrium toward synthesis is to replace water with an organic solvent. The organic solvent suppresses the ionization of the starting materials and also reduces the concentration of water. Other common ways to shift the equilibrium are to increase the concentrations of the starting materials or to choose protective groups that promote precipitation of the product.

In kinetically controlled syntheses, researchers start with an activated carboxyl component, usually an ester. The ester reacts with the enzyme to form an acyl enzyme intermediate, which then reacts either with an amine to form the desired amide, or with water to form a carboxylic acid. Because the starting material is an activated carboxyl component, reactions are faster in the kinetically-controlled approach than in the thermodynamically-controlled approach. Because the kinetically-controlled approach requires an acyl enzyme intermediate, only serine hydrolases (e.g., subtilisin, lipases) are suitable. Metallo proteases such as thermolysin work only in thermodynamically-controlled syntheses. Kinetically-controlled syntheses are more common.

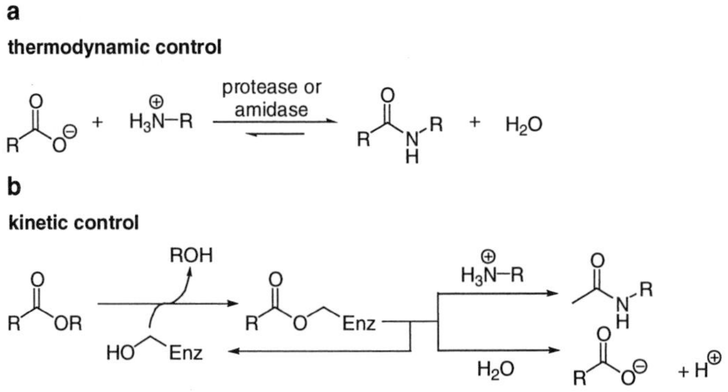

Fig. 25. Synthesis of amide bonds using proteases and amidases. **a** Thermodynamic control shifts the equilibrium toward synthesis by changing the reaction conditions. For example, researchers add organic solvents to reduce the concentration of water and suppress ionization of the starting materials. **b** Kinetic control starts with an activated carboxyl component (e.g., an ester) and forms an acyl enzyme intermediate. The acyl enzyme intermediate then reacts with an amine to form the amide. In a competing side reaction, water may react with the acyl enzyme intermediate. The kinetic control approach requires the formation of an acyl enzyme intermediate; thus, serine hydrolases and cysteine hydrolases are suitable, but not metallo proteases.

Protease-catalyzed peptide synthesis was first reported in 1901 (Savjalov, 1901). Until the end of the 1930's researchers believed that biosynthesis of proteins involved the reverse action of proteases. In the late 1970's synthetic chemists began to use proteases to simplify peptide synthesis and it continues to be an active area of research. Several excellent reviews on the synthesis of amide links using proteases and amidases are available (Kullmann, 1987; Schellenberger and Jakubke, 1991; Wong and Whitesides, 1994; Drauz and Waldmann, 1995).

The advantages of a hydrolase-catalyzed peptide synthesis over a chemical synthesis are mild conditions, no racemization, minimal need for protective groups, and high regio- and enantioselectivity.

The largest scale application (hundreds to thousands of tons) of protease-catalyzed peptide synthesis is the thermolysin catalyzed synthesis of aspartame, a low calorie sweetener (Fig. 26) (Isowa et al., 1979; Oyama, 1992). Precipitation of the product drives this thermodynamically-controlled synthesis. The high regioselectivity of thermolysin ensures that only the α-carboxyl group in aspartate reacts. Thus, there is no need to protect the β-carboxylate. The high enantioselectivity allows Tosoh to use racemic amino acids; only the L-enantiomer reacts.

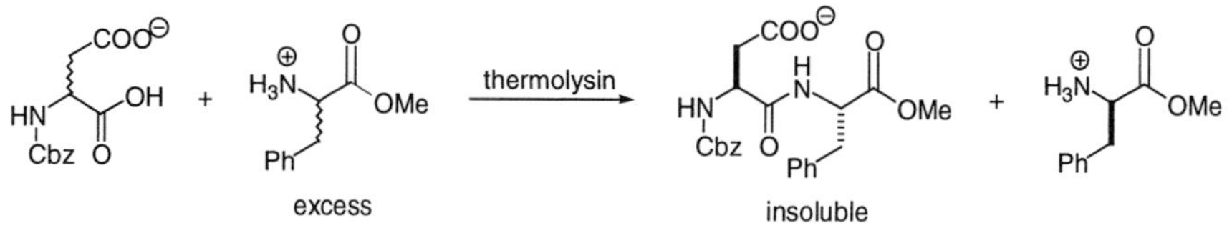

Fig. 26.   Commercial process for the production of aspartame (α-L-aspartyl-L-phenylalanine methyl ester) by Tosoh Corporation (Japan). Thermolysin catalyzes the coupling of an *N*-Cbz protected aspartic acid with phenylalanine methyl ester. The product forms an insoluble salt with excess phenylalanine methyl ester. This precipitation drives this thermodynamically-controlled peptide synthesis. The high regioselectivity of thermolysin for the α-carboxylate allows Tosoh to leave the β-carboxylate in aspartate unprotected. The enantioselectivity of thermolysin allows Tosoh to use racemic starting materials.

An example of a kinetically-controlled peptide synthesis is the α-CT-catalyzed production of kyotorphin (Tyr-Arg) (Fig. 27) (Fischer et al., 1994). To minimize the hydrolysis of the acyl enzyme intermediate, Fischer et al. (1994) used high concentrations of the nucleophile. The charged maleyl protective group increased the solubility of the carboxyl component.

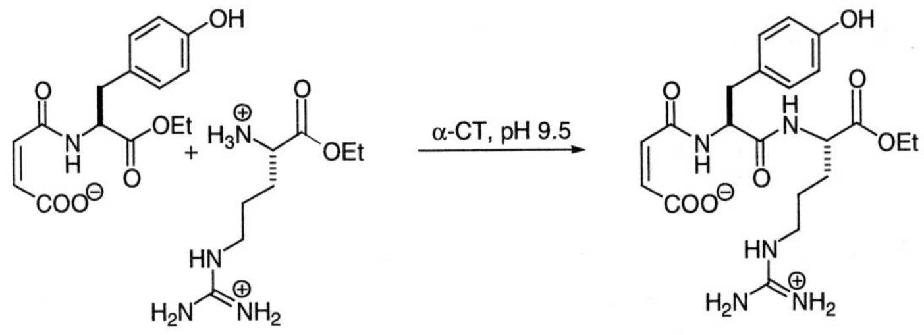

maleyl-Tyr-Arg-OEt

Fig. 27.   Large scale synthesis of a dipeptide, kyotorphin. α-CT catalyzes the coupling of the two amino acid derivatives in a concentrated aqueous solution (1.5 M in each). The N-maleyl protective group on the tyrosine moiety increases its solubility. The high concentration of nucleophile minimized the competing hydrolysis of the acyl enzyme intermediate and increased the yield. The remaining N-maleyl and –OEt protective groups were removed by a subsequent treatment with acid.

Subtilisin accepts a broader range of substrates than other proteases, so researchers usually use subtilisin for amide couplings involving unnatural substrates (Moree et al., 1997). When coupling a D-amino acid, it is best to use it as the nucleophile, not the carboxyl donor because subtilisin is more tolerant of changes in the nucleophile than in the carboxyl group.

Researchers also coupled larger peptides using proteases. For example, subtilisin cleaved a protein such as lysozyme and RNAse into several peptides (Vogel et al., 1996; Witte et al., 1997). Next, addition of an organic solvent shifted the equilibrium toward peptide synthesis and the same subtilisin now reassembled the protein. In the future, this condensation may create proteins containing unnatural amino acids or sugars. Manufacturers convert porcine insulin to human insulin by a similar process (Morihara and Oka, 1983).

Acylases and amidases also catalyze the formation of peptide bonds. For example, Justiz et al. (1997) coupled a side chain to 7-aminocephalosporanic acid (7-ACA) in the key step of an antibiotic synthesis (Fig. 28). They used the kinetically-controlled approach and obtained a 98 % yield in aqueous solution. Amino acid acylases are metallo proteases, so only the thermodynamically-controlled approach can be used.

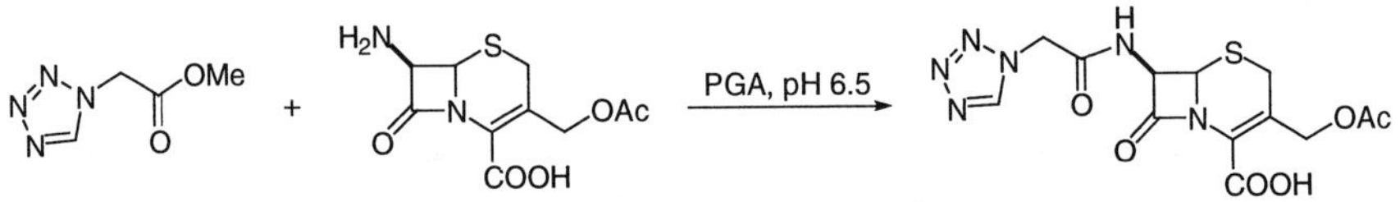

7-ACA   98%

Fig. 28.   The penicillin acylase (PGA)-catalyzed coupling of a side chain to a cephalosporin nucleus.

4.3   Other Reaction Media

4.3.1   Reverse Micelles

Reverse micelles are the simplest way to run enzyme-catalyzed reactions in almost pure organic solvent, but researchers rarely use reverse micelles for preparative reactions due to difficulties in workup related to the surfactant. Reverse micelles consist of a bulk organic phase containing aqueous droplets stabilized by surfactant. The biocatalyst remains soluble and active in the water, while substrates and products dissolve in the organic phase (Fig. 29). The amount of aqueous phase is small, so that e.g., lipases can catalyze transesterification and ester synthesis reactions under these conditions. A second advantage is a large interfacial area between the micelles and the organic phase, which eliminates mass transfer limitations. These advantages simplify the kinetic analysis of lipases (Han et al., 1987; Walde et al., 1993; Stamatis et al., 1995) and other enzymes (Bommarius et al., 1995). Reverse micelles are also transparent and therefore suitable for spectrophotometric studies. Holmberg (1994) and Ballesteros et al. (1995) reviewed enzymic reactions in microemulsions.

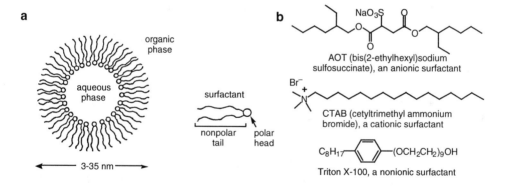

Fig. 29.   Reverse micelles or water in oil microemulsions. **a** Reverse micelle contain an inner aqueous phase stabilized by surfactant in a bulk organic phase. **b** Surfactants used to stabilize reverse micelles include anionic (e.g., AOT), cationic (e.g., CTAB) and nonionic (e.g., Triton X-100) surfactants.

Anionic surfactants, in particular AOT, are best for lipase-catalyzed reactions because nonionic surfactants can inhibit lipases and can also react in transesterification reactions when the surfactant contains a free hydroxyl group (Skagerlind et al., 1992). A cationic surfactant (CTAB) decreased the maximal rate of a ROL-catalyzed hydrolysis of triolein by a factor of 50 compared to AOT (Valis et al., 1992).

For preparative work, reverse micelles have several disadvantages. First, recovery of products from surfactant-containing organic solvent can be difficult. To simplify recovery, researchers added gelatin to the aqueous phase. Simple filtration recovers the lipase-containing aqueous phase. Several groups used CVL to make simple esters (Rees et al., 1991, 1993, 1995; Uemasu and Hinze, 1994; Backlund et al., 1995) and CVL or *Pseudomonas* sp. lipase (Genzyme) to resolve secondary alcohols (de Jesus et al., 1995). CRL was inactive under these conditions. Interestingly, CVL within the gel remained active at −20 °C (Rees et al., 1991). Another method to recover products from reverse micelles is to disrupt the emulsion with a temperature change into an oil-rich and a water-rich phase (Larsson et al., 1990).

Another disadvantage is that esterifications and transesterifications in reverse micelles often have lower yields than in other systems. For example, Borzeix et al. (1992) compared the RML-catalyzed synthesis of butyl butyrate in hexane, in a two-phase mixture of water-hexane, and in AOT-stabilized reverse micelles in hexane. The rate of ester synthesis was similar in all three systems, but the yield was significantly lower in the reverse micelles. Lipid modification in reverse micelles, especially synthesis of monoglycerides, also yielded less product than other reaction systems (Holmberg et al., 1989b; Hayes and Gulari, 1991; Chang et al., 1991; Singh et al., 1994a, b; Bornscheuer et al., 1994b). In addition, the surfactants used to stabilize reverse micelles can also denature lipases, but optimizing the water-to-surfactant ratio can minimize denaturation (Fletcher et al., 1985; Han and Rhee, 1986; Kim and Chung, 1989; Valis et al., 1992).

Although some enzymes show 'superactivity' and changes in selectivity in reverse micelles (Martinek et al., 1982), lipases show only small changes in selectivity. Bello et al. (1987) noted that CRL, which normally shows little fatty acid chain length selectivity, favored longer chain lengths in the transesterification of triglycerides in reverse micelles. Hedström et al. (1993) reported significantly increased enantioselectivity of the CRL-catalyzed esterification of ibuprofen in reverse micelles as compared to hexane ($E > 100$ vs. 3) (Eq. 9). Many other treatments and reaction conditions also increase the enantioselectivity of this reaction (see Sect. 5.2.2).

$$\text{(9)}$$

Lipase from *Penicillium simplicissimum* showed low selectivity (relative initial rates of 6–7) toward menthol enantiomers in reverse micelles (Stamatis et al., 1995).

4.3.2   Supercritical Fluids

As the temperature and pressure of a liquid are raised above the critical point, separate phases of liquid and gas disappear into a single phase called a supercritical fluid. Supercritical fluids have densities and dissolving powers near those of a liquid, but the viscosities near that of a gas. The advantages of supercritical fluids are rapid mass transfer due to the low viscosity, simple downstream processing by evaporation, the elimination of organic solvents, and the ability to change solvation properties by changing the pressure. The disadvantages of supercritical fluids are higher equipment costs and more com-plex reaction engineering.

Of the several possible supercritical fluids, most researchers use carbon dioxide be-cause it is nonflammable, non-toxic, cheap, and reaches the supercritical state at low temperature (31.1 °C). Moreover, its solvating properties are comparable to acetone. To dissolve polar substrates in supercritical carbon dioxide (SCCO$_2$), researchers either added a small amount of polar solvent such as dichloromethane, acetone, or t-butanol (Capewell et al., 1996) or they used techniques developed previously for organic sol-vents, e.g., complexation of fructose with phenyl boronic acid or immobilization of glyc-erol on silica gel (Castillo et al., 1994). Kamat et al. (1995) suggested that fluoroform may be a better supercritical fluid for enzyme-catalyzed reaction because carbon dioxide reacts with the lysine residues on an enzyme to make carbamates. In addition, supercriti-cal fluoroform is a better solvent than SCCO$_2$.

Nakamura et al. (1986) first showed that lipases remain active in supercritical fluids. ROL catalyzed the interesterification of triolein and stearic acid to 8 % conversion in SCCO$_2$. Since then researchers examined a wide range of reactions, especially lipid modifications (Tab. 8). The observed changes in conversion, enantioselectivity or lipase stability are similar to those in organic solvents. Several reviews of enzyme-catalyzed reactions in supercritical fluids have appeared (Hammond et al., 1985; Nakamura, 1990; Aaltonen and Rantakylae, 1991; Ballesteros et al., 1995).

Most reactions in Tab. 8 used immobilized RML in SCCO$_2$ at ~40 °C and ~150 bar in batch systems and the research focused on optimizing the activity and stability of the lipase. For example, Marty et al. (1992) varied the water content to maximize the rate of esterification of oleic acid with ethanol. The optimum water amount increased upon ad-dition of a small amount of ethanol to the reaction. As expected, Marty et al. (1992) observed no diffusion limitations, nor did Miller et al. (1990) in a similar reaction, but Bernard and Barth (1995) observed a partial diffusion-limitation. Conversion and resid-ual activity of PCL were improved by adding molecular sieves to the reaction (Capewell et al., 1996), maximum conversion were influenced by pressure and temperature (Nakamura et al., 1986; Chi et al., 1988), initial rates were twice higher in SCCO$_2$ com-pared to n-hexane, which was attributed in part due to different solubility of the sub-strates in the two solvents (Marty et al., 1990, 1992).

Most comparisons suggest that the enantioselectivity of lipases in SCCO$_2$ is similar or slightly lower than in organic solvents (Martins et al., 1992; Michor et al., 1996; Cape-well et al., 1996).

Pressure changes the solvating power of a supercritical fluid and several groups found that pressure changes the selectivity of a lipase. Ikushima et al. (1995) found that the enantioselectivity of a CRL-catalyzed acetylation of (±)-citronellol in SCCO$_2$ varied with pressure and suggested that pressure changes may change the conformation of the lipase. On the other hand, Rantakylae and Aaltonen (1994) found no changes in enantioselectiv-ity for the RML-catalyzed esterification of ibuprofen with n-propanol. Chaudhary et al. (1995) controlled the molecular weight of polyester formed in a PPL-catalyzed trans-esterification of 1,4-butanediol and bis(2,2,2-trichloroethyl)adipate by changing the pressure of supercritical fluoroform. As the pressure increased, supercritical fluoroform dissolved longer polymer chains and the molecular weight of the product increased.

Batch supercritical reactors allow analysis only at the end of the reaction. To monitor while the reaction is in progress, Marty et al. (1990, 1992) used a reactor with a sampling loop and a saphire window for visual monitoring. To avoid taking samples, Bornscheuer et al. (1996) monitored formation of acetaldehyde in the acylation of a 3-hydroxy ester with vinyl acetate through a high-pressure flow-through cell at 320 nm (Fig. 30). The online data agreed with off-line values up to 60 % conversion.

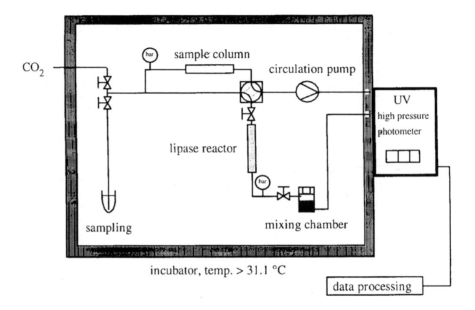

Fig. 30.   Schematic diagram of a supercritical CO_2 reactor with a high pressure flow through cell for online measurement of the formation of acetaldehyde (Bornscheuer et al., 1996).

Tab. 8.   Examples of Lipase-Catalyzed Reactions in Supercritical Carbon Dioxide.

Lipase	Reaction	Process conditions[a]	References
ROL	triolein/stearic acid	B, 35–50°C, 150 bar	Chi et al. (1988); Nakamura et al. (1986)
RML	ethyl acetate/iso-amyl alcohol	C, 35–80°C, 80–140 bar, IPI	Doddema (1990); Janssens et al. (1992)
ROL	trilaurin/palmitic acid	B, 40°C, 90–290 bar	Erickson et al. (1990)
RML	oleic acid/ethanol	C, 33–50°C,110–170 bar,	Marty et al. (1990, 1992)
RML	myristic acid/ethanol	B, 50°C, 150 bar	Dumont et al. (1992)
ROL	trilaurin/myristic acid	S–C, 35°C, 79–107 bar	Miller et al. (1990)
RML	trilaurin/oleic acid methylester	C, 40°C, 100 bar, IPI	Adshiri et al. (1992)
RML	myristic acid/ethanol	B, 50°C, 125 bar	Bernard and Barth (1995)
RML, ANL	oleic acid/oleyl alcohol	B, 40°C, 84–167 bar	Knez and Habulin (1992)
RML	myristic acid/ethanol	B, 50°C, 125 bar	Bernard et al. (1992)
PPL	butyric acid/glycidol	B, 35°C, 140 bar	Martins et al. (1992)
RML	ethyl acetate/nonanol	C, 60°C, 125–200 bar	Vermuë et al. (1992)
several	methacrylate/2-ethyl hexanol	B, 40–50°C, 107 bar[b]	Kamat et al. (1992, 1993)
RML	propyl acetate/geraniol	B, 40°C, 140 bar	Chulalaksananukul et al. (1993)
RML	e.g., fructose-PBA/oleic acid	B, 40°C, 150 bar	Castillo et al. (1994)
PPL	adipate/1,4-butanediol	B, 50°C, 60–340 bar[c]	Chaudhary et al. (1995)
CRL	oleic acid/citronellol	B, 31–40°C, 76–193 bar	Ikushima et al. (1993)
PCL	3-hydroxyester/vinyl acetate	B, 36–70°C, 100–170 bar	Bornscheuer et al. (1996); Capewell et al. (1996)
PCL	2°-alcohols/acylation	B, 40°C, 200 bar	Cernia et al. (1994a, b)
CRL, PME	menthol or citronellol /IA, TA	B, 35–70°C, 150 bar	Michor et al. (1996)
RML	ibuprofen	B, 50°C, 100–150 bar	Rantakylae and Aaltonen (1994)
CAL-B	randomization of e.g., palm olein	B, 65°C, 200–345 bar	Jackson et al. (1997)
CAL-B	soybean oil/glycerol or MeOH	B, 40–70°C, 200–345 bar	Jackson and King (1997)

[a] B: batch reactor; C: continuous reactor; S-C: semi-continuous reactor; IPI: integrated product isolation; PBA: phenyl boronic acid; IA: isopropenyl acetate; TA: triacetylglycerol. [b] Ethane, ethylene, fluoroform, propane, sulfur hexafluoride also. [c] Fluoroform also.

4.4   Useful Techniques

4.4.1   Assays for Hydrolase Activity

4.4.1.1 Lipase Assays

One unit of lipase activity is defined as the amount of enzyme liberating 1 μmol fatty acid per min under assay conditions. Common methods for the determination of lipase activity are either based on a titrimetric (e.g., pH-stat) or spectrophotometrical (e.g., hydrolysis of *p*-nitrophenyl ester) assay. However, no standard protocol has defined so far and as a consequence each research group or supplier of lipases uses a different procedure. A number of lipase assays including detailed protocols are described by Vorderwülbecke et al. (1992), who compared the activity of a wide range of commercial lipases. Care should be taken when comparing lipases from commercial suppliers, because most enzymes are sold as crude preparations, which might contain several hydrolytic enzymes including isozymes of lipases. In addition, different protein content, additives or carriers for immobilization affect specific activity, which is usually expressed as U/mg protein (Vorderwülbecke et al., 1992; Soumanou et al., 1997). Representative procedures for the determination of lipase activity are:

pH-Stat assay: Triglyceride (5 % v/v) and gum arabic (2 % w/v) (or other appropriate detergents) are emulsified with water using an ultraturrax for 3 min. 20 ml of this emulsion are added to the reaction chamber of the pH-stat equipment, thermostated to 37 °C and pH is adjusted to pH 7.0. Then a known amount of lipase (solid or dissolved in buffer) is added to the vigorously stirred emulsion and pH is kept constant by continuous addition of NaOH solution. Specific activity is then calculated from the initial rate of NaOH consumption. If no pH-stat is available, one can perform an end-point titration by addition of a stop solution (EtOH:acetone, 1:1) after 10–20 min followed by titration of the total amount of released fatty acids with NaOH. However, this method is less accurate due to a significant pH-drop during hydrolysis causing enzyme inactivation and inhibition of lipase by the fatty acids released.

Spectrophotometric assay: Common substrates are *p*-nitrophenyl esters such as *p*-nitrophenyl palmitate (pNPP) or acetate (pNPA). 100 μl pNPP (10 mM dissolved in DMSO or isopropanol) solution and 900 μl lipase solution are mixed in a cuvette and the amount of *p*-nitrophenol released is determined by recording the absorbance at 410 nm at 25 °C using a spectrophotometer. One unit is defined as the amount of lipase releasing 1 μmol *p*-nitrophenol per min under assay conditions.

More sophisticated assays take into account the phenomenon of interfacial activation (see Sect. 2.1.7) by e.g., monolayer techniques using Langmuir-Blodget film balance, however these methods require pure enzyme and are not of practical importance. Other assays allow the determination of the regioselectivity of lipases (Jaeger et al., 1994;

Farias et al., 1997) by using artificial triglycerides or the direct measurement of enantio-selectivity (Janes and Kazlauskas, 1997b; Janes et al., 1998) (see Sect. 3.1.1).

Due to the increased use of lipases in organic synthesis, also assays based on (trans)esterification are employed, because they also cover stability and activity of lipases in organic solvents. For instance, Novo (Denmark) determines activity of RML by measuring the amount of decanoic acid incorporated into sn-1 and sn-3 positions of triglycerides of high oleic sunflower oil. Boehringer characterizes PCL by acylation of α-phenylethanol with vinyl acetate in n-hexane followed by chiral analysis of the reaction products.

Zymogram: This activity staining allows identification, calculation of molecular weight and purity of lipases and esterases on polyacrylamide gels. Proteins are first denatured with sodium dodecyl sulfate (SDS) at 90 °C to obtain monomeric proteins, followed by separation using polyacrylamide gel electrophoresis. SDS is removed by incubation of the gel in a detergent solution (e.g., 0.5 % w/v Triton-X-100 in Tris/HCl buffer (100 mM, pH 7.5)) and then stained by adding a 1:1 mixture of solutions A and B (see below). The gel is then incubated until a red-colored band indicates proteins with hydrolytic activity due to formation of a complex between α-naphthol and Fast Red. Solution A: 20 mg α-naphhtyl acetate (dissolved in 5 ml acetone and 45 ml Tris/HCl buffer (100 mM, pH 7.5)), solution B: 50 mg Fast Red (Sigma) in 50 ml of the same buffer. Both solutions must be prepared directly prior to use.

Fig. 31 shows an SDS-PAGE of samples taken during cultivation of *E. coli* harboring the gene encoding esterase from *Pseudomonas fluorescens* (PFE). Inductions was performed by adding L-rhamnose. Proteins were stained with coomassie brillant blue (left gel) or solutions A and B (right gel).

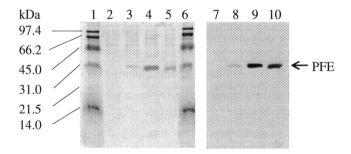

Fig. 31.   Separation of proteins (*left*) and identification of the esterase by activity staining with α-naphthyl acetate and Fast Red (*right*). Lanes 1 and 6: low molecular weight standard, lanes 2 and 7: before induction, lanes 3 and 8: 1 h after induction, lanes 4 and 9: 3.5 h after induction, lanes 5 and 10: crude cell extract after cultivation (Krebsfänger et al., 1998).

Screening of new microbial lipases or esterases can be performed by an agar plate assay. Tributyrin (1 % v/v) is added to the agar nutrient media before sterilization and mixed with an Ultraturrax. Lipase or esterase producing microorganims are identified by the formation of clear zones surrounding the colonies caused by release of free fatty acids. The assay becomes more sensitive by addition of rhodamin B creating a fluorescent complex.

4.4.1.2   Esterase Assays

In principle, esterase activity is determined using the same assays as described above for lipases. However, substrates are short-chain triglycerides (e.g., triacetin, tributyrin), p-nitrophenyl acetate or simple aliphatic esters (e.g., ethyl acetate). To distinguish between lipases and esterases, one uses long-chain triglycerides or pNPP, which are not hydrolyzed by esterases.

4.4.1.3   Protease Assays

The determination of protease activity is usually performed using p-nitrophenyl derivatives of amino acids. Representative chromogenic substrates are shown in Fig. 32 for trypsin, elastase, chymotrypsin and subtilisin. A detailed procedure for assaying chymotrypsin or subtilisin activity is:

100 µl N-Succinyl-L-Ala-Ala-Pro-Phe-p-nitrophenyl anilide (0.5 mM) dissolved in HEPES buffer (5 mM, pH 8.0) containing 9 % DMF and 900 µl protease solution are mixed in a cuvette and the amount of p-nitrophenyl anilide released is determined by recording the absorbance at 410 nm at 25 °C using a spectrophotometer. One unit is defined as the amount of protease releasing 1 µmol p-nitrophenyl anilide per min under assay conditions (Graham et al., 1993).

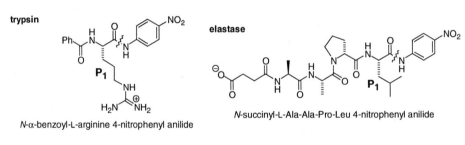

Fig. 32.   A chromogenic substrates often used to assay subtilisin or chymotrypsin. A hydrophobic residue, Phe, lies at the P_1 position. The cleaved position is indicated by a wavy bond.

4.4.2 Immobilization

4.4.2.1 Increasing the Surface Area to Increase Catalytic Activity

Enzyme powders are insoluble in organic solvents and can be recovered by simple filtration at the end of the reaction. Unfortunately, even after optimizing the solvent and the water content, catalysis is often thousands of times slower than in water or water-organic solvent mixtures. One reason for the drop in activity is diffusional limitations, that is, the substrate cannot reach the enzyme molecules in the center of the particle. Another reason for the drop in activity is denaturation of the biocatalyst during lyophilization. The simplest solution to both these problems is adsorbing the enzyme on an insoluble support such as Celite. Adsorption both increases the surface area and also avoids lyophilization of the biocatalyst. Most examples given below deal with the immobilization of lipases. However, the principles and most findings also hold true for other hydrolases.

Adsorption and entrapment. In a typical procedure, PCL (0.4 g) was dissolved in buffer (15 mL), mixed with insoluble support (4.0 g) and dried at room temperature (Bianchi et al., 1988c; Inagaki et al., 1992; Kanerva and Sundholm, 1993a). Reactions catalyzed by PPL absorbed on Celite were seven to twenty times faster than for crude PPL (Banfi et al., 1995). Sugars added to the buffer further increased the activity of immobilized PCL by a factor of 2–3. Dabulis and Klibanov (1993) also found similar rate increases for PCL, while Sanchez-Montero et al. (1991) found that the rate of heptyl oleate formation catalyzed by CRL depended on the type of carbohydrate. Rates increased upon adding lactose, but decreased upon adding fructose, glucose, sucrose or sorbitol. Sanchez-Montero et al. (1991) suggested these differences may be due to the ability of sugars to change the activity of water. Indeed, many commercial samples of lipases contain large amounts of inert materials such as sugars or Celite (sometimes > 95 wt %), so the available surface area is already large. CRL adsorbed on Celite showed increased stability toward acetaldehyde, a product of esterifications with vinyl esters (Kaga et al., 1994). Adsorption of PCL on an acrylic resin (Amberlite XAD-8) also increased the catalytic activity > 200-fold (Hsu et al., 1990). Several adsorption-immobilized lipases are commercially-available from Boehringer Mannheim: CAL-B immobilized by adsorption on macroporous acrylic resin and RML immobilized by adsorption on microporous phenolic anion exchange resin. Adsorption-immobilized lipases may desorb from the support in water; thus, covalently immobilized lipases (see below) should be used in water.

Because lipases often show higher catalytic activity in the presence of insoluble organic substrates, several groups have adsorbed or entrapped lipases in hydrophobic matrices. For example, hydrolysis of mixtures of tetramethoxysilane and alkyltrimethoxysilanes, $RSi(OCH_3)_3$, in the presence of lipases forms sol gel-entrapped lipases. Lipases entrapped in hydrophobic sol-gels also showed up to 100-fold increased activity in organic solvents (Reetz et al., 1995, 1996a, b; Reetz, 1997). Enantioselectivities remained unchanged in most cases. The sol-gel entrapped lipases are also easily recovered and reused with no loss in activity. Fluka awarded the Reagent-of-the-year-1997 prize to Manfred Reetz for his discovery of the sol-gel immobilized lipases. Other workers found smaller improvements in similar systems (Sato et al., 1994; Kawakami and Yoshida, 1995). Sol-gel immobilized lipases also work in aqueous solutions (Reetz et al., 1995). The mechanism for activation is probably lid opening as suggested for the lipid- or surfactant-coated lipases (Sect. 4.4.2.2).

Covalent Immobilization. Covalent immobilization creates a more stable link between the enzyme and the support, but requires more effort than adsorption (for reviews see Balcao et al., 1996; Akita, 1996). The most common method is the cross-linking of adsorbed biocatalysts with glutaraldehyde. Boehringer Mannheim sells several lipases immobilized by this method. Other examples include linking to an epoxy-containing resin (Berger and Faber, 1991), to polystyrene via a cysteinyl-*S*-ethyl spacer (Stranix and Darling, 1995) or via a poly(ethylene glycol) linker (Ampon et al., 1994), or entrapping the lipase in urethane prepolymers (Koshiro et al., 1985). Covalent immobilization often increases the thermal or operational stability of the enzyme, but does not activate it.

Cross-Linked Enzyme Crystals – CLEC's. Cross-linked enzyme crystals are crystals (typically 1–100 μm diameter) of pure enzyme cross-linked with glutaraldehyde (St.Clair and Navia, 1992; Margolin, 1996). Two lipase CLECs are commercially-available from Altus Biologics Inc. (Cambridge, MA, USA): lipase from *Candida rugosa* (CLEC-CR) and *Pseudomonas cepacia* (CLEC-PC) (Lalonde, 1995; Lalonde et al., 1995). Lipase CLEC remain active and insoluble in organic solvents and, unlike adsorbed lipases, in water-organic solvent mixtures. In addition, CLEC lipases are more stable to high temperatures and retain their activity over many cycles of use. For example, Lalonde et al. (1995) reused CLEC-CR eighteen times (recovering 20 % of the activity), whereas crude CRL lost virtually all activity after one cycle. Mechanical losses and lipase inactivation contributed approximately equally to the loss in CLEC-CR activity. CLEC-CR also showed improved enantioselectivity toward 2-arylpropionic acids; for ketoprofen the enantioselectivity increased from 5.2 to 64 (Lalonde et al., 1995; Persichetti et al., 1996). Lalonde et al. (1995) attributed the increase to the removal of a nonselective esterase during purification, but a conformational change may also contribute. CLEC's prepared from different conformations of CRL (open vs. closed) also differed in their enantioselectivity. Coating the crystals with surfactants increased the activity of the lipase in organic solvents (Khalaf et al., 1996). After taking into account the protein content, the activity was 2–90 times greater than crude lipase. The surfactant may maintain the water balance or it may facilitate transfer of hydrophobic substrates through the tightly bound layer of water. Two protease CLEC's are also avaibable from Altus Biologics Inc.: CLEC-BL, derived from subtilisin A and CLEC-TR, derived from thermolysin.

Covalently-Modified Lipases Soluble in Organic Solvents. Another way to increase the activity in organic solvents is to modify the lipase so that it dissolves in the organic solvent. Kikkawa et al. (1989) coupled polyethylene glycol (PEG) to the free amino groups of PCL and lipase from *Pseudomonas fragi*. The modified lipases were soluble in benzene, toluene and chlorinated hydrocarbons and catalyzed the formation of lactones from ethyl 16-hydroxyhexadecanoate and the resolution of 2-phenylethanol. Addition of hexane precipitated the PEG-lipase thereby simplifying recovery (Kodera et al., 1994). Using the same principle, Kodera et al. (1994) coupled PFL to a comb-shaped polymer yielding a more stable and more active lipase. PEG-modified CRL was significantly more stable in organic solvents than unmodified CRL (Basri et al., 1995; Hernáiz et al., 1996).

4.4.2.2   Lipid- or Surfactant-Coated Lipases

One of the simplest ways to increase the activity of a lipase in organic solvent is to coat the lipase with a lipid or surfactant before lyophilization (Tab. 9; Fig. 33). For example, a water-insoluble complex containing approximately 10 wt % protein formed upon mixing aqueous solutions of ROL and a nonionic amphiphile (didodecyl N-D-glucono-L-glutamate) (Okahata and Ijiro, 1988, 1992). Researchers estimated that 150 ± 30 amphiphile molecules surrounded each lipase molecule. The modified lipase was soluble in most organic solvents and was > 100 times more active than suspended enzyme in the synthesis of di- and triglycerides from lauric acid and monolaurin. Other researchers reported similar increases in reaction rate with other lipases and other lipids or surfactants.

Adding a hydrophobic interface before lyophilization may prevent the denaturation which often accompanies lyophilization. Further, the hydrophobic interface may open the lid of the lipase (Mingarro et al., 1995, 1996). The open form of lipases is more active due to a more accessible active site and the more-suitable orientation of the oxyanion-stabilizing residues. The open form is then 'frozen' removing the water by lyophilization or precipitation. Consistent with this explanation, surfactant treatment increased the catalytic activity of lipases RML, GCL, and CRL, all of which contain lids, but not cutinase, which lacks a lid. Surfactants also activated cross-linked crystals of lipases (Khalaf et al., 1996), but the mechanism is unclear since the cross-linking presumably prevents significant conformational changes.

Other workers used imprinting to increase enantioselectivity as well as reaction rate. Lyophilization of CRL with (R)-α-phenylethanol followed by lipid coating increased the enantioselectivity from 5.5 to 77. Heating or storage for a few days abolished the increased enantioselectivity (Okahata et al., 1995b).

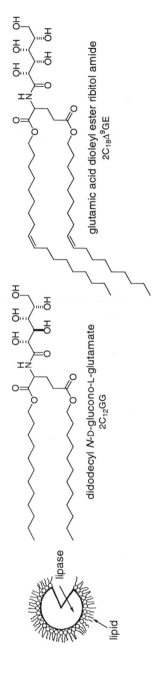

didodecyl *N*-D-glucono-L-glutamate
2C₁₂GG

glutamic acid dioleyl ester ribitol amide
2C₁₈Δ⁹GE

Fig. 33.   Schematic of lipid-coated lipase and structure of lipids used for the coating.

Tab. 9.   Lipid- or Surfactant-Coated Lipases.

Coating	Reaction	Lipase	Reference
2C$_{12}$GG	synthesis of triglycerides	ROL, P. fragiL, PCL	Okahata and Ijiro (1988, 1992)
2C$_{12}$GG	hydrolysis of triglycerides	P. fragiL, PCL	Tsuzuki et al. (1991, 1995)
Dialkyl-ether PL	resolution of diltiazem precursor	PCL	Akita et al. (1991)
Synthetic PL	resolution of α-hydroxy ester	CRL	Akita et al. (1993)
Linoleic or oleic acids	synthesis of benzyl alcohol ester	PCL	Balkenhohl et al. (1993a)
2C$_{18}$Δ^{9}GE	synthesis of benzyl alcohol ester	PCL, ROL, CRL	Goto et al. (1994)
2C$_{18}$Δ^{9}GE	tripalmitin/oleic acid interesterific.	PCL, ROL, CRL	Goto et al. (1995)
2C$_{18}$Δ^{9}GE	resolution of menthol	CRL	Kamiya et al. (1995)
2C$_{12}$GG	resolution of α-phenylethanol	CRL, P. fragiL	Okahata et al. (1995a, b)
Sorbitan monostearate	tripalmitin/stearic acid interesterific.	ROL	Basheer et al. (1995a, b)
Sorbitan monostearate	tripalmitin/stearic acid interesterific.	PCL	Isono et al. (1995)
n-Octyl-β-D-glucose[a]	tripalmitin/oleic acid interesterific.	PPL, CRL, ROL, ANL, PCL	Mingarro et al. (1995, 1996)

[a] Researchers washed the lipase-surfactant complex with anhydrous benzene/ethanol (90:10) to remove the surfactant.

5 Survey of Enantioselective Lipase-Catalyzed Reactions

5.1 Alcohols

A number of recent reviews also include surveys of lipase enantioselectivity (Chen and Sih, 1989; Sih and Wu, 1989; Klibanov, 1990; Xie, 1991; Kazlauskas et al., 1991; Boland et al., 1991; Faber and Riva, 1992; Santaniello et al., 1992; Margolin, 1993a; Gais and Elferink, 1995; Theil, 1995; Mori, 1995; Schoffers et al., 1996). The survey below includes only representative examples to give the reader a feel for the type and range of molecules that undergo enantioselective reactions with lipases.

5.1.1 Secondary Alcohols

5.1.1.1 Overview and Models

Although lipases show high enantioselectivity toward a wide range of substrates, the most common substrates are secondary alcohols and their derivatives. Researchers have resolved hundreds of secondary alcohols using lipases. Selected examples, including asymmetric syntheses of secondary alcohols, are collected below.

Based on the observed enantioselectivity of lipases, researchers proposed a rule to predict which enantiomer reacts faster in lipase-catalyzed reactions (Fig. 34; Tab. 10). This rule is based on the size of the substituents and suggest that lipases distinguish between enantiomeric secondary alcohols primarily by comparing the sizes of the two substituents. Indeed, a number of researchers increased the enantioselectivity of lipase-catalyzed reactions by modifying the substrate to increase the size of the large substituent (for examples see Scilimati et al., 1988; Johnson et al., 1991; Kazlauskas et al., 1991; Kim and Choi, 1992; Gupta and Kazlauskas, 1993; Adam et al., 1997b; Rotticci et al., 1997). Similarly, Shimizu et al. (1992a) reversed the enantioselectivity by converting the medium substituent into the large one.

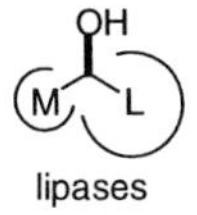

lipases

Fig. 34. An empirical rule to predict which enantiomer of a secondary alcohol reacts faster in lipase-catalyzed reactions. M, medium-sized substituent, e.g., methyl. L, large substituent, e.g., phenyl. In acylation reactions, the enantiomer shown reacts faster; in hydrolysis reactions, the ester of the enantiomer shown reacts faster.

To add more detail to this model, many groups tried to more precisely define the size limits of the medium and large substituents (PFL: Burgess and Jennings, 1991; Naemura et al., 1994, 1995; lipase QL: Naemura et al., 1996; PCL: Theil et al., 1995; Lemke et al., 1997), while other have tried to include electronic effects (PCL: Hönig et al., 1994).

Tab. 10.   Sized-Based Rules Similar to Those in Fig. 34 Proposed for Different Lipases.

Lipase	Reference	Comments
CAL-B	Orrenius et al. (1995b)	8 substrates
CRL	Kazlauskas et al. (1991)	86 substrates; reliable for cyclic, but not acyclic, substrates
PAL	Kim and Cho (1992)	28 substrates
PCL	Laumen (1987)	tried to also include primary alcohols and acids
PCL	Xie et al. (1990)	6 substrates
PCL	Kazlauskas et al. (1991)	64 substrates
PFL	Burgess and Jennings (1991)	31 substrates
PFL	Naemura et al. (1993b, 1995)	27 substrates
PPL	Janssen et al. (1991b)	23 substrates
PPL	Lutz et al. (1992)	21 substrates
RML	Roberts (1989)	6 substrates
CE	Kazlauskas et al. (1991)	15 substrates
lipase QL	Naemura et al. (1996)	27 substrates

Although steric effects are the most important determinant of lipase enantiopreference, electronic effects also contribute. For example, the CAL-B shows high enantioselectivity toward 3-nonanol (E > 300), but low enantioselectivity toward 1-bromo-2-octanol (E = 7.6) under the same conditions (Fig. 35). Both an ethyl and a –CH_2Br group are similar in size, so the difference suggests that an electronic effect lowers the enantioselectivity.

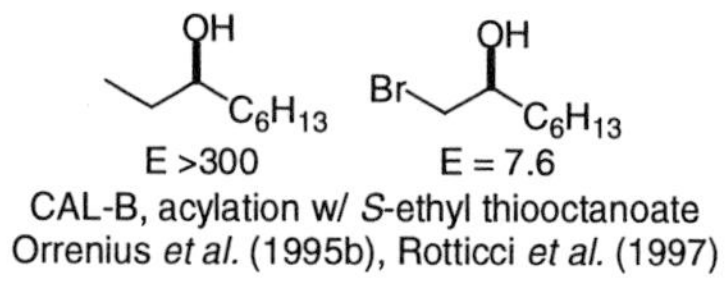

E >300   E = 7.6
CAL-B, acylation w/ S-ethyl thiooctanoate
Orrenius et al. (1995b), Rotticci et al. (1997)

Fig. 35.   Electronic effects also change enantioselectivity.

X-ray structures of transition state analogs containing a secondary alcohol, menthol, bound to CRL identified the alcohol binding pocket (Cygler et al., 1994). This pocket indeed resembled the empirical rule: a large hydrophobic pocket and a smaller pocket for the medium-sized substituent (Fig. 36). A comparison of the structures of the fast- and slow-reacting enantiomers of menthol showed that in both cases the large substituent binds in the large hydrophobic pocket and the medium substituent binds in the smaller pocket. In the slow enantiomer the alcohol oxygen points away from the catalytic histidine, thus, this transition state analog lacks a key hydrogen bond. This observation suggests that enantiomers differ mainly in their rate of reaction, not in their relative affinity to the lipase. Consistent with this idea, Nishizawa et al. (1997) measured the kinetic constants with PCL for two enantiomers of a secondary alcohol and found similar values for the apparent K_m, but very different values for k_{cat}. However, modeling of the transition state for ester hydrolysis in CAL-B suggested another explanation. Uppenberg et al. (1995) found that transition states for hydrolysis of the enantiomers of 1-phenylethanol showed different binding. In the fast enantiomer the large substituent binds in the large pocket, and medium in the medium pocket, but in the slow enantiomer, the large substituent binds in the medium pocket and medium substituent in the large pocket.

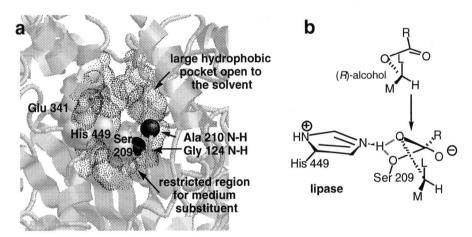

Fig. 36.   Proposed binding site for secondary alcohols in CRL. **a** X-ray structure of CRL highlighted to show the catalytic machinery (Ser 209, His 449, Glu 341 and the N-H groups of Ala 210 and Gly 124) and the alcohol binding site. **b** Schematic of the first step of hydrolysis of an ester of a secondary alcohol. The alcohol oxygen orients to form a hydrogen bond with the catalytic His, while the large and medium substituents orient in their respective pockets.

Further support for the proposed alcohol binding site comes from variations in the amino acids within the pocket for the medium substituent (Tab. 11). These variations are consistent with differences in the selectivity of lipases. For example, smaller amino acids line the M region of CRL (Glu, Ser, Gly) than in PCL and CVL (=PGL) (His, Leu, Gly). If the backbone lies in the same place for both lipases, then the smaller side chains in CRL create a larger binding site. Consistent with this suggestion, CRL catalyzes the hydrolysis of esters of large alcohols (esters of norborneols (Oberhauser et al., 1987)) and esters of tertiary alcohols (O'Hagan and Zaidi, 1992), while the *Pseudomonas* lipases do not. Using substrate mapping Exl et al. (1992) found that CRL had a larger alcohol binding site than PCL. Further, CRL shows low enantioselectivity toward esters of primary alcohols, while the *Pseudomonas* lipases show moderate enantioselectivity. All of these characteristics are consistent with a larger binding site in CRL.

Because the same size rule works for all lipases, Cygler et al. (1994) suggested that structures common to all lipases cause this enantiopreference. Indeed, all lipases follow the α/β-hydrolase fold and have a similar catalytic machinery. On the basis of x-ray crystal structures of chiral transition state analogs bound to the active site of CRL, Cygler et al. (1994) proposed that the loops that assemble the catalytic machinery also assemble an alcohol binding site that is similar to the rule in Fig. 34. A large hydrophobic pocket open to the solvent can bind the large substituent, while a restricted pocket near the catalytic machinery can bind the medium substituent (Fig. 36).

Although all lipases favored the same enantiomer of a secondary alcohol, subtilisin favors the opposite enantiomer and the enantioselectivity is usually lower (Kazlauskas and Weissfloch, 1997). One way to reverse the enantiopreference of CAL-B toward secondary alcohols is to place the alcohol in a different place within the binding site. For example, aminolysis of an allyl carbonate derivative (Eq. 10), replaces the allyl group (Pozo and Gotor, 1993b). In this reaction, RCHMeOC(O)– behaves as the acid portion of an ester; thus, the alcohol stereocenter probably binds in the acid binding site. Of course, the secondary alcohol rule no longer applies to this reaction. Pozo and Gotor (1993b) found that CAL-B favored the alcohol enantiomer opposite to the one predicted in Fig. 34.

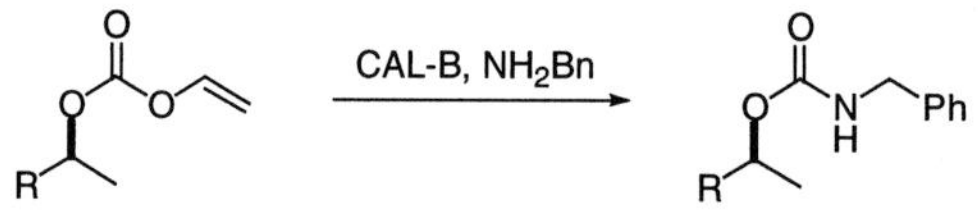

R = *n*-hexyl, Ph, Et; E >50

(10)

Tab. 11. Amino Acid Residues in 11 Lipases Showing the Catalytic Triad (**Bold**), in the Oxyanion Hole (***Bold Italics***), and in the Proposed M Binding Site (*Italics*).

Lipase	Consensus sequence near nucleophilic Ser Gly–X–Ser –X–Gly	Catalytic His	Catalytic Asp /Glu	Oxyanion Hole[a]
CVL, PCL	–Gly85–His86–**Ser87**–*Gln88*–Gly89–	**His285**–*Leu286*–	–**Asp263**–	–*Leu17–Ala18*–
RML	–Gly142–His143–**Ser144**–*Leu145*–Gly146–	**His257**–*Leu258*–	–**Asp203**–	–*Gly81–Ser82*–
HLL	–Gly144–His145–**Ser146**–*Leu147*–Gly148–	**His258**–*Leu259*–	–**Asp201**–	–*Gly82–Ser83*–
PcamL	–Gly143–His144–**Ser145**–*Leu146*–Gly147–	**His259**–*Ile260*–	–**Asp199**–	–*Gly83–Ser84*–
ROL	–Gly143–His144–**Ser145**–*Leu146*–Gly147–	**His257**–*Leu258*–	–**Asp204**–	–*Gly82–Thr83*–
PPL[b]	–Gly150–His151–**Ser152**–*Leu153*–Gly154–	**His263**–*Leu264*–	–**Asp176**–	–*Gly76–Phe77*–
CAL-B	–Thr103–Trp104–**Ser105**–*Gln106*–Gly107–	**His224**–*Ala225*–	–**Asp187**–	–*Gly39–Thr40*–
CRL (lip1)	–Gly207–Glu208–**Ser209**–*Ala210*–Gly211	**His449**–*Ser450*–[c]	–**Glu341**–	–*Gly123–Gly124*–
GCL (lipII)	–Gly215–Glu216–**Ser217**–*Ala218*–Gly219	**His463**–*Gly464*–	–**Glu354**–	–*Gly131–Ala132*–

[a] All lipases use two hydrogen bonds from the amide N–H of the two residues shown in bold italics. In addition, structures of transition state analogs show that RML stabilizes the oxyanion with a third hydrogen bond from the hydroxyl group of Ser or Thr. Other lipases with a Ser or Thr at this position may also use this third hydrogen bond. [b] The corresponding residues in the humane enzyme are identical. [c] The minor isozymes of CRL contain either Gly or Ala in place of Ser 450.

5.1.1.2   *Candida antarctica* **Lipase B**

CAL-B and PCL are usually the most enantioselective lipases toward secondary alcohols, see Figs. 38–40 for CAL-B, Figs. 46–48 for PCL. All examples follow the empirical rule in Fig. 34. CAL-B is more enantioselective toward secondary alcohols where the medium-sized substituent (M in Fig. 34) is relatively small, e.g., methyl, ethyl, –C≡CH, –CH=CH₂. Reactions are slower, but still highly enantioselective when M is *n*-propyl or –CH₂OCH₃, but no reaction occurs when M = *i*-propyl or *t*-butyl as in the substrates in Fig. 37 (Orrenius et al., 1995b). In contrast, PCL accepts longer *n*-alkyl chains as the M substituent (Fig. 47).

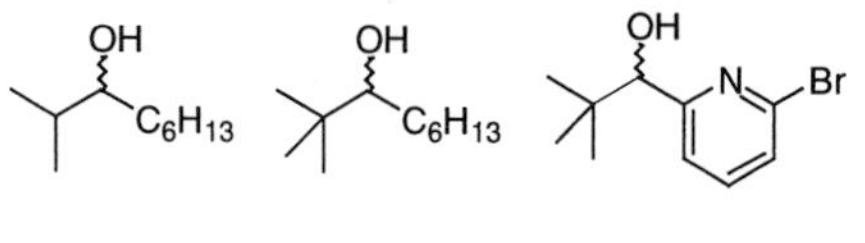

Fig. 37.   No reaction with CAL-B.

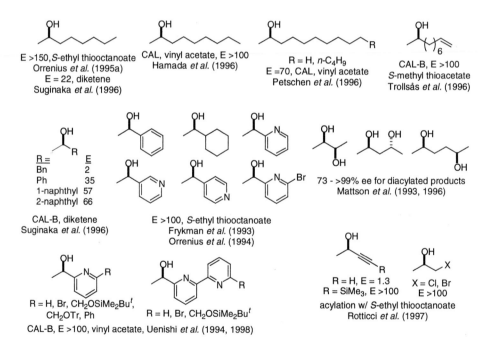

Fig. 38.   Selected examples of 2-alkanols resolved by CAL-B.

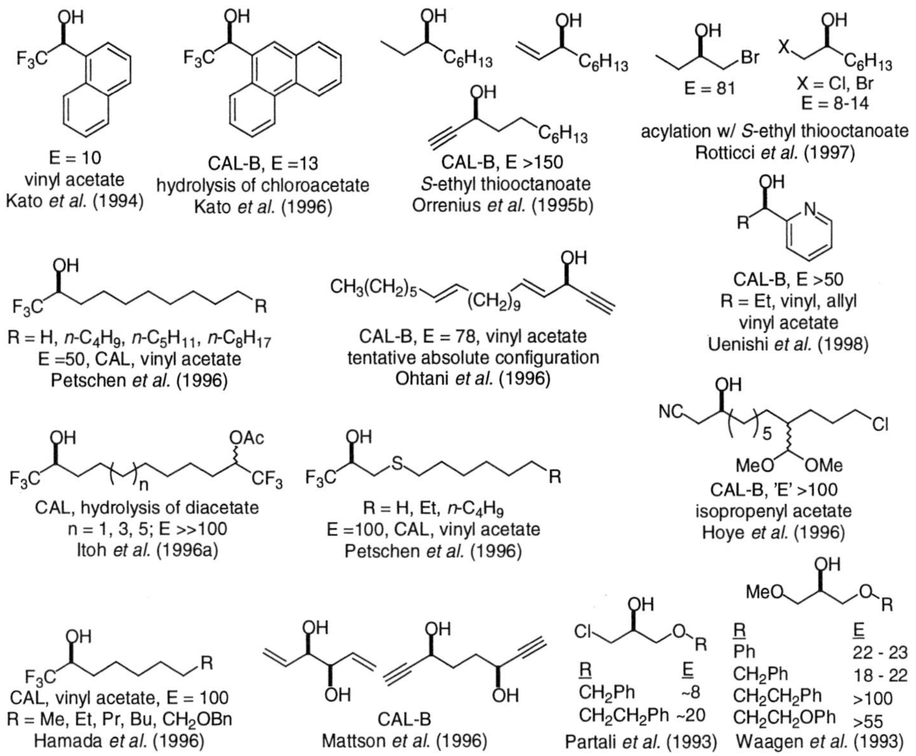

Fig. 39.   Selected examples of acyclic secondary alcohols resolved by CAL-B.

CAL-B also resolves a number of 2-substituted cycloalkanols (Fig. 40). Many *trans*-oriented substituents gave high enantioselectivity, but a *cis* hydroxyl gave low enantio-selectivity.

The x-ray crystal structure of CAL-B shows a deeply buried M-pocket (or alcohol stereoselectivity pocket) large enough to accommodate a methyl or ethyl group without changing the conformation of the protein (see Fig. 5).

Ohtani et al. (1998) recently compared the enantioselectivity of CAL-B toward twelve secondary alcohols with that of CRL and several *Pseudomonas* lipases (PCL, PFL, PAL). In general, CAL-B showed the highest enantioselectivity followed by the *Pseudomonas* lipases while CRL showed the lowest enantioselectivity. However, PCL showed the highest enantioselectivity for one alcohol, in another case, CRL showed higher enantioselectivity than PCL.

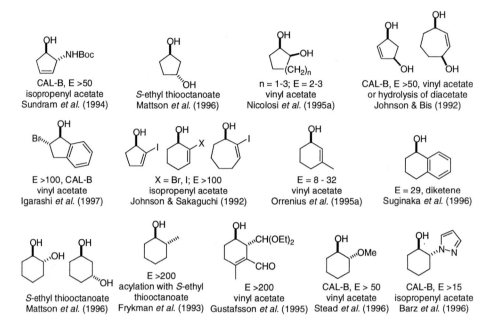

Fig. 40.   Selected examples of cyclic secondary alcohols resolved by CAL-B.

5.1.1.3   *Candida rugosa* Lipase

In a previous survey of secondary alcohols resolved by CRL, Kazlauskas et al. (1991) found that the secondary alcohol rule in Fig. 34 is *not* reliable for acyclic alcohols. Out of 31 examples, only 14 followed the rule. Since there are only two choices in predicting the fast reacting enantiomer, even guessing yields 50 % correct predictions. Thus the rule is little better than guessing. More recent examples (Fig. 41), include seven examples that follow the rule, four that do not and two with either an uncertain absolute configuration or large and small substituents with similar sizes. Compared to other lipases (especially CAL-B, PCL and RML) CRL accepts larger substrates and the x-ray crystal structures show a larger active site. This larger active site may allow acyclic substrates to react in several productive conformations. Some of these conformations may favor the opposite enantiomers.

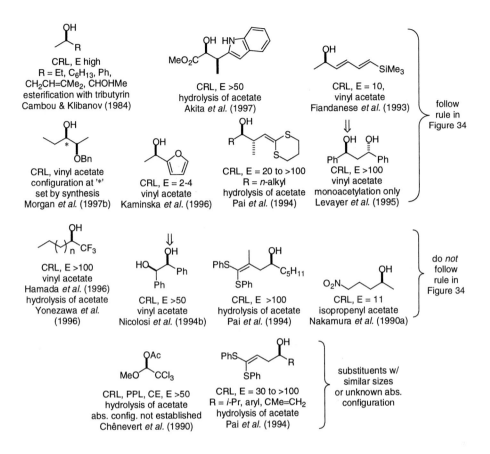

Fig. 41. Selected examples of acyclic secondary alcohols resolved by CRL. Note that several examples do not follow the empirical rule in Fig. 34. This rule is not reliable for CRL-catalyzed reactions of *acyclic* secondary alcohols.

In contrast, cyclic secondary alcohols reliably follow the rule in Fig. 34. Of the 55 substrates in an earlier survey, 51 followed the rule (Kazlauskas et al., 1991). All but two of the selected examples in Fig. 42 follow the rule.

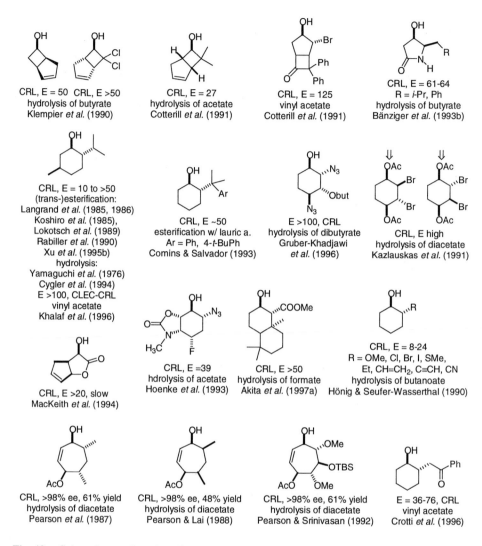

CRL, E = 50 CRL, E >50
hydrolysis of butyrate
Klempier et al. (1990)

CRL, E = 27
hydrolysis of acetate
Cotterill et al. (1991)

CRL, E = 125
vinyl acetate
Cotterill et al. (1991)

CRL, E = 61-64
R = i-Pr, Ph
hydrolysis of butyrate
Bänziger et al. (1993b)

CRL, E = 10 to >50
(trans-)esterification:
Langrand et al. (1985, 1986)
Koshiro et al. (1985),
Lokotsch et al. (1989)
Rabiller et al. (1990)
Xu et al. (1995b)
hydrolysis:
Yamaguchi et al. (1976)
Cygler et al. (1994)
E >100, CLEC-CRL
vinyl acetate
Khalaf et al. (1996)

CRL, E ~50
esterification w/ lauric a.
Ar = Ph, 4-t-BuPh
Comins & Salvador (1993)

E >100, CRL
hydrolysis of dibutyrate
Gruber-Khadjawi
et al. (1996)

CRL, E high
hydrolysis of diacetate
Kazlauskas et al. (1991)

CRL, E >20, slow
MacKeith et al. (1994)

CRL, E =39
hdrolysis of acetate
Hoenke et al. (1993)

CRL, E >50
hydrolysis of formate
Akita et al. (1997a)

CRL, E = 8-24
R = OMe, Cl, Br, I, SMe,
Et, CH=CH₂, C≡CH, CN
hydrolysis of butanoate
Hönig & Seufer-Wasserthal (1990)

CRL, >98% ee, 61% yield
hydrolysis of diacetate
Pearson et al. (1987)

CRL, >98% ee, 48% yield
hydrolysis of diacetate
Pearson & Lai (1988)

CRL, >98% ee, 61% yield
hydrolysis of diacetate
Pearson & Srinivasan (1992)

E = 36-76, CRL
vinyl acetate
Crotti et al. (1996)

Fig. 42.   Selected examples of cyclic secondary alcohols resolved by CRL. Two examples do not follow the rule in Fig. 34: MacKeith et al. (1994) and Hoenke et al. (1993).

5.1.1.4   Porcine Pancreatic Lipase

All examples of PPL-catalyzed reactions of secondary alcohols follow the rule in Fig. 34. The examples are divided into 2-alkanols in Fig. 43, other acyclic secondary alcohols in Fig. 44, and cyclic secondary alcohols in Fig. 45. Note in Fig. 43 that the *cis* vs. *trans* configuration of the double bond in the large substituent strongly influenced the enantio-selectivity.

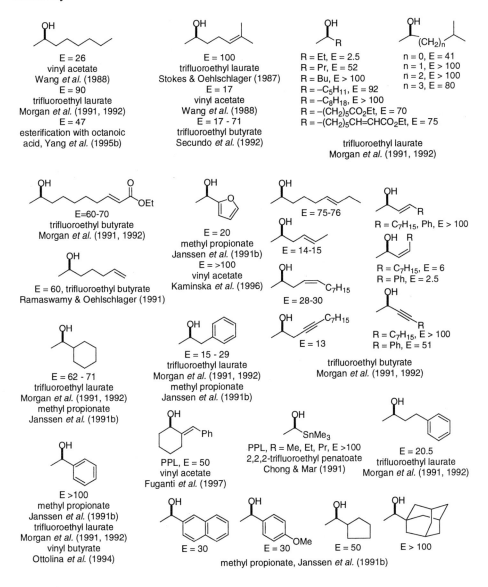

Fig. 43.   Examples of 2-alkanols resolved by porcine pancreatic lipase.

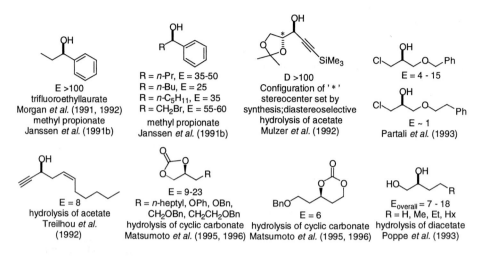

Fig. 44.   Examples of other acyclic secondary alcohols resolved by porcine pancreatic lipase.

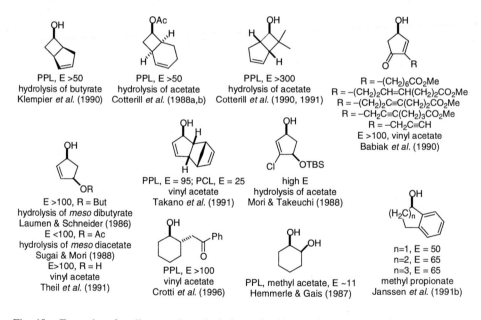

Fig. 45.   Examples of cyclic secondary alcohols resolved by porcine pancreatic lipase.

5.1.1.5   *Pseudomonas* Lipases

The *Pseudomonas* lipases show high enantioselectivity toward a wide range of secondary alcohols, Figs. 46–48. A previous survey also includes 64 secondary alcohols (Kazlauskas et al., 1991).

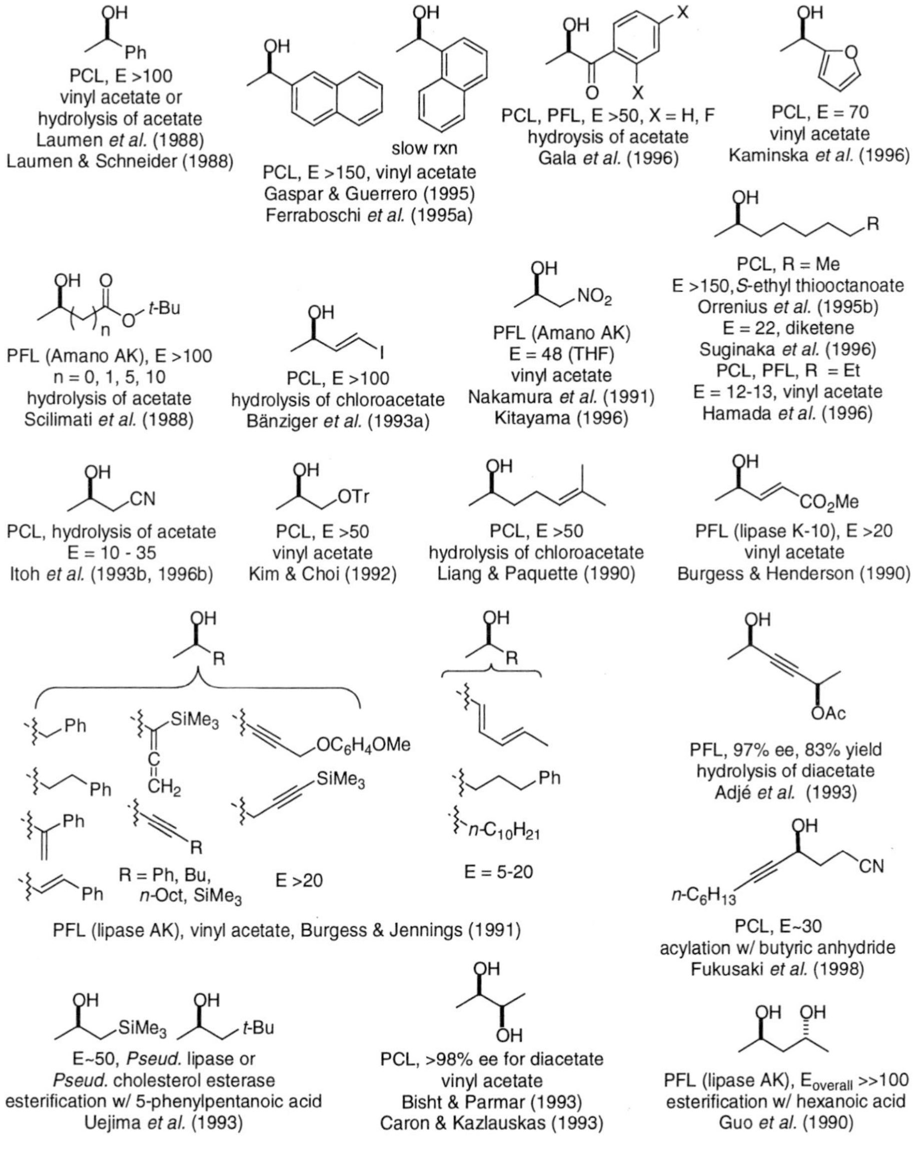

OH
Ph
PCL, E >100
vinyl acetate or
hydrolysis of acetate
Laumen *et al.* (1988)
Laumen & Schneider (1988)

OH

slow rxn
PCL, E >150, vinyl acetate
Gaspar & Guerrero (1995)
Ferraboschi *et al.* (1995a)

OH
X
O   X
PCL, PFL, E >50, X = H, F
hydroysis of acetate
Gala *et al.* (1996)

OH
O
PCL, E = 70
vinyl acetate
Kaminska *et al.* (1996)

OH
R
PCL, R = Me
E >150, *S*-ethyl thiooctanoate
Orrenius *et al.* (1995b)
E = 22, diketene
Suginaka *et al.* (1996)
PCL, PFL, R = Et
E = 12-13, vinyl acetate
Hamada *et al.* (1996)

OH   O
n   O   *t*-Bu
PFL (Amano AK), E >100
n = 0, 1, 5, 10
hydrolysis of acetate
Scilimati *et al.* (1988)

OH
I
PCL, E >100
hydrolysis of chloroacetate
Bänziger *et al.* (1993a)

OH
NO₂
PFL (Amano AK)
E = 48 (THF)
vinyl acetate
Nakamura *et al.* (1991)
Kitayama (1996)

OH
CN
PCL, hydrolysis of acetate
E = 10 - 35
Itoh *et al.* (1993b, 1996b)

OH
OTr
PCL, E >50
vinyl acetate
Kim & Choi (1992)

OH
PCL, E >50
hydrolysis of chloroacetate
Liang & Paquette (1990)

OH
CO₂Me
PFL (lipase K-10), E >20
vinyl acetate
Burgess & Henderson (1990)

OH
R

Ph   SiMe₃   OC₆H₄OMe
C
Ph   CH₂   SiMe₃
Ph
R
R = Ph, Bu,   E >20
n-Oct, SiMe₃
PFL (lipase AK), vinyl acetate, Burgess & Jennings (1991)

OH
R

Ph
n-C₁₀H₂₁
E = 5-20

OH
OAc
PFL, 97% ee, 83% yield
hydrolysis of diacetate
Adjé *et al.* (1993)

OH
CN
n-C₆H₁₃
PCL, E~30
acylation w/ butyric anhydride
Fukusaki *et al.* (1998)

OH
SiMe₃
E~50, *Pseud.* lipase or
Pseud. cholesterol esterase
esterification w/ 5-phenylpentanoic acid
Uejima *et al.* (1993)

OH
t-Bu
OH
PCL, >98% ee for diacetate
vinyl acetate
Bisht & Parmar (1993)
Caron & Kazlauskas (1993)

OH   OH
PFL (lipase AK), E_overall >>100
esterification w/ hexanoic acid
Guo *et al.* (1990)

Fig. 46.   Selected example of 2-alkanols resolved by *Pseudomonas* lipases.

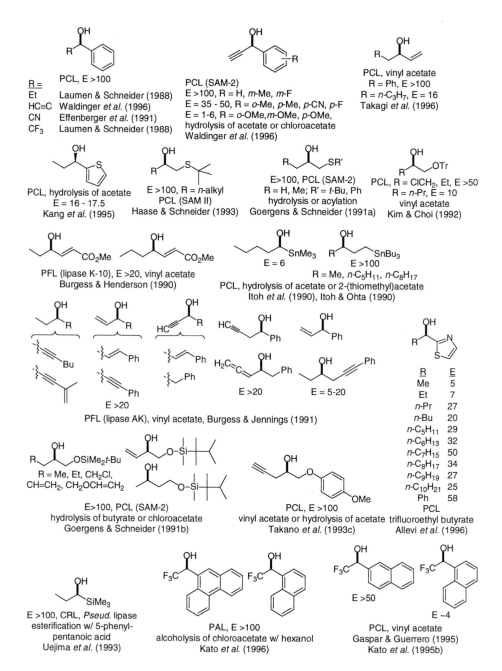

Fig. 47.   Selected examples of enantioselective reactions of *Pseudomonas* lipases with other racemic acyclic alcohols. Note that substituent type and substituent location in the aromatic ring strongly influences the enantioselectivity in some cases.

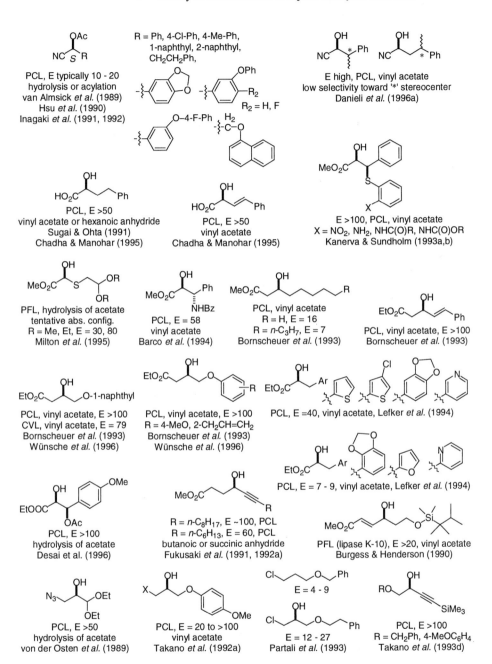

Fig. 47.   Selected example of 2-alkanols resolved by *Pseudomonas* lipases (continued).

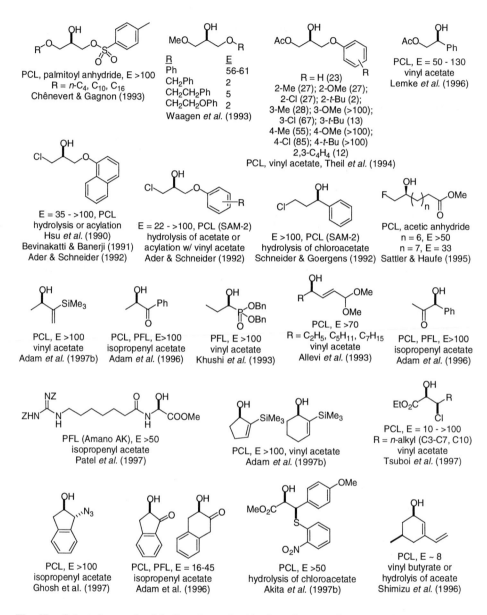

PCL, palmitoyl anhydride, E >100
R = n-C₄, C₁₀, C₁₆
Chênevert & Gagnon (1993)

R	E
Ph	56-61
CH₂Ph	2
CH₂CH₂Ph	5
CH₂CH₂OPh	2

Waagen et al. (1993)

R = H (23)
2-Me (27); 2-OMe (27);
2-Cl (27); 2-t-Bu (2);
3-Me (28); 3-OMe (>100);
3-Cl (67); 3-t-Bu (13)
4-Me (55); 4-OMe (>100);
4-Cl (85); 4-t-Bu (>100)
2,3-C₄H₄ (12)
PCL, vinyl acetate, Theil et al. (1994)

PCL, E = 50 - 130
vinyl acetate
Lemke et al. (1996)

E = 35 - >100, PCL
hydrolysis or acylation
Hsu et al. (1990)
Bevinakatti & Banerji (1991)
Ader & Schneider (1992)

E = 22 - >100, PCL (SAM-2)
hydrolysis of acetate or
acylation w/ vinyl acetate
Ader & Schneider (1992)

E >100, PCL (SAM-2)
hydrolysis of chloroacetate
Schneider & Goergens (1992)

PCL, acetic anhydride
n = 6, E >50
n = 7, E = 33
Sattler & Haufe (1995)

PCL, E >100
vinyl acetate
Adam et al. (1997b)

PCL, PFL, E>100
isopropenyl acetate
Adam et al. (1996)

PFL, E >100
vinyl acetate
Khushi et al. (1993)

PCL, E >70
R = C₂H₅, C₅H₁₁, C₇H₁₅
vinyl acetate
Allevi et al. (1993)

PCL, PFL, E>100
isopropenyl acetate
Adam et al. (1996)

PFL (Amano AK), E >50
isopropenyl acetate
Patel et al. (1997)

PCL, E >100, vinyl acetate
Adam et al. (1997b)

PCL, E = 10 - >100
R = n-alkyl (C3-C7, C10)
vinyl acetate
Tsuboi et al. (1997)

PCL, E >100
isopropenyl acetate
Ghosh et al. (1997)

PCL, PFL, E = 16-45
isopropenyl acetate
Adam et al. (1996)

PCL, E >50
hydrolysis of chloroacetate
Akita et al. (1997b)

PCL, E ~ 8
vinyl butyrate or
hydrolyis of aceate
Shimizu et al. (1996)

Fig. 47.   Selected example of 2-alkanols resolved by *Pseudomonas* lipases (continued).

racemic cyclic secondary alcohols

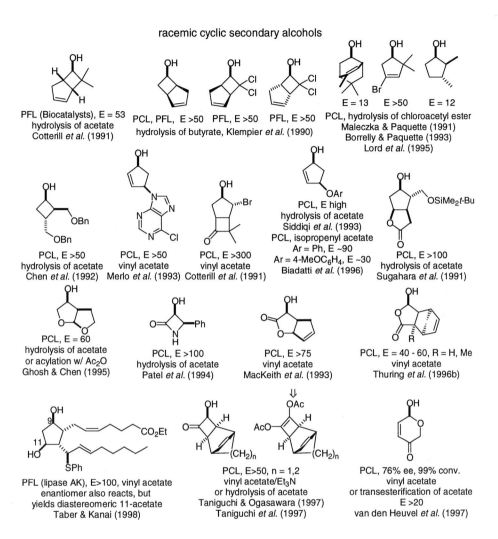

PFL (Biocatalysts), E = 53
hydrolysis of acetate
Cotterill *et al.* (1991)

PCL, PFL, E >50
hydrolysis of butyrate, Klempier *et al.* (1990)

PFL, E >50

PFL, E >50

E = 13   E >50   E = 12
PCL, hydrolysis of chloroacetyl ester
Maleczka & Paquette (1991)
Borrelly & Paquette (1993)
Lord *et al.* (1995)

PCL, E >50
hydrolysis of acetate
Chen *et al.* (1992)

PCL, E >50
vinyl acetate
Merlo *et al.* (1993)

PCL, E >300
vinyl acetate
Cotterill *et al.* (1991)

PCL, E high
hydrolysis of acetate
Siddiqi *et al.* (1993)
PCL, isopropenyl acetate
Ar = Ph, E ~90
Ar = 4-MeOC$_6$H$_4$, E ~30
Biadatti *et al.* (1996)

PCL, E >100
hydrolysis of acetate
Sugahara *et al.* (1991)

PCL, E = 60
hydrolysis of acetate
or acylation w/ Ac$_2$O
Ghosh & Chen (1995)

PCL, E >100
hydrolysis of acetate
Patel *et al.* (1994)

PCL, E >75
vinyl acetate
MacKeith *et al.* (1993)

PCL, E = 40 - 60, R = H, Me
vinyl acetate
Thuring *et al.* (1996b)

PFL (lipase AK), E>100, vinyl acetate
enantiomer also reacts, but
yields diastereomeric 11-acetate
Taber & Kanai (1998)

PCL, E>50, n = 1,2
vinyl acetate/Et$_3$N
or hydrolysis of acetate
Taniguchi & Ogasawara (1997)
Taniguchi *et al.* (1997)

PCL, 76% ee, 99% conv.
vinyl acetate
or transesterification of acetate
E >20
van den Heuvel *et al.* (1997)

Fig. 48.   Selected examples of enantioselective reactions of cyclic secondary alcohols catalyzed by *Pseudomonas* lipases.

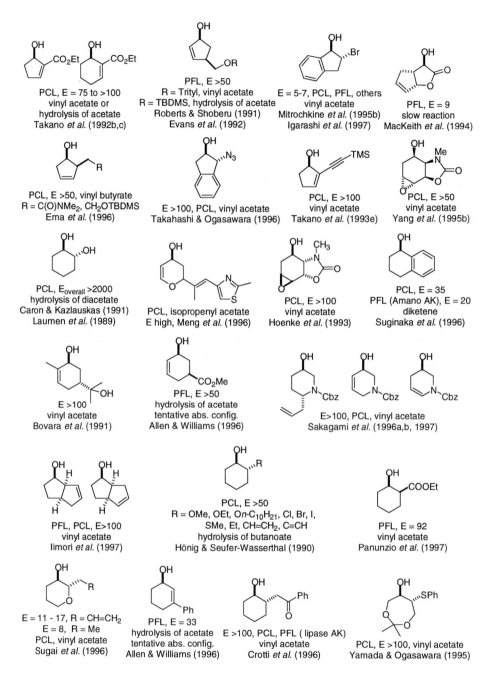

Fig. 48. Selected examples of enantioselective reactions of cyclic secondary alcohols catalyzed by *Pseudomonas* lipases (continued).

meso cyclic secondary alcohols

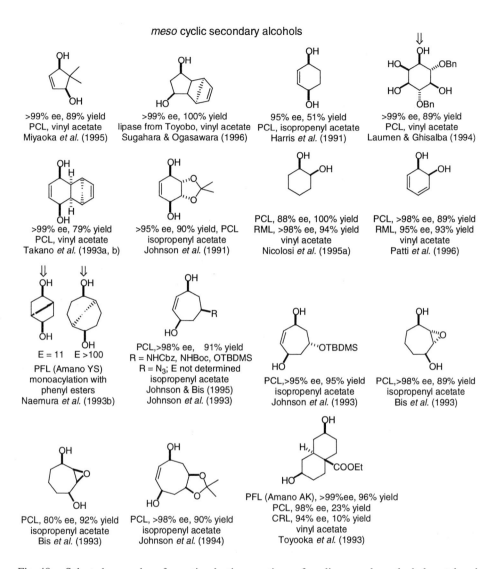

>99% ee, 89% yield
PCL, vinyl acetate
Miyaoka *et al.* (1995)

>99% ee, 100% yield
lipase from Toyobo, vinyl acetate
Sugahara & Ogasawara (1996)

95% ee, 51% yield
PCL, isopropenyl acetate
Harris *et al.* (1991)

>99% ee, 89% yield
PCL, vinyl acetate
Laumen & Ghisalba (1994)

>99% ee, 79% yield
PCL, vinyl acetate
Takano *et al.* (1993a, b)

>95% ee, 90% yield, PCL
isopropenyl acetate
Johnson *et al.* (1991)

PCL, 88% ee, 100% yield
RML, >98% ee, 94% yield
vinyl acetate
Nicolosi *et al.* (1995a)

PCL, >98% ee, 89% yield
RML, 95% ee, 93% yield
vinyl acetate
Patti *et al.* (1996)

E = 11   E >100

PFL (Amano YS)
monoacylation with
phenyl esters
Naemura *et al.* (1993b)

PCL,>98% ee,   91% yield
R = NHCbz, NHBoc, OTBDMS
R = N$_3$; E not determined
isopropenyl acetate
Johnson & Bis (1995)
Johnson *et al.* (1993)

PCL,>95% ee, 95% yield
isopropenyl acetate
Johnson *et al.* (1993)

PCL,>98% ee, 89% yield
isopropenyl acetate
Bis *et al.* (1993)

PCL, 80% ee, 92% yield
isopropenyl acetate
Bis *et al.* (1993)

PCL, >98% ee, 90% yield
isopropenyl acetate
Johnson *et al.* (1994)

PFL (Amano AK), >99%ee, 96% yield
PCL, 98% ee, 23% yield
CRL, 94% ee, 10% yield
vinyl acetate
Toyooka *et al.* (1993)

Fig. 48.   Selected examples of enantioselective reactions of cyclic secondary alcohols catalyzed by *Pseudomonas* lipases (continued).

Except for increasing the difference in size of the substituents (see Sect. 5.1.1), or lowering the temperature (for an example see Sakai et al., 1997), no general method exists to increase the enantioselectivity of PCL-catalyzed reactions. Longer esters or acylating agents (butyrates and above) sometimes give higher enantioselectivity than acetates. Hydrolyses of β-(phenylthio)- or β-(methylthio)acetoxy groups increased enantioselectivity ten-fold (e.g., from 6 to 55) as compared to hydrolyses of acetates or valerates (Itoh et al., 1991). Thiocrown ethers (e.g., 1,4,8,11-tetrathiacyclotetradecane) increased the enantioselectivity of PCL four-fold (from 9 to 27–37 and from 100 to 400) in the resolution of several secondary alcohols (Takagi et al., 1996; Itoh et al., 1996b). Itoh et al. (1993b) observed smaller increases with simple crown ethers.

5.1.1.6 *Rhizomucor* Lipases

Researchers have resolved fewer secondary alcohols using RML (Fig. 49). In triglycerides, RML selectively hydrolyzes esters at the primary alcohol positions (see Sect. 6.2.1), but the examples below show that RML can also hydrolyze esters of secondary alcohols.

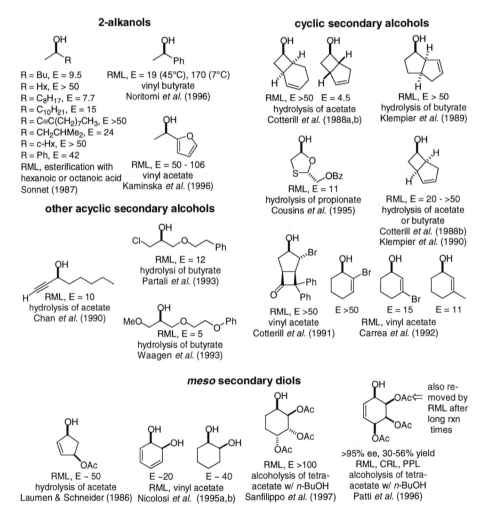

Fig. 49. Examples of *Rhizomucor miehei* lipase-catalyzed enantioselective reactions of secondary alcohols.

Noritomi et al. (1996) increased the enantioselectivity of an RML-catalyzed acylation of α-phenylethanol by lowering the temperature. Acylation with vinyl butyrate in dioxane gave E = 19 at 45 °C, but E = 170 at 7 °C. In other solvents (pyridine, THF, triethyl-amine) temperature did not affect E.

5.1.1.7   Other Lipases

Other lipases are also enantioselective toward secondary alcohols and usually follow the rule in Fig. 34. Selected examples are in Fig. 50.

Naemura et al. (1996) surveyed the enantioselectivity of a lipase from *Alcaligenes* sp. and found good enantioselectivity toward a range of secondary alcohols (27 examples, mostly MeCH(OH)aryl). The favored enantiomer was the one predicted by the empirical rule in Fig. 34.

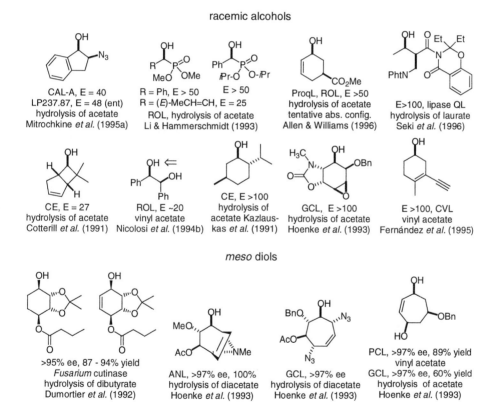

Fig. 50.   Selected examples of enantioselective reactions catalyzed by other lipases.

5.1.1.8   Choosing the Best Route

The best route to a particular compound is rarely a straightforward choice. In addition to several lipase-catalyzed routes, researchers consider other chemical and biochemical routes. The 'best' is often an individual decision which depends on the intended next steps and available starting materials. The two examples below summarize only the lipase-catalyzed routes to these targets.

Inositols. Researchers have found a number of different routes to enantiomerically pure inositol derivatives. Starting from the achiral *myo*-inositol, researchers added protective groups to increase the size of one substituent. Protection yields either a racemate or a *meso* derivative. Several different lipases showed excellent enantioselectivity. The asymmetric synthesis starting from *meso* derivatives is probably the best route since it gives both high yield and high enantioselectivity. However, in some cases another route may fit better with subsequent synthetic steps (Fig. 51).

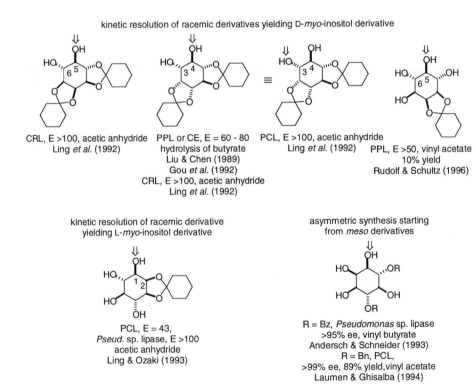

Fig. 51.   Lipase-catalyzed enantioselective reactions of *myo*-inositol derivatives.

β-Blockers. Most β-adrenergic antagonists (β-blockers), used for the treatment of hypertension and angina pectoris, are 3-aryloxy-2-propanolamines which contain a secondary alcohol stereocenter. The (*S*)-enantiomer is usually more active than (*R*), e.g., one hundred times more active in the case of propranolol. For this reason, chemists have developed routes to the (*S*)-enantiomer, including lipase-catalyzed reactions (Fig. 52) (reviewed by Kloosterman et al., 1988; Sheldon, 1993). None of these routes have been commercialized. In most examples in Fig. 52 the aryl group is the large substituent, thus the (*R*)-enantiomer reacts faster. Resolution by acylation of the alcohol is preferred over hydrolysis of the acetate because acylation yields the desired (*S*)-alcohol as the unreacted starting material. A resolution by hydrolysis requires an extra step to hydrolyze the unreacted ester.

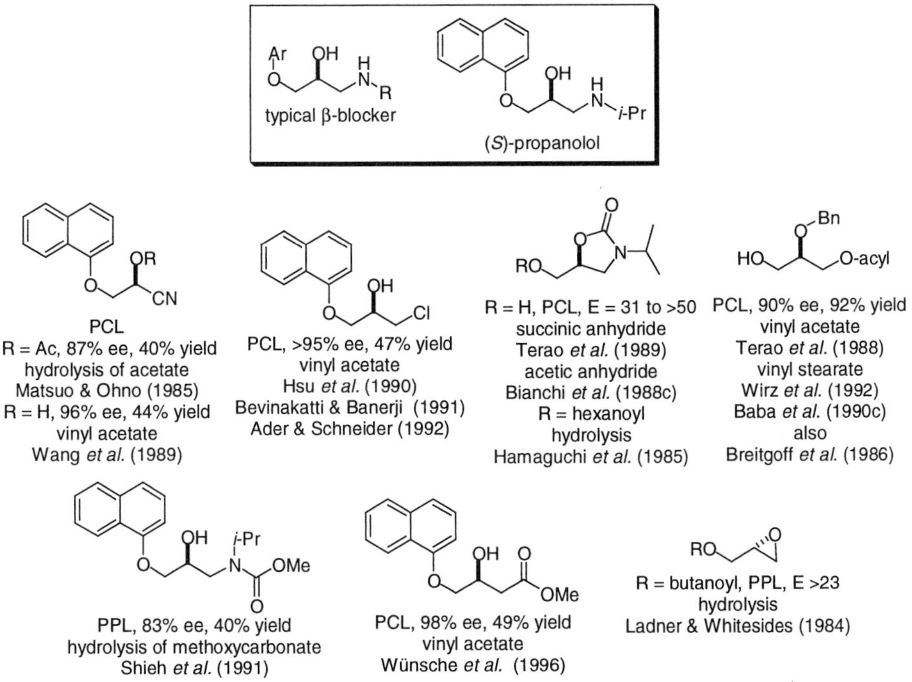

typical β-blocker

(*S*)-propanolol

PCL
R = Ac, 87% ee, 40% yield
hydrolysis of acetate
Matsuo & Ohno (1985)
R = H, 96% ee, 44% yield
vinyl acetate
Wang *et al.* (1989)

PCL, >95% ee, 47% yield
vinyl acetate
Hsu *et al.* (1990)
Bevinakatti & Banerji (1991)
Ader & Schneider (1992)

R = H, PCL, E = 31 to >50
succinic anhydride
Terao *et al.* (1989)
acetic anhydride
Bianchi *et al.* (1988c)
R = hexanoyl
hydrolysis
Hamaguchi *et al.* (1985)

PCL, 90% ee, 92% yield
vinyl acetate
Terao *et al.* (1988)
vinyl stearate
Wirz *et al.* (1992)
Baba *et al.* (1990c)
also
Breitgoff *et al.* (1986)

PPL, 83% ee, 40% yield
hydrolysis of methoxycarbonate
Shieh *et al.* (1991)

PCL, 98% ee, 49% yield
vinyl acetate
Wünsche *et al.* (1996)

R = butanoyl, PPL, E >23
hydrolysis
Ladner & Whitesides (1984)

Fig. 52.   Lipase-catalyzed routes to enantiomerically pure precursors of propranolol.

5.1.2   Primary Alcohols

Lipases usually show lower enantioselectivity toward primary alcohols than toward secondary alcohols. Only PPL and PCL show high enantioselectivity toward a wide range of primary alcohols.

5.1.2.1   *Pseudomonas* Lipases

An empirical rule can predict some of the enantiopreference of PCL toward primary alcohols (Weissfloch and Kazlauskas, 1995). Like the secondary alcohol rule above, the primary alcohol rule is based on the size of the substituents, but, surprisingly the sense of enantiopreference toward it is opposite. That is, the –OH of secondary alcohols and the –CH$_2$OH of primary alcohols point in opposite directions. One way to rationalize this opposite enantiopreference is to assume the extra CH$_2$ in primary alcohols introduces a kink between the stereocenter and the oxygen as suggested in Fig. 53. In this manner the large and medium substituents bind in the same enzyme pockets in both cases. Another possibility is a different binding mode for primary alcohols. Indeed, modeling suggests that the large substituent of primary alcohols does not bind in the same pocket as secondary alcohols (Tuomi and Kazlauskas, 1999).

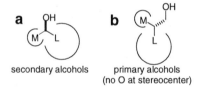

secondary alcohols   primary alcohols
                            (no O at stereocenter)

Fig. 53.   Empirical rules summarize the enantiopreference of PCL toward primary and secondary alcohols. **a** Shape of the favored enantiomer of secondary alcohols. **b** Shape of the favored enantiomer of primary alcohols. This rule is reliable only when there is no oxygen directly bonded at the stereocenter. Computer modeling (and the drawing above) suggest that the large substituent, L, binds to different regions of the active site. Note that PCL shows an opposite enantiopreference toward these two classes of alcohols.

Not all primary alcohols fit this rule. In particular, primary alcohols that have an oxygen at the stereocenter (e.g., glycerol derivatives) do not fit this rule. Of the remaining primary alcohol examples, the rule showed an 89 % reliability (correct for 54 of 61 examples).

An example of how the empirical rule does not apply to primary alcohols with an oxygen at the stereocenter are the γ-butyrolactones in Fig. 54. For the *trans* isomer, PCL favors one enantiomer, but for the *cis* isomer PCL favors the other (Ha et al., 1998).

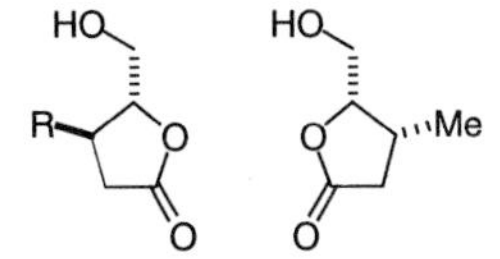

R = H, E = 30   E >100
R = Me, E = 38
R = Et, E = 16

PCL, hydrolysis of acetate
Ha *et al.* (1998)

Fig. 54.   Examples which do not follow the empirical rule for primary alcohols.

For secondary alcohols, increasing the difference in the size of the substituents often increases the enantioselectivity of PCL and other lipases. Indeed, researchers often introduce a large protective group to increase the enantioselectivity (see Sect. 5.1.1.1). However, for primary alcohols this strategy is not reliable. Upon adding large substituents the enantioselectivity sometimes increased (Lampe et al., 1996), sometimes decreased, and sometimes remained unchanged (Weissfloch and Kazlauskas, 1995)

Selected examples of primary alcohols that are resolved or desymmetrized by PCL are shown in Fig. 55. More examples are found in a recent survey (Weissfloch and Kazlauskas, 1995).

One group of popular substrates are the *meso*-1,3-diols which can be asymmetrized either by hydrolysis of the diester or acylation of the diol. For hydrolysis reactions, Liu et al. (1990) noted that acyl migration in the monoester was fast enough in aqueous solution at pH 7 and above to lower the enantiomeric purity of the product. Acyl migration slows considerably in organic solvents (benzene, chloroform, THF). For this reason, asymmetrization of a diol by acylation can give higher enantiomeric purity (see also Sect. 6.2.1.2 for discussion of acyl migration in the similar glycerides).

racemic primary alcohols

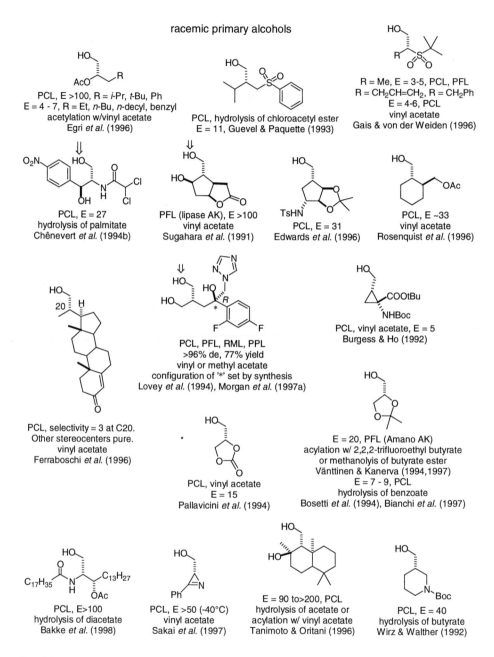

PCL, E >100, R = *i*-Pr, *t*-Bu, Ph
E = 4 - 7, R = Et, *n*-Bu, *n*-decyl, benzyl
acetylation w/vinyl acetate
Egri *et al.* (1996)

PCL, hydrolysis of chloroacetyl ester
E = 11, Guevel & Paquette (1993)

R = Me, E = 3-5, PCL, PFL
R = CH₂CH=CH₂, R = CH₂Ph
E = 4-6, PCL
vinyl acetate
Gais & von der Weiden (1996)

PCL, E = 27
hydrolysis of palmitate
Chênevert *et al.* (1994b)

PFL (lipase AK), E >100
vinyl acetate
Sugahara *et al.* (1991)

PCL, E = 31
Edwards *et al.* (1996)

PCL, E ~33
vinyl acetate
Rosenquist *et al.* (1996)

PCL, selectivity = 3 at C20.
Other stereocenters pure.
vinyl acetate
Ferraboschi *et al.* (1996)

PCL, PFL, RML, PPL
>96% de, 77% yield
vinyl or methyl acetate
configuration of '*' set by synthesis
Lovey *et al.* (1994), Morgan *et al.* (1997a)

PCL, vinyl acetate, E = 5
Burgess & Ho (1992)

PCL, vinyl acetate
E = 15
Pallavicini *et al.* (1994)

E = 20, PFL (Amano AK)
acylation w/ 2,2,2-trifluoroethyl butyrate
or methanolyis of butyrate ester
Vänttinen & Kanerva (1994,1997)
E = 7 - 9, PCL
hydrolysis of benzoate
Bosetti *et al.* (1994), Bianchi *et al.* (1997)

PCL, E>100
hydrolysis of diacetate
Bakke *et al.* (1998)

PCL, E >50 (-40°C)
vinyl acetate
Sakai *et al.* (1997)

E = 90 to>200, PCL
hydrolysis of acetate or
acylation w/ vinyl acetate
Tanimoto & Oritani (1996)

PCL, E = 40
hydrolysis of butyrate
Wirz & Walther (1992)

Fig. 55.   Selected examples of PCL-catalyzed enantioselective reactions of primary alcohols.

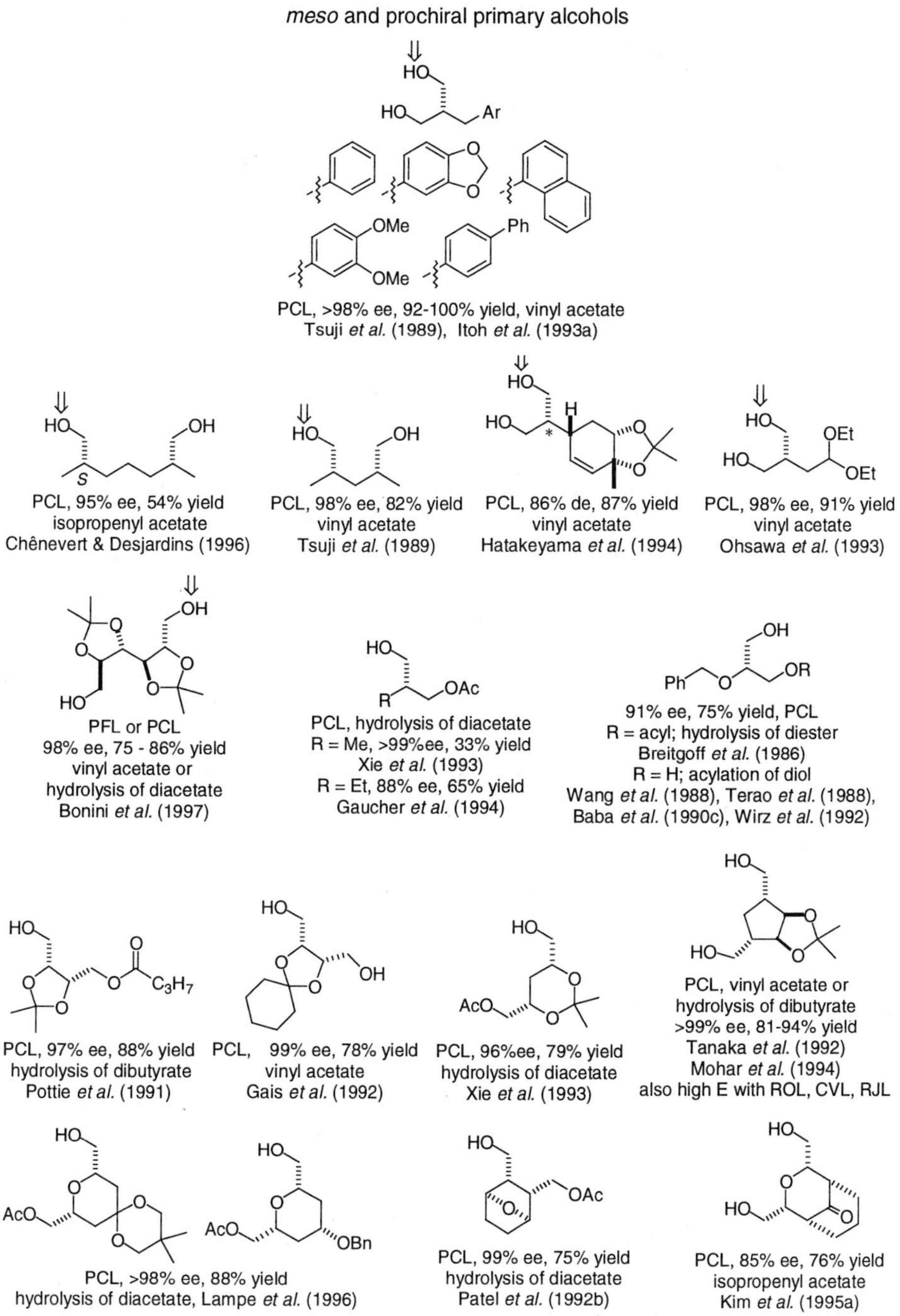

meso and prochiral primary alcohols

PCL, >98% ee, 92-100% yield, vinyl acetate
Tsuji *et al.* (1989), Itoh *et al.* (1993a)

PCL, 95% ee, 54% yield
isopropenyl acetate
Chênevert & Desjardins (1996)

PCL, 98% ee, 82% yield
vinyl acetate
Tsuji *et al.* (1989)

PCL, 86% de, 87% yield
vinyl acetate
Hatakeyama *et al.* (1994)

PCL, 98% ee, 91% yield
vinyl acetate
Ohsawa *et al.* (1993)

PFL or PCL
98% ee, 75 - 86% yield
vinyl acetate or
hydrolysis of diacetate
Bonini *et al.* (1997)

PCL, hydrolysis of diacetate
R = Me, >99%ee, 33% yield
Xie *et al.* (1993)
R = Et, 88% ee, 65% yield
Gaucher *et al.* (1994)

91% ee, 75% yield, PCL
R = acyl; hydrolysis of diester
Breitgoff *et al.* (1986)
R = H; acylation of diol
Wang *et al.* (1988), Terao *et al.* (1988),
Baba *et al.* (1990c), Wirz *et al.* (1992)

PCL, 97% ee, 88% yield
hydrolysis of dibutyrate
Pottie *et al.* (1991)

PCL, 99% ee, 78% yield
vinyl acetate
Gais *et al.* (1992)

PCL, 96%ee, 79% yield
hydrolysis of diacetate
Xie *et al.* (1993)

PCL, vinyl acetate or
hydrolysis of dibutyrate
>99% ee, 81-94% yield
Tanaka *et al.* (1992)
Mohar *et al.* (1994)
also high E with ROL, CVL, RJL

PCL, >98% ee, 88% yield
hydrolysis of diacetate, Lampe *et al.* (1996)

PCL, 99% ee, 75% yield
hydrolysis of diacetate
Patel *et al.* (1992b)

PCL, 85% ee, 76% yield
isopropenyl acetate
Kim *et al.* (1995a)

Fig. 55. Selected examples of PCL-catalyzed enantioselective reactions of primary alcohols (continued).

5.1.2.2   Porcine Pancreatic Lipase

No generally applicable rule to predict the fast-reacting enantiomer in PPL catalyzed reactions of primary alcohols exists. Researchers have proposed several rules, but none are satisfactory (Ehrler and Seebach, 1990; Wimmer, 1992; Guanti et al., 1992; Hultin and Jones, 1992). Two rules even predict opposite enantiomers. An example of the difficulties is shown in Fig. 56. Enantiopreference of PPL reversed upon changing from a *trans* to a *cis* configuration of the double bond in the 2-substituted 1,3-propandiol derivatives below. PPL favored the (*S*)-enantiomer with high enantioselectivity for the *trans* isomer, the (*S*) enantiomer with moderate enantioselectivity for the saturated analog, but the (*R*)-enantiomer with low to moderate enantioselectivity for the *cis* isomer.

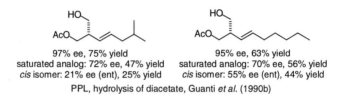

97% ee, 75% yield   95% ee, 63% yield
saturated analog: 72% ee, 47% yield   saturated analog: 70% ee, 56% yield
cis isomer: 21% ee (ent), 25% yield   *cis* isomer: 55% ee (ent), 44% yield
PPL, hydrolysis of diacetate, Guanti *et al.* (1990b)

Fig. 56.   The presence and configuration of a double bond change and even reverse the enantioselectivity.

This reversal is difficult to explain using only the relative sizes of the substituents. Note that for secondary alcohols, the enantioselectivity also varied with the configuration of double bonds in the large substituent, but the enantiopreference remained the same (Morgan et al., 1992). Selected examples of PPL-catalyzed resolutions of primary alcohols are shown in Fig. 57.

kinetic resolution of racemic primary alcohols

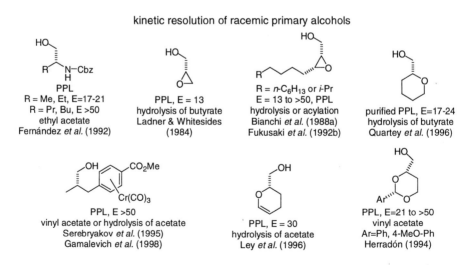

Fig. 57.   Selected examples of PPL-catalyzed enantioselective reactions of primary alcohols.

Although there is no general method to increase the enantioselectivity of PPL-catalyzed reactions of primary alcohols, Serebryakov et al. (1995) dramatically increased the
enantioselectivity of PPL toward 3-aryl-2-methyl propanol from E = 2 to E > 50 by
coordinating a chromium tricarbonyl group to the aryl. After resolution, oxidation with
iodine removed the chromium tricarbonyl group. Large differences in the size of the
substituents are not always necessary. Ley et al. (1996) efficiently resolved pyrans having
minimal differences in the sizes of the substituents (Fig. 57).

asymmetric syntheses starting from *meso* or prochiral primary alcohols

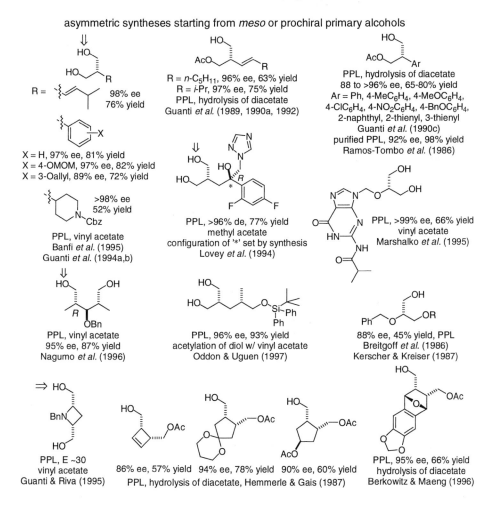

Fig. 57.   Selected examples of PPL-catalyzed enantioselective reactions of primary alcohols
(continued).

5.1.2.3   Other Lipases

Only a handful of primary alcohols have been either resolved or desymmetrized by other lipases. Selected examples are shown in Fig. 58.

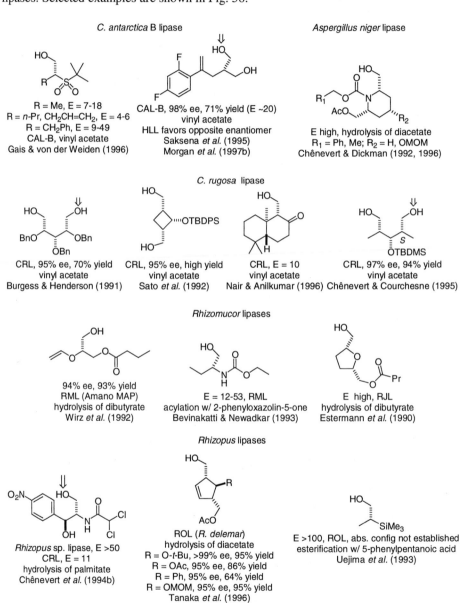

Fig. 58.   Selected examples of enantioselective reactions of primary alcohols catalyzed by other lipases.

5.1.2.4   Enantioselectivity of Lipases Toward Triglycerides

Triglycerides are presumably the natural substrates of lipases, so many researchers have investigated the enantioselectivity of lipases toward triglycerides. Triglycerides with identical functional groups at the two primary alcohols (*sn*-1 and *sn*-3) are prochiral. The stereochemical numbering for triglycerides and other glycerol derivatives starts with a Fischer projection of the triglyceride with the central hydroxyl group positioned to the left. Numbering the carbons *sn*-1, *sn*-2, and *sn*-3 from top to bottom uniquely identifies each position. Enantioselectivity towards triglycerides is based on the ability of a lipase to discriminate between the *sn*-1 and *sn*-3 position. Enantiomers thus formed are 1,2- or 2,3-diglycerides or 1- or 3- monoglycerides (Fig. 59).

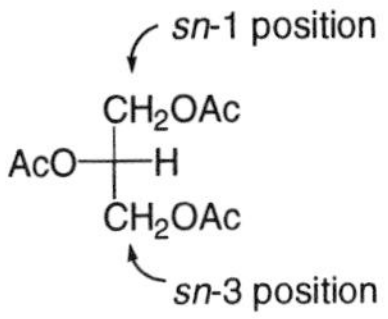

$$
\begin{array}{c}
\nearrow \text{ } sn\text{-}1 \text{ position} \\
\text{CH}_2\text{OAc} \\
\text{AcO} \longrightarrow \text{H} \\
\text{CH}_2\text{OAc} \\
\searrow \text{ } sn\text{-}3 \text{ position}
\end{array}
$$

Fig. 59.   Stereochemical numbering of triglycerides.

Rogalska et al. (1993) surveyed the enantioselectivity of 25 lipases in triglyceride monolayers. Some lipases (e.g., from *Pseudomonas* sp., RML, CRL) showed high selectivity toward the *sn*-1 position, but the stereoselectivity of other lipases varied with the triglyceride: CAL-B showed *sn*-3 selectivity with trioctanoin, but *sn*-1 selectivity with triolein. Selectivity also varied with interfacial tension. Stadler et al. (1995) found large changes as well as reversals in selectivity using analogs of triglycerides with ether or alkyl groups at the *sn*-2 position. For ROL, Holzwarth et al. (1997) rationalized these changes using computer modeling. In recent publications it was proposed that the flexibility of the substituent in *sn*-2 position as well as the geometry of the substrate docked to the active site of ROL determines enantioselectivity. If the torsion angle $\Phi_{O3\text{-}C3}$ of glycerol is > 150°, the substrate will be hydrolyzed preferentially in *sn*-1 position, if $\Phi_{O3\text{-}C3}$ is < 150°, ROL is *sn*-3 selective. Moreover, enantioselectivity could be altered by site-directed mutagenesis (Scheib et al., 1998, 1999).

5.1.3   Other Alcohols, Amines, and Alcohol Analogs

5.1.3.1   Tertiary Alcohols and Other Quaternary Stereocenters

Lipase-catalyzed reactions involving tertiary alcohols are slow, presumably due to steric hindrance. O'Hagan and Zaidi (1992, 1994b) resolved several acetylenic alcohols with CRL (Fig. 60). O'Hagan and Zaidi (1992, 1994b) suggested that the acetylenic moiety in these tertiary alcohols occupies the same site as a hydrogen in secondary alcohols because CRL showed low enantioselectivity toward alcohols containing both a hydrogen and a –C≡CH substituent at the stereocenter. No CRL-catalyzed hydrolysis occurred when O'Hagan and Zaidi (1992, 1994b) replaced the acetylenic moiety with Me, vinyl, or nitrile. The ability of CRL to catalyze reactions involving tertiary alcohols is consistent with a large alcohol binding site as suggested above in Sect. 5.1.1. Surprisingly, RML also catalyzed hydrolysis of an acetate of a tertiary alcohol.

Brackenridge et al. (1993) used an oxalate ester to introduce a less hindered ester group. PPL-catalyzed hydrolysis of the mixed oxalate ester showed moderate enantiose-lectivity, although the actual site of reaction was not determined.

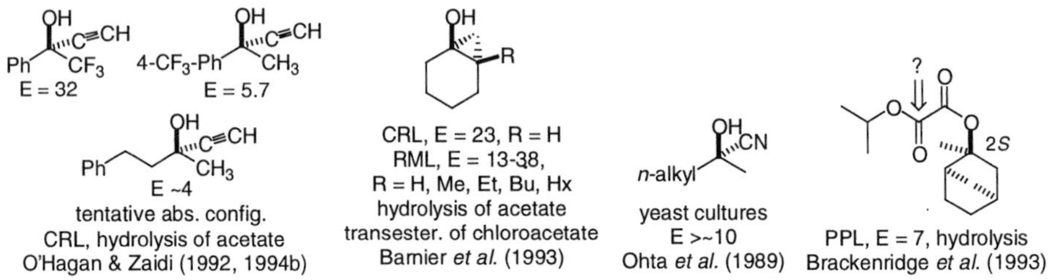

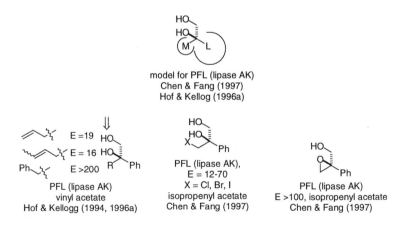

Fig. 60.   Lipase catalyzed enantioselective reactions of tertiary alcohols.

In primary alcohols with quaternary stereocenter, the hindered stereocenter lies further from the reactive hydroxyl, so lipase-catalyzed reactions remain fast. For diols of the type RR'C(OH)CH₂OH, two groups proposed models to predict the fast reacting enantiomer in PFL-catalyzed reactions (Fig. 61) (Hof and Kellog, 1996a; Chen and Fang, 1997). Hof and Kellog (1996a) proposed a flat pocket for one substituent, while Chen and Fang proposed a simplified version. Fig. 62 shows more examples of primary alcohols with quaternary stereocenters. Many of these also follow the model, even though the reactions use lipases other than PFL. Other examples of quaternary stereocenters are in Sect. 5.2.4 on acids and Sect. 5.1.3.2 on alcohols with remote stereocenters.

Fig. 61.   Model and several examples of primary alcohols with a quaternary stereocenter resolved by PFL (lipase AK).

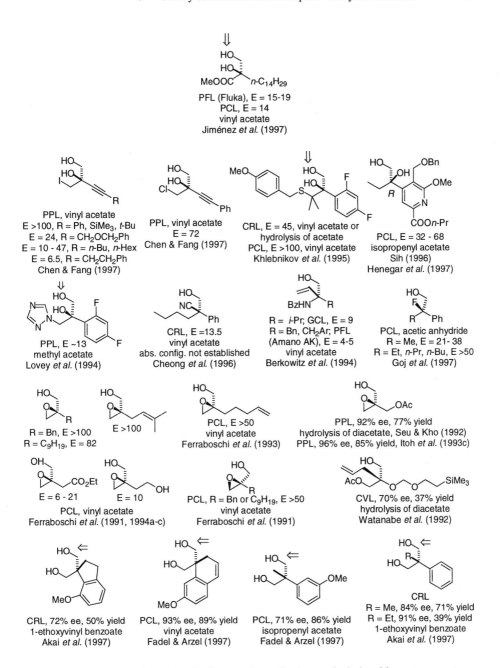

Fig. 62.   Lipase-catalyzed enantioselective reactions of primary alcohols with quaternary stereo-centers.

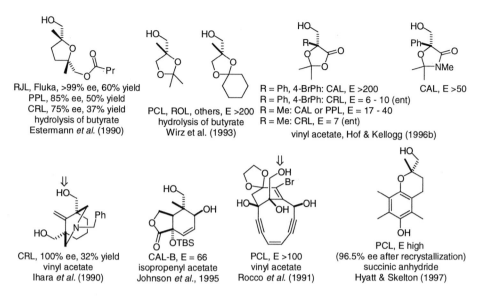

Fig. 62.   Lipase-catalyzed enantioselective reactions of primary alcohols with quaternary stereo-centers (continued).

5.1.3.2   Alcohols with Axial Chirality or Remote Stereocenters

Pure enantiomers of axially-disymmetric and spiro compounds are often difficult to make using traditional chemical methods, so lipase-catalyzed reactions are often the best route to these compounds (Fig. 63).

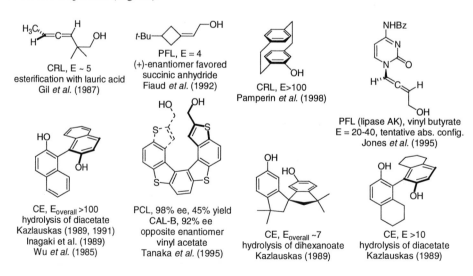

Fig. 63.   Selected examples of lipase-catalyzed enantioselective reactions of axially-disymmetric and spiro alcohols.

Other difficult-to-resolve compounds are those with a stereocenter remote from the reaction site. Nevertheless, lipases often showed high enantioselectivity toward these compounds (reviewed by Mizuguchi et al., 1994). Selected examples involving alcohols are summarized in Fig. 64.

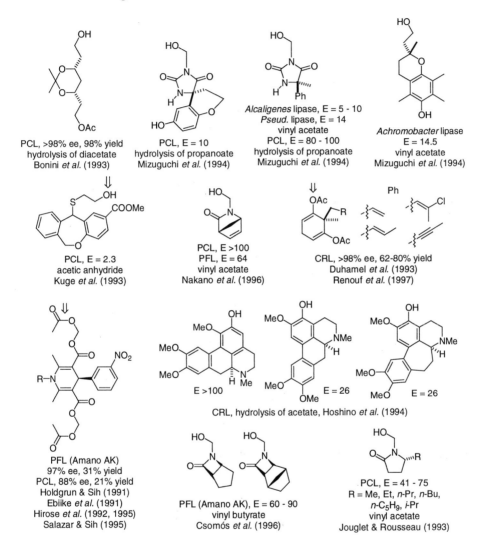

Fig. 64.   Lipase-catalyzed enantioselective reactions of alcohols with remote stereocenters.

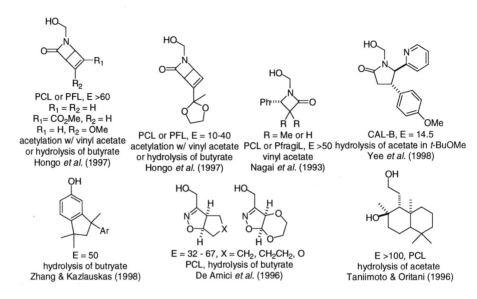

Fig. 64.   Lipase-catalyzed enantioselective reactions of alcohols with remote stereocenters (continued).

The prochiral dihydropyridine (Fig. 64, first example in the last row), is a chiral acid, but is included with the chiral alcohols due to the acetyloxymethyl ester. Lipases do not catalyze hydrolysis of simple esters of these dihydropyridines presumably due to a combination of steric hindrance and lower reactivity (this carbonyl is a vinylogous carbamate). Researchers used the acetyloxymethyl group to introduce a more reactive and more accessible carbonyl group. This strategy places the stereocenter to the alcohol portion of the ester.

5.1.3.3   Alcohols with Non-Carbon Stereocenters

The first examples of lipase-catalyzed resolutions involved secondary alcohols where the organometallic was the large substituent (Wang et al., 1988; Boaz, 1989; Chong and Mar, 1991; Izumi et al., 1992). Acylation in organic solvent was crucial to the success of the resolution of the ferrocenyl derivatives because the corresponding acetate reacts readily with water (Fig. 65).

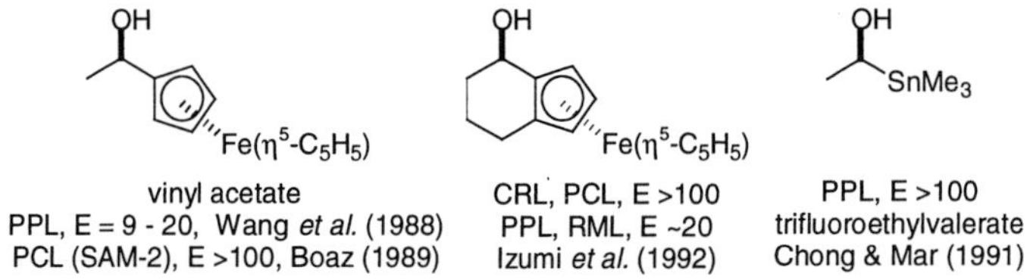

Fig. 65.   Secondary alcohols containing an organometallic substituent.

Later examples of lipase-catalyzed resolutions involved organometallics with planar chirality. For the (arene)chromium tricarbonyl complexes (Fig. 66), the *Pseudomonas* lipases were the most enantioselective. CRL showed opposite, but low E (Uemura et al., 1994). PCL also resolved several other metal carbonyl complexes (Fig. 67), and ferrocenes (Fig. 68). Surprisingly, the shape of the favored enantiomers in PCL catalyzed reactions is similar for all the (arene)chromium tricarbonyl complexes, but is opposite for most of the ferrocenes.

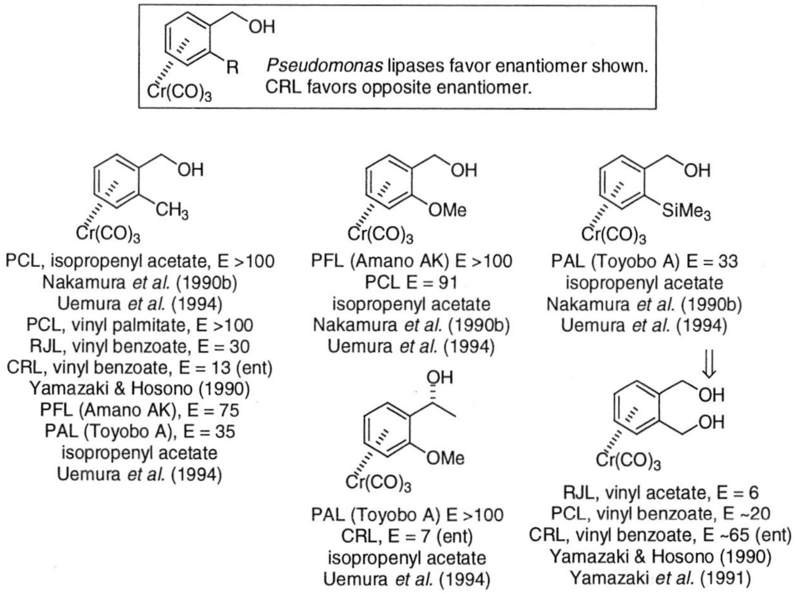

OH
R
Cr(CO)3
Pseudomonas lipases favor enantiomer shown.
CRL favors opposite enantiomer.
OH
CH3
Cr(CO)3
PCL, isopropenyl acetate, E >100
Nakamura et al. (1990b)
Uemura et al. (1994)
PCL, vinyl palmitate, E >100
RJL, vinyl benzoate, E = 30
CRL, vinyl benzoate, E = 13 (ent)
Yamazaki & Hosono (1990)
PFL (Amano AK), E = 75
PAL (Toyobo A), E = 35
isopropenyl acetate
Uemura et al. (1994)
OH
OMe
Cr(CO)3
PFL (Amano AK) E >100
PCL E = 91
isopropenyl acetate
Nakamura et al. (1990b)
Uemura et al. (1994)
OH
OMe
Cr(CO)3
PAL (Toyobo A) E >100
CRL, E = 7 (ent)
isopropenyl acetate
Uemura et al. (1994)
OH
SiMe3
Cr(CO)3
PAL (Toyobo A) E = 33
isopropenyl acetate
Nakamura et al. (1990b)
Uemura et al. (1994)
OH
OH
Cr(CO)3
RJL, vinyl acetate, E = 6
PCL, vinyl benzoate, E ~20
CRL, vinyl benzoate, E ~65 (ent)
Yamazaki & Hosono (1990)
Yamazaki et al. (1991)

Fig. 66. Favored enantiomer in the lipase-catalyzed reaction of *ortho*-substituted hydroxymethyl benzene chromium (0) tricarbonyl complexes. The *Pseudomonas* lipases favor the enantiomer with the general structure shown for five examples, while CRL favors the opposite enantiomer in three examples.

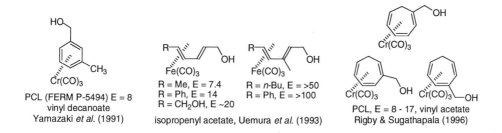

HO
CH3
Cr(CO)3
PCL (FERM P-5494) E = 8
vinyl decanoate
Yamazaki et al. (1991)
R
OH
Fe(CO)3
R = Me, E = 7.4
R = Ph, E = 14
R = CH2OH, E ~20
R
OH
Fe(CO)3
R = n-Bu, E = >50
R = Ph, E = >100
isopropenyl acetate, Uemura et al. (1993)
OH
Cr(CO)3
OH
Cr(CO)3
Cr(CO)3
OH
PCL, E = 8 - 17, vinyl acetate
Rigby & Sugathapala (1996)

Fig. 67. Other examples of metal carbonyl complexes resolved by lipases.

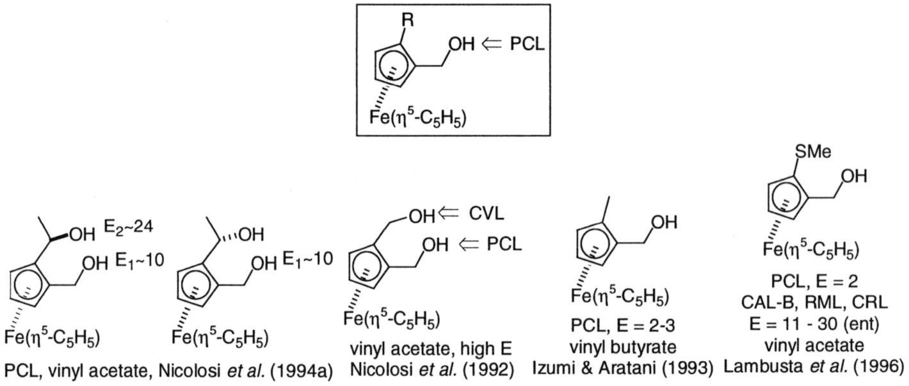

Fig. 68.   Favored enantiomer in the lipase-catalyzed reaction of 1,2-disubstituted ferrocenes. For the hydroxymethyl substituted ferrocenes, PCL favors the general structure shown. Note the absolute configuration of the favored enantiomer of the ferrocenes differs from the benzene tricarbonyls in Fig. 66.

CRL and CE were the most enantioselective lipases toward phenols containing sulfur or phosphorus stereocenters (Fig. 69). Examples of acids containing phosphorus or sulfur stereocenters are summarized in Sect. 5.2.4.

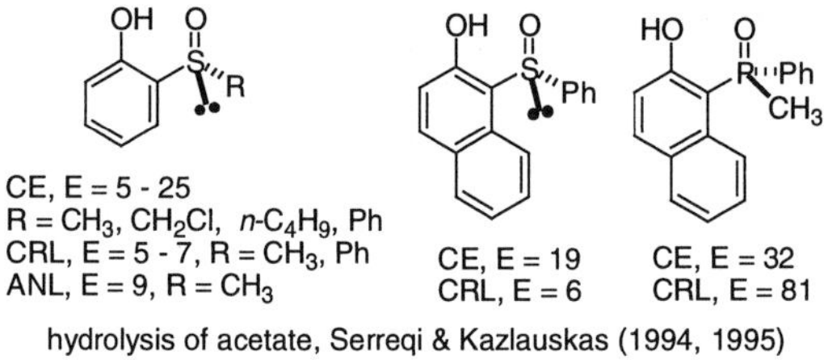

Fig. 69.   Lipase-catalyzed enantioselective reactions of alcohols containing phosphorus or sulfur stereocenters.

CRL and CVL catalyzed the enantioselective acylation of prochiral 2-sila-1,3-pro-panediols (Fig. 70), but favored opposite enantiomers (Djerourou and Blanco, 1991). Unfortunately, several closely related compounds reacted very slowly and with low enantioselectivity (Aouf et al., 1994).

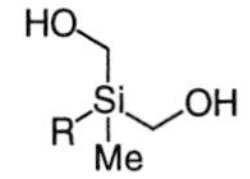

CRL or CVL (favor opposite enantiomers)
R = Ph, 70% ee, 50-80% yield
R = *n*-octyl, 75% ee, 63-70% yield
acylation with isobutyrate esters
absolute configuration not established
Djerourou & Blanco (1991)

Fig. 70.   CRL and CVL catalyzed enantioselective acylation of prochiral silanes.

5.1.3.4   Analogs of Alcohols: Amines, Thiols, and Hydroperoxides

Amines. Lipases also catalyze the enantioselective acylation of amines, although reactions are slower than for alcohols. CAL-B is the most popular lipase, but PCL and lipase from *Pseudomonas aeruginosa* also showed high enantioselectivity. Most researchers used less reactive acylating agents (e.g., esters or carbonates) to avoid chemical acylation of the more nucleophilic amines. Primary amines of the type NH_2CHRR' are isosteric with secondary alcohols. Smidt et al. (1996) proposed extending the secondary alcohol rule to primary amines for CAL-B and indeed all of the amines below fit this rule (Fig. 71).

BASF AG commercialized the resolution of primary amines by a *Pseudomonas* lipase-catalyzed acylation (Balkenhohl et al., 1997). A key discovery was the acylation reagent ethyl methoxyacetate. Activated acyl donors react chemically, while lipase-catalyzed reactions with simple esters or carbonates are usually slow. Acylation with ethyl methoxyacetate proceeds at least 100 times faster than acylation with ethyl butyrate. The reason for this acceleration is not known, but one possibility is that the oxygen may help deprotonate the amine. The hydrolysis of the corresponding methoxyacetate esters of amines also proceeded significantly faster compared to hydrolysis of simple acetates, however here CAL-B was superior compared to *Pseudomonas* sp. lipases (Wagegg et al., 1998).

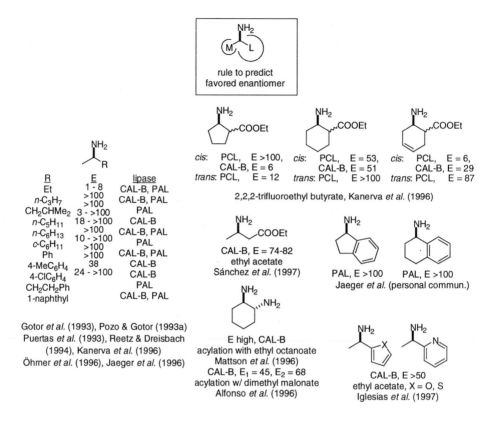

Fig. 71.   Examples of lipase-catalyzed resolutions of amines.

Researchers also resolved several amines which do not resemble secondary alcohols. Orsat et al. (1996) resolved a secondary amine with CRL and Yang et al. (1995a) resolved a primary amine that is isosteric with a primary alcohol (Fig. 72).

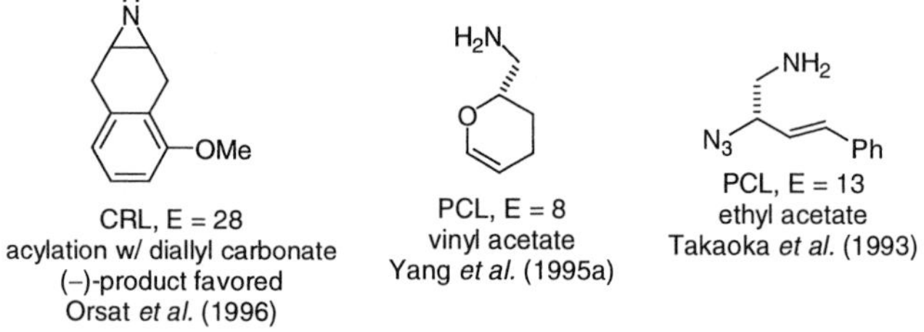

Fig. 72.   Other amines resolved by lipases.

Lipases usually do not catalyze hydrolysis of amides. One exception is the CAL-B-catalyzed hydrolysis of N-acetyl 1-arylethylamines, but reaction times were a week or longer (Smidt et al., 1996). Chapman et al. (1996) found an indirect way to resolve amides using oxalamic esters. CAL-B catalyzed hydrolysis of the ester group and showed high enantioselectivity toward the remote stereocenter. The stereocenter now lies in the acid portion of the reacting ester, so the rule above no longer applies. Coincidentally, this reaction also favors the same amine enantiomer predicted above (Fig. 73).

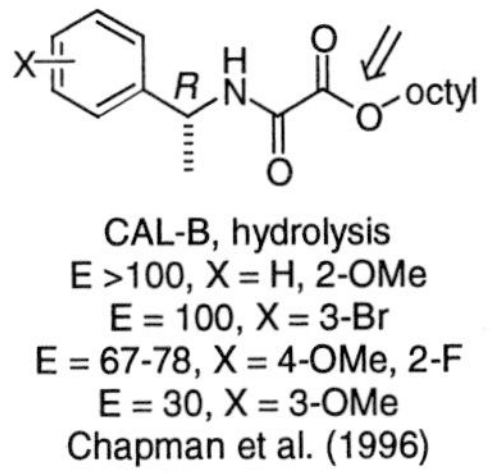

CAL-B, hydrolysis
E >100, X = H, 2-OMe
E = 100, X = 3-Br
E = 67-78, X = 4-OMe, 2-F
E = 30, X = 3-OMe
Chapman et al. (1996)

Fig. 73.   Resolution of amines as oxalamic esters.

Thiols. Researchers resolved several thiols, which are the simplest analogs of alcohols (Fig. 74). Resolutions used either hydrolysis or alcoholysis of thiol esters and examples include primary and secondary thiols and even an axially chiral thiol. Enantioselectivities were similar to those for the corresponding alcohols. Baba et al. (1990a) and Öhrner et al. (1996) noted that acylation of thiols gave no reaction. The acyl enzyme intermediate may not be a strong enough acyl donor to acylate thiols.

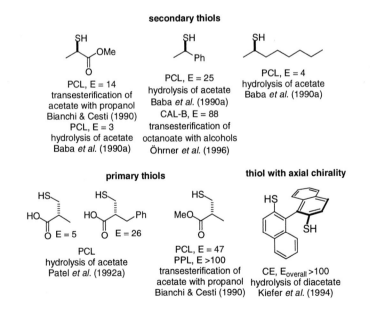

secondary thiols

PCL, E = 14
transesterification of
acetate with propanol
Bianchi & Cesti (1990)
PCL, E = 3
hydrolysis of acetate
Baba et al. (1990a)

PCL, E = 25
hydrolysis of acetate
Baba et al. (1990a)
CAL-B, E = 88
transesterification of
octanoate with alcohols
Öhrner et al. (1996)

PCL, E = 4
hydrolysis of acetate
Baba et al. (1990a)

primary thiols

E = 5   E = 26

PCL
hydrolysis of acetate
Patel et al. (1992a)

PCL, E = 47
PPL, E >100
transesterification of
acetate with propanol
Bianchi & Cesti (1990)

thiol with axial chirality

CE, E$_{overall}$ >100
hydrolysis of diacetate
Kiefer et al. (1994)

Fig. 74.   Selected examples of lipase-catalyzed enantioselective reactions of thiols.

Peroxides. Lipases also discriminate between enantiomers of alkyl peroxides, which resemble primary alcohols. Enantioselective acylation of alkyl peroxides yielded unreacted starting material in high enantiomeric excess, but the produced peroxyesters decomposed under the reaction conditions to ketones. The structures in Fig. 75 show the fast-reacting enantiomer.

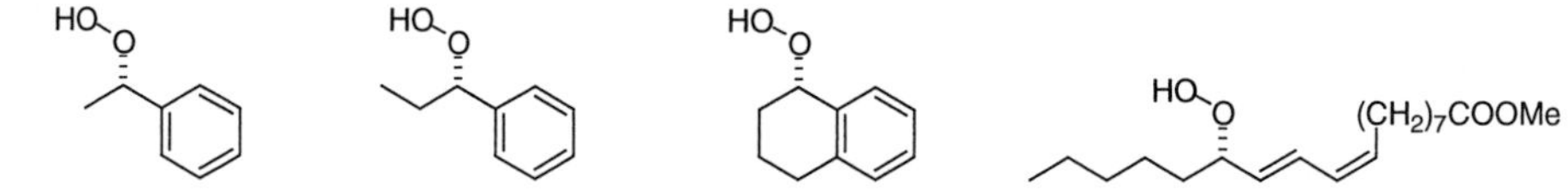

PCL, E = 29	PCL, E = 4	PPL, E = 4	PCl, E = 4
isopropenyl acetate	isopropenyl acetate	isopropenyl acetate	vinyl acetate
Baba *et al.* (1988)	Baba *et al.* (1988)	Höft *et al.* (1995)	Baba *et al.* (1990b)

Fig. 75.   Alkyl peroxides resolved by PCL-catalyzed acylations.

5.2   Carboxylic Acids

5.2.1   General Considerations

There are fewer examples of lipase-catalyzed enantioselective reactions of carboxylic acids (reviewed by Haraldsson, 1992). In water, lipases catalyze hydrolyses of various carboxylic acid esters, while in organic solvents, lipases catalyze esterification of acids, transesterification of esters and aminolysis of esters. Reactions of chiral anhydrides and lactones are discussed below in Sects. 5.2.6 and 5.3.

5.2.2   Carboxylic Acids with a Stereocenter at the α-Position

5.2.2.1   *Candida antarctica* Lipase B

In contrast to its high enantioselectivity toward alcohols, CAL-B usually shows low to moderate enantioselectivity toward carboxylic acids (Fig. 76). Preparation of enantiomerically-pure 2-arylpropionic acids, a class of non-steroidal anti-inflammatory drugs, required two sequential resolutions (Morrone et al., 1995; Trani et al., 1995). Starting from 300 g of racemic ibuprofen (Ar = 4-i-BuC$_6$H$_4$) Trani et al. (1995) used two sequential esterifications to make 38 g of (S)-ibuprofen with 97.5 %ee.

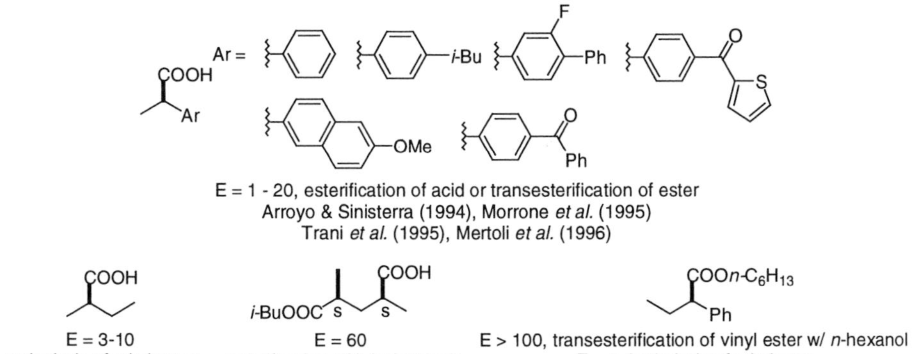

Fig. 76.   CAL-B-catalyzed enantioselective reaction of carboxylic acids with a stereocenter at the α-position.

The acyl binding site of CAL-B is a shallow crevice. It is likely that the lower enantioselectivity toward stereocenters in the acyl part of an ester stems from fewer and/or weaker contacts between the acyl part and its binding site. In contrast, the alcohol binding site appears to engulf the alcohol.

5.2.2.2   *Candida rugosa* Lipase

In contrast to CAL-B, CRL shows high enantioselectivity toward many carboxylic acids (Fig. 77), and a rule can predict the enantiopreference of CRL-catalyzed reactions of carboxylic acids with a stereocenter at the α-position. Recent reviews also contain a few more examples of acids resolved by CRL (Ahmed et al., 1994; Franssen et al., 1996).

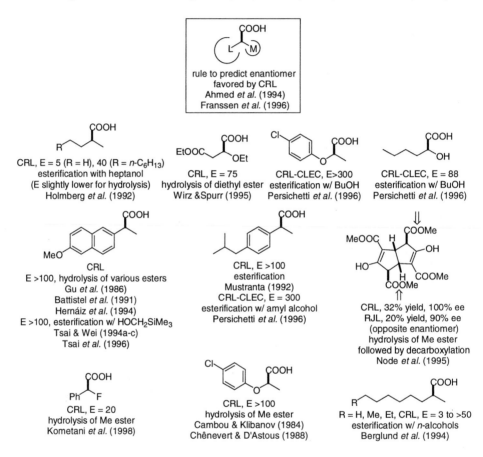

Fig. 77.   Selected examples of CRL-catalyzed enantioselective reactions of carboxylic acids with a stereocenter at the α-carbon.

One important commercial target are pure (*S*)-enantiomers of 2-arylpropionic acids. Although CRL shows high enantioselectivity toward these acids, reaction rates are too slow for commercial use. Chirotech in the UK developed a resolution of naproxen using a *Bacillus* esterase (Quax and Broekhuizen, 1994, see also Sect. 9.4.1). This process produced 13 tons of (*S*)-naproxen in 1996 (Stinson, 1997).

Comparing the above empirical rule to the X-ray structure of CRL suggests that the large substituent, L, binds in a tunnel, while the stereocenter lies at the mouth of this tunnel. Indeed, molecular modeling supports this proposal (Holmquist et al., 1996; Botta et al., 1997). Further modeling rationalized some known exceptions to the empirical rule (Holmquist et al., 1996). When the large substituent is extensively branched, it no longer fits in the tunnel. An alternate binding mode with the substrate outside the tunnel favors the opposite enantiomer.

CRL, E = 2 - 26
transester. of Me ester w/ BuOH
Kanerva & Sundholm (1993b)

CRL, E = 12 - 83
hydrolysis of methyl ester
Bhaskar Rao *et al.* (1994)

R = *i*-Pr, CRL, PPL, E >100
R = *t*-Bu, CRL, E = 36
abs. config. not established
transester. of $-OCH_2CF_3$ ester w/ BuOH
Martres *et al.* (1994)

R = H, E>100
R = Me, E = 13
R = Ph, E not reported
R = COOMe, E = 11
hydrolysis of Me of Et ester
Gu & Li (1992)

CRL, E >50
hydrolysis of Bu ester
Csuk & Dorr (1994)

CRL, E >100
transester. of Meester w/ BuOH
Franssen *et al.* (1996)

CRL, E = 13
hydrolysis of Me ester
Schueller *et al.* (1996)

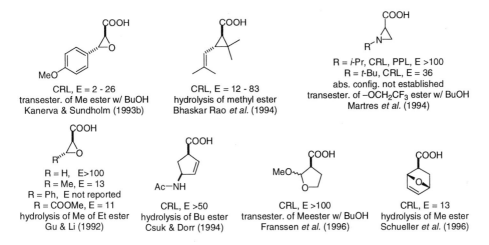

Fig. 77. Selected examples of CRL-catalyzed enantioselective reactions of carboxylic acids with a stereocenter at the α-carbon (continued).

Researchers increased the enantioselectivity of CRL-catalyzed resolution of chiral acids using a number of different methods. Most examples involve either 2-arylpropionic acids or 2-aryloxypropionic acids, a class of herbicides. For example, Guo and Sih increased the enantioselectivity of a CRL-catalyzed hydrolysis of the 2-chloroethyl ester of 2-(3-benzoyl)phenylpropanoic acid by adding a chiral amine, dextromethorphan (Guo and Sih, 1989; Sih et al., 1992). Kinetic analysis showed that dextromethorphan increased the enantioselectivity by inhibiting the hydrolysis of the slow-reacting enantiomer. In another example, Colton et al. (1995) increased the enantioselectivity of a CRL-catalyzed hydrolysis of the methyl ester of 2-(4-chloro)phenoxypropanoic acid by a purification procedure that involved treating the enzyme with isopropanol (Fig. 78).

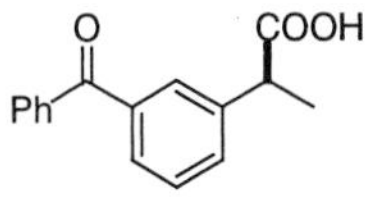

crude, E = 4
added dextromethorphan, E = 42
hydrolysis of 2-chloroethyl ester
Guo & Sih (1989)

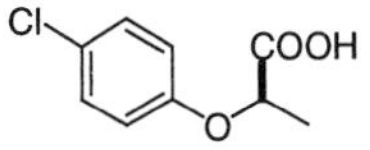

crude, E = 2.3-17
isopropanol-treated, E >100
hydrolysis of methyl ester
Colton *et al.* (1995)

Fig. 78. Several methods increase the enantioselectivity of CRL toward acids.

Other researchers have increased the enantioselectivity of CRL toward 2-aryl- or 2-aryloxypropionic acids by changing the solvent (Miyazawa et al., 1992), temperature (Yasufuku and Ueji, 1995) or pH, by carrying out the reaction in a microemulsion (Hedström et al., 1993), by adding (*S*)-2-amino-4-methylthio-1-butanol (Itoh et al., 1991) or Triton X-100 (a surfactant) (Bhaskar-Rao et al., 1994), by linking the ε-amino group of lysine residues to a solid support (Sinisterra et al., 1994), by nitration of tyrosyl residues (Gu and Sih, 1992), by purification and cross-linking of crystals of CRL (Lalonde et al., 1995; Persichetti et al., 1996), by purification (Wu et al., 1990; Allenmark and Ohlsson, 1992a, b), and by careful addition of water to avoid clumping in organic solvents (Tsai and Dordick, 1996).

Chemists do not know how these changes increase enantioselectivity on a molecular level, but two possibilities are most reasonable. First, the treatments may remove or inactivate a contaminating hydrolase with low or opposite enantioselectivity. Indeed, Lalonde et al. (1995) reported a contaminating esterase in commercial CRL. Second, the treatments may change the conformation of CRL. Crystallographers solved the structures for both an 'open' and a 'closed' form of CRL which differed in the orientation of the lipase lid (a surface helical region). Indeed, Lalonde et al. (1995) found that crystals of the 'open' and 'closed' forms differed in their enantioselectivity. These two possibilities do not exclude each other, so both effects may contribute to the increased enantioselectivity.

5.2.2.3 *Pseudomonas* Lipases

PCL also catalyzes enantioselective reactions of acids (Fig. 79). The relative sizes of the substituents cannot account for the enantiopreference, but note that all but one example have a similar orientation of an electron-withdrawing group.

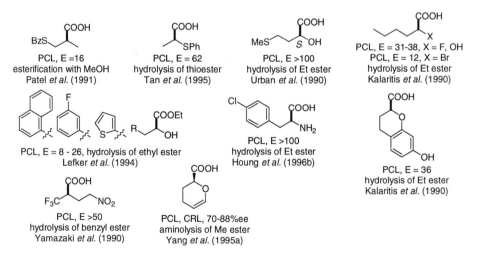

Fig. 79.   Selected examples of PCL-catalyzed enantioselective reactions of carboxylic acids with a stereocenter at the α-carbon.

O'Hagan and Rzepa (1994) suggested that the high enantioselectivity of PCL toward acids with a fluorine substituent at the α-position may be due to a stereoelectronic effect. In nonenzymic reactions, nucleophilic attack at a carbonyl favors an *anti* orientation of an electron withdrawing substituent at the α-position. A similar preference in the active site of PCL may also account for the observed enantioselectivity.

5.2.2.4 Other Lipases

Fig. 80 summarizes selected examples of enantioselective reactions involving other lipases.

The PPL-catalyzed resolution of amino acid esters (Houng et al., 1996a, b) used crude enzyme which contains protease contaminants. Researchers observed high enantioselectivity (E > 100) only for amino acids where the alkyl group is –CH$_2$–aryl. These amino acid esters are good substrates for chymotrypsin, thus chymotrypsin, a likely contaminant, may contribute to the observed selectivity.

A survey of the enantioselectivity of ANL toward carboxylic acids identified α-amino acids as the best resolved class of carboxylic acids (Janes and Kazlauskas, 1997a). Replacement of the positively charged –NH$_3^+$ substituent with –OH or –CH$_3$ lowered the enantioselectivity drastically. Note that CRL and RML favor opposite enantiomers of 2-arylpropionic acids.

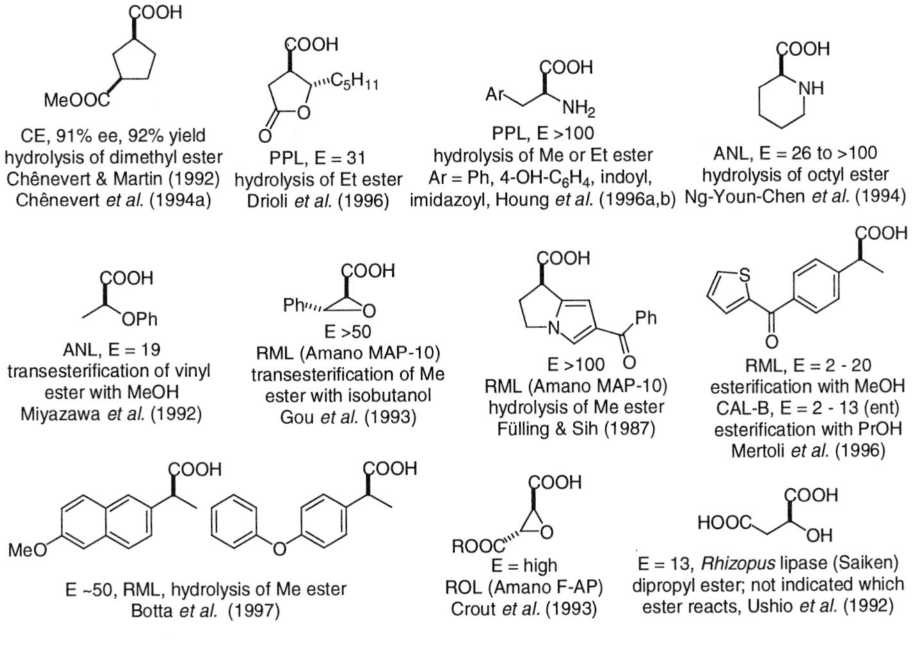

Fig. 80.   Selected examples of lipase-catalyzed enantioselective reactions of carboxylic acids with a stereocenter at the α-carbon.

5.2.3   Carboxylic Acids with a Stereocenter at the β-Position

Even though the stereocenter is further away, a number of lipases also catalyze enantio-selective reactions of carboxylic acids with a stereocenter at the β-position (Fig. 81).

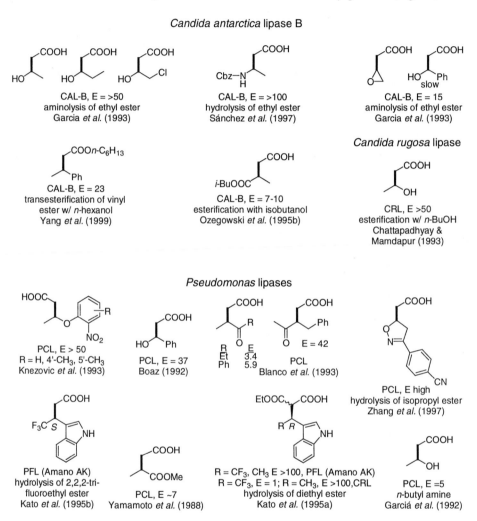

Fig. 81.   Selected examples of lipase-catalyzed enantioselective reactions of carboxylic acids with a stereocenter at the β-position.

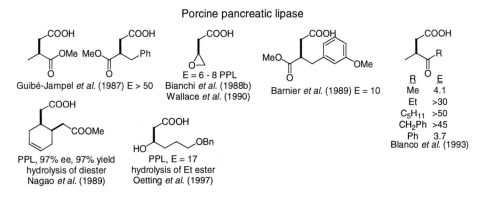

Porcine pancreatic lipase

Guibé-Jampel *et al.* (1987) E > 50

Bianchi *et al.* (1988b)
Wallace *et al.* (1990)
E = 6 - 8 PPL

Barnier *et al.* (1989) E = 10

R	E
Me	4.1
Et	>30
C$_5$H$_{11}$	>50
CH$_2$Ph	>45
Ph	3.7

Blanco *et al.* (1993)

PPL, 97% ee, 97% yield
hydrolysis of diester
Nagao *et al.* (1989)

PPL, E = 17
hydrolysis of Et ester
Oetting *et al.* (1997)

Fig. 81.   Selected examples of lipase-catalyzed enantioselective reactions of carboxylic acids with a stereocenter at the β-position (continued).

5.2.4   Other Carboxylic Acids

5.2.4.1   Quaternary Stereocenters

Several examples are shown in Fig. 82.

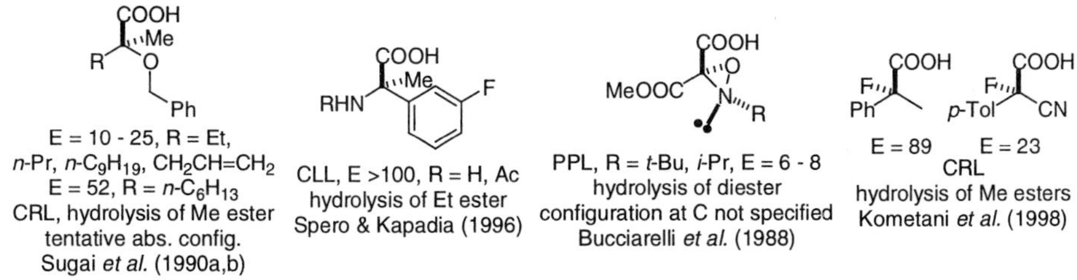

E = 10 - 25, R = Et,
n-Pr, *n*-C$_9$H$_{19}$, CH$_2$CH=CH$_2$
E = 52, R = *n*-C$_6$H$_{13}$
CRL, hydrolysis of Me ester
tentative abs. config.
Sugai *et al.* (1990a,b)

CLL, E >100, R = H, Ac
hydrolysis of Et ester
Spero & Kapadia (1996)

PPL, R = *t*-Bu, *i*-Pr, E = 6 - 8
hydrolysis of diester
configuration at C not specified
Bucciarelli *et al.* (1988)

E = 89   E = 23
CRL
hydrolysis of Me esters
Kometani *et al.* (1998)

Fig. 82.   Selected examples of lipase-catalyzed enantioselective reactions of carboxylic acids with quaternary stereocenters.

5.2.4.2 Sulfur Stereocenters

Several examples are shown in Fig. 83.

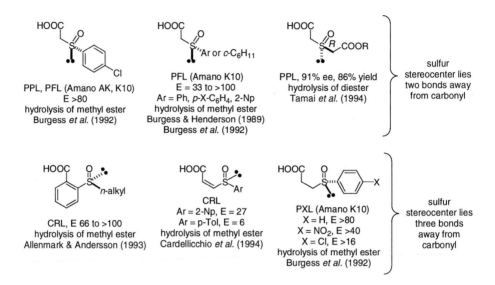

Fig. 83.   Selected examples of lipase-catalyzed enantioselective reactions of carboxylic acids with sulfur stereocenters.

5.2.4.3 Remote Stereocenters

Lipases occasionally show high enantioselectivity toward carboxylic acids with stereocenters far from the carbonyl (Fig. 84). For example, researchers at Merck Research Laboratories (Rahway, NJ, USA) used *Pseudomonas* lipases to enantioselectively hydrolyze the *pro-R* ester in the dithioacetal (Fig. 84) yielding the (*S*)-monoester in enantiomeric purity even though the sterecenter lies four bonds from the carbonyl (Hughes et al., 1989, 1990, 1993; Smith et al., 1992; Chartrain et al., 1993). The enantioselectivity dropped for analogues where the stereocenter lies either three or five bonds from the carbonyl. Bhalerao et al. (1991) found that CRL showed surprisingly high enantioselectivity toward several carboxylic acids where the stereocenter lies eight or nine bonds (but not seven bonds) from the carbonyl. The x-ray crystal structures of CRL suggest that the carboxylic acid binds in a tunnel which bends approximately at C-9 of a fatty acid. The observed enantioselectivity toward stereocenters in this region may be due to this bend. Researchers also resolved several other synthetic intermediates with remote stereocenters.

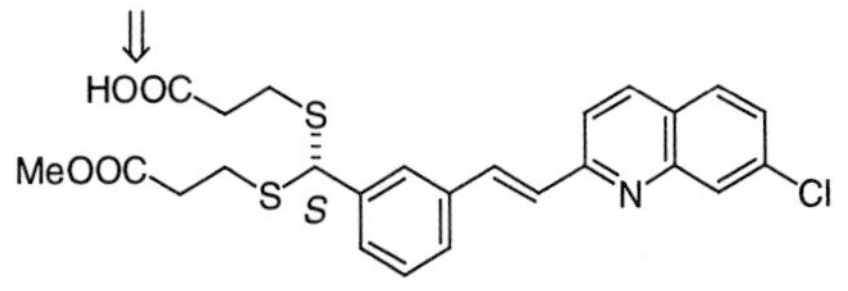

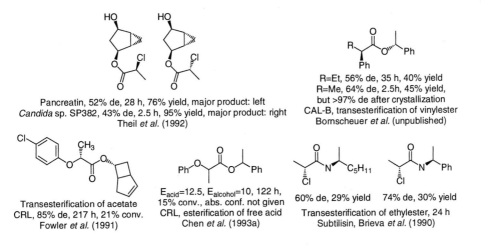

>98% ee, 95% yield, hydrolysis of dimethyl ester
PCL, Hughes *et al.* (1989, 1990, 1993), Smith *et al.* (1992)
P. aeruginosa lipase, Chartrain *et al.* (1993)

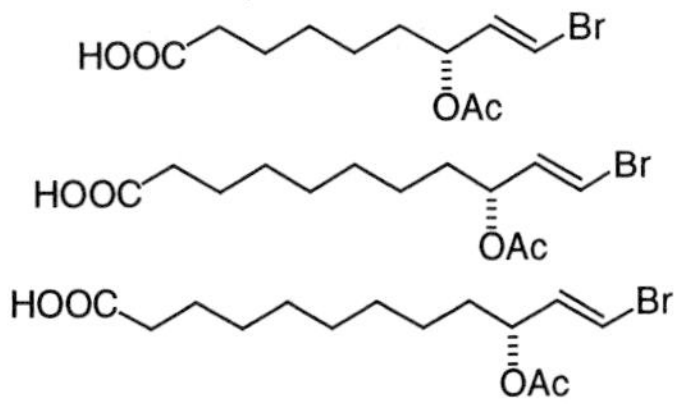

CRL, hydrolysis of methyl or butyl ester
E = 4, >100, >100, Bhalerao *et al.* (1991)

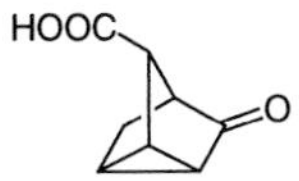

CAL-A, E >100
hydrolysis of methyl ester
Kingerywood & Johnson (1996)

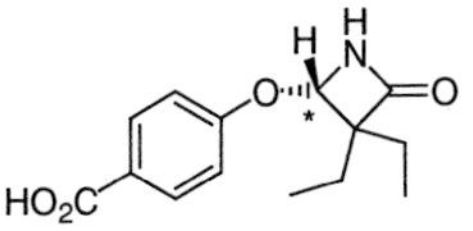

PCL, E = 37, PAL, E ~20
hydrolysis of benzyl ester
Cvetovich *et al.* (1996)

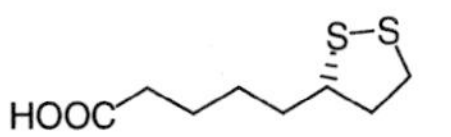

CRL, E = 7.5
esterification w/ 1-hexanol
Fadnavis & Koteshwar (1997)

Fig. 84.   Selected examples of lipase-catalyzed enantioselective reactions of carboxylic acids with remote stereocenters.

5.2.5   Double Enantioselection

The term double enantioselection refers to the synthesis of esters from racemic (or prostereogenic) carboxylic acids and alcohols. However, in order to achieve an efficient reaction, the following criteria must be fulfilled: (1) the enzyme should be highly enantioselective towards both substrates and (2), the reaction equilibrium must be shifted towards ester synthesis in order to obtain high conversion of both enantiomers. The first problem might be solved by investigating different hydrolases, whereas the low reaction rate might be enhanced by using activated acyl donors (see Sect. 4.2.3).

Pancreatin, 52% de, 28 h, 76% yield, major product: left
Candida sp. SP382, 43% de, 2.5 h, 95% yield, major product: right
Theil *et al.* (1992)

R=Et, 56% de, 35 h, 40% yield
R=Me, 64% de, 2.5h, 45% yield,
but >97% de after crystallization
CAL-B, transesterification of vinylester
Bornscheuer *et al.* (unpublished)

Transesterification of acetate
CRL, 85% de, 217 h, 21% conv.
Fowler *et al.* (1991)

E_{acid}=12.5, $E_{alcohol}$=10, 122 h,
15% conv., abs. conf. not given
CRL, esterification of free acid
Chen *et al.* (1993a)

60% de, 29% yield   74% de, 30% yield

Transesterification of ethylester, 24 h
Subtilisin, Brieva *et al.* (1990)

Fig. 85.   Examples for lipase- or subtilisin-catalyzed double enantioselections.

To date, double enantioselection has been investigated only to a limited extent (Fig. 85). For instance, Theil and coworkers used *rac*-2,2,2-trifluoroethyl-2-chloropropanoate in the resolution of a *meso*-diol in THF. However, the reaction proceeded only with moderate selectivity and best results were achieved using a lipase from *Candida* sp. yielding a diastereomeric excess of 52 %de (Theil et al., 1992). Direct esterification of e.g., *rac*-α-phenoxypropionic acid with *rac*-α-phenylethanol in *n*-hexane was much slower (only 15 % conversion after 122 h) and enantioselectivity was very low (E = 10.4 (alcohol) and 12.5 (acid) (Chen et al., 1993a).

5.2.6 Anhydrides

PCL catalyzed the regioselective ring opening of 2-substituted succinic and glutaric acid anhydrides, but without enantioselectivity (Hiratake et al., 1989). Reaction occurred at the less hindered carbonyl (Fig. 86).

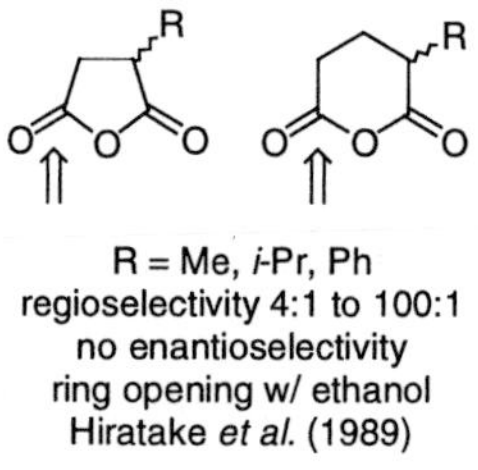
R
R
O
O
O
O
O
O
R = Me, *i*-Pr, Ph
regioselectivity 4:1 to 100:1
no enantioselectivity
ring opening w/ ethanol
Hiratake *et al.* (1989)

Fig. 86. Regioselective ring opening of anhydrides occurred at the less hindered carbonyl.

However, CAL-B was both regio- and enantioselective (Ozegowski et al., 1994b, 1995b). For 2-methyl glutaric acid anhydride, the (*R*)-enantiomer reacted at the more hindered carbonyl, while the (*S*)-enantiomer reacted at the less hindered carbonyl (Eq. 11). Similarly, the (2*R*)-enantiomer of 2,3-dimethylglutarates reacted at the more hindered carbonyl while the (2*S*)-enantiomer reacted at the less-hindered carbonyl (Eq. 12 and 13) (Ozegowski et al., 1996). A similar reaction with the five-membered succinic anhydride was less enantioselective (not shown).

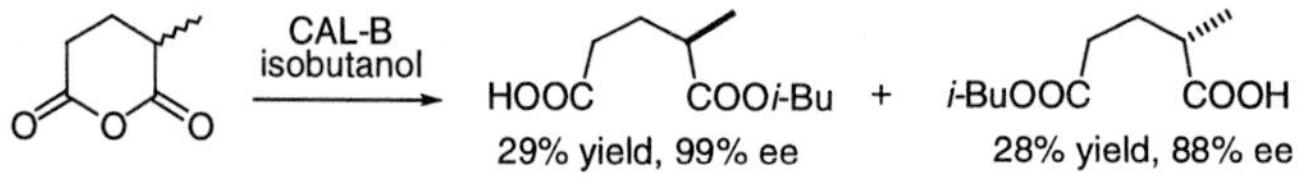
O
O
O
CAL-B
isobutanol
HOOC
COO*i*-Bu
+
i-BuOOC
COOH
29% yield, 99% ee
28% yield, 88% ee

(11)

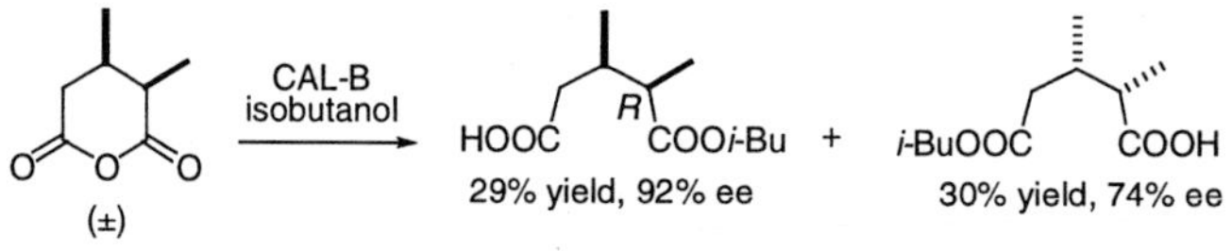
O
O
O
(±)
CAL-B
isobutanol
HOOC
R
COO*i*-Bu
+
i-BuOOC
COOH
29% yield, 92% ee
30% yield, 74% ee

(12)

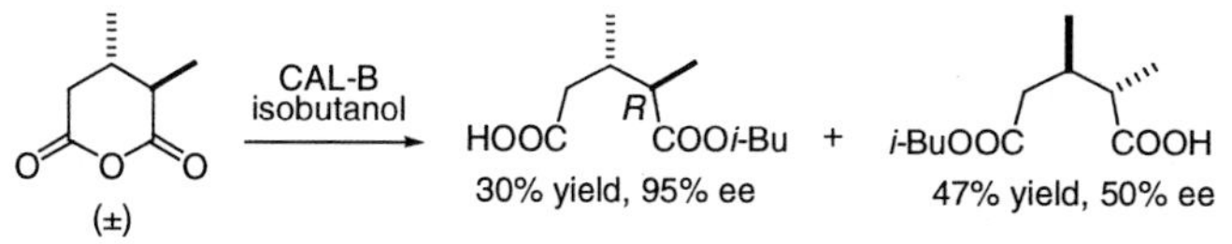

$$\text{(13)}$$

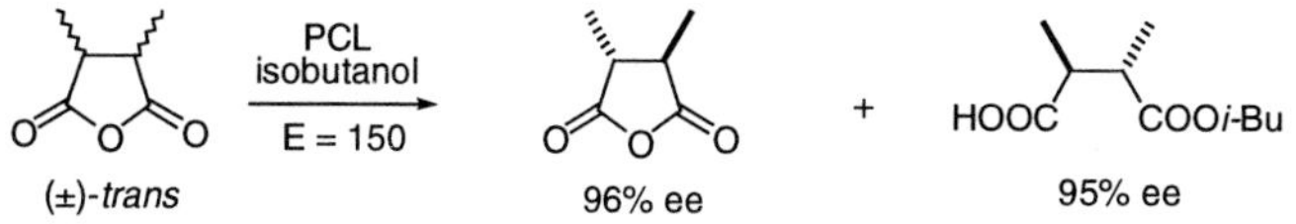

CAL-B
isobutanol

HOOC $\overset{R}{\underset{}{}}$ COO*i*-Bu   +   *i*-BuOOC $\overset{}{\underset{}{}}$ COOH

(±)   30% yield, 95% ee   47% yield, 50% ee

PCL
isobutanol
E = 150

(±)-*trans*   96% ee   +   HOOC $\overset{}{\underset{}{}}$ COO*i*-Bu

95% ee

$$\text{(14)}$$

For resolution of *syn*-2,3-dimethylbutandioic anhydride, PCL was the most enantioselective enzyme (Eq. 14) (Ozegowski et al., 1995a).

Lipase-catalyzed ring-opening of prochiral and *meso* anhydrides also proceeded with good enantioselectivity (Fig. 87).

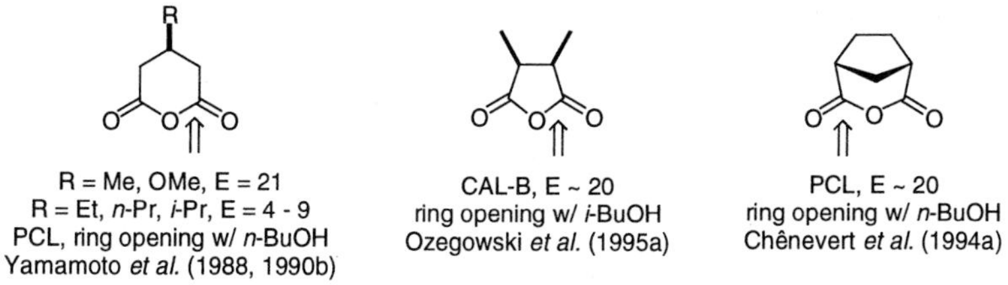

R = Me, OMe, E = 21
R = Et, *n*-Pr, *i*-Pr, E = 4 - 9
PCL, ring opening w/ *n*-BuOH
Yamamoto *et al.* (1988, 1990b)

CAL-B, E ~ 20
ring opening w/ *i*-BuOH
Ozegowski *et al.* (1995a)

PCL, E ~ 20
ring opening w/ *n*-BuOH
Chênevert *et al.* (1994a)

Fig. 87.   Lipase-catalyzed ring opening of prochiral and *meso* anhydrides.

5.3   Lactones

Lactones are important flavor compounds and synthetic intermediates. Researchers have used lipases for the synthesis of enantiomerically pure lactones either directly using reactions involving the lactone link, or indirectly using other reactions that eventually lead to enantiomerically pure lactones. Another application of lipases is the selective formation of macrolides or diolides (cyclic dimers) from hydroxyacids. Without a lipase catalyst, oligomers are the major product.

Enantioselective reactions involving the lactone link are summarized in Fig. 88. PPL catalyzed the enantioselective lactonization of a wide range of γ-hydroxyesters to the five-membered γ-lactones (Eq. 15). Researchers used hydroxy esters, not hydroxy acids, as the starting materials to avoid spontaneous lactonization. The enantioselectivity was moderate to good, but the reaction times were often several days. Hydrolysis of the γ-lactones was less enantioselective than lactonization. Resolutions using esterases are covered in Sect. 9.4.2.

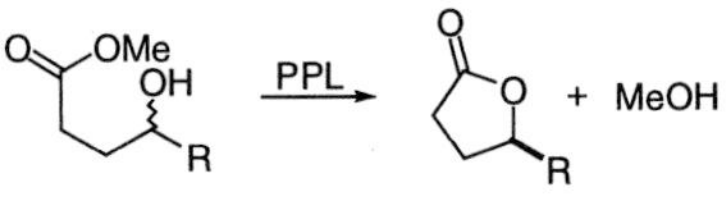
OMe
OH
R
PPL
+ MeOH
R

(15)

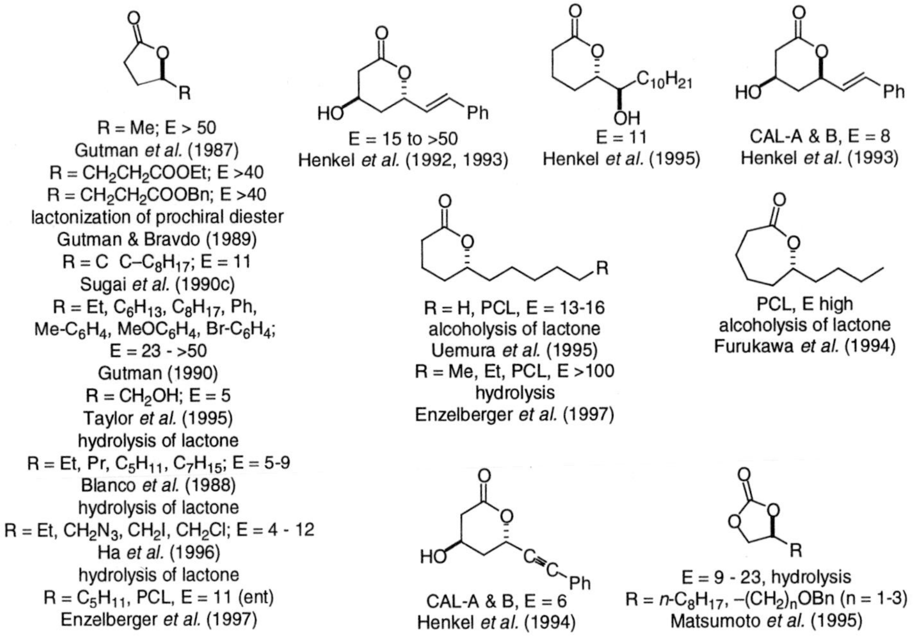
R
R = Me; E > 50
Gutman et al. (1987)
R = CH2CH2COOEt; E >40
R = CH2CH2COOBn; E >40
lactonization of prochiral diester
Gutman & Bravdo (1989)
R = C C–C8H17; E = 11
Sugai et al. (1990c)
R = Et, C6H13, C8H17, Ph,
Me-C6H4, MeOC6H4, Br-C6H4;
E = 23 - >50
Gutman (1990)
R = CH2OH; E = 5
Taylor et al. (1995)
hydrolysis of lactone
R = Et, Pr, C5H11, C7H15; E = 5-9
Blanco et al. (1988)
hydrolysis of lactone
R = Et, CH2N3, CH2I, CH2Cl; E = 4 - 12
Ha et al. (1996)
hydrolysis of lactone
R = C5H11, PCL, E = 11 (ent)
Enzelberger et al. (1997)
HO
Ph
E = 15 to >50
Henkel et al. (1992, 1993)
C10H21
OH
E = 11
Henkel et al. (1995)
HO
Ph
CAL-A & B, E = 8
Henkel et al. (1993)
R
R = H, PCL, E = 13-16
alcoholysis of lactone
Uemura et al. (1995)
R = Me, Et, PCL, E >100
hydrolysis
Enzelberger et al. (1997)
PCL, E high
alcoholysis of lactone
Furukawa et al. (1994)
HO
Ph
CAL-A & B, E = 6
Henkel et al. (1994)
R
E = 9 - 23, hydrolysis
R = n-C8H17, –(CH2)nOBn (n = 1-3)
Matsumoto et al. (1995)

Fig. 88. Lipase-catalyzed enantioselective reactions involving the lactone ring. Unless otherwise noted, all reactions refer to the lactonization of the ester catalyzed by PPL. In all cases, the faster-reacting enantiomer is shown. The cyclic carbonate resembles a lactone so it is included in this list.

PPL also catalyzed the formation of δ-lactones, but with lower enantioselectivity. Surprisingly, the fast-reacting enantiomer differed for the γ- and δ-lactones. Although the alcohol portion of γ- and δ-lactones is a secondary alcohol, the secondary alcohol rule cannot be used here because the stereocenter lies in a different position as shown in Fig. 89.

Attempts to form four- or seven-membered lactones yielded oligomers and polymers as discussed in Sect. 6.3. In some cases, oligomeric side products also formed during the lactonization of six-membered rings. However, ring-opening alcoholysis can resolve seven- (Furukawa et al., 1994) or four-membered lactones (Xu et al., 1996; Adam et al., 1997a) (Fig. 90).

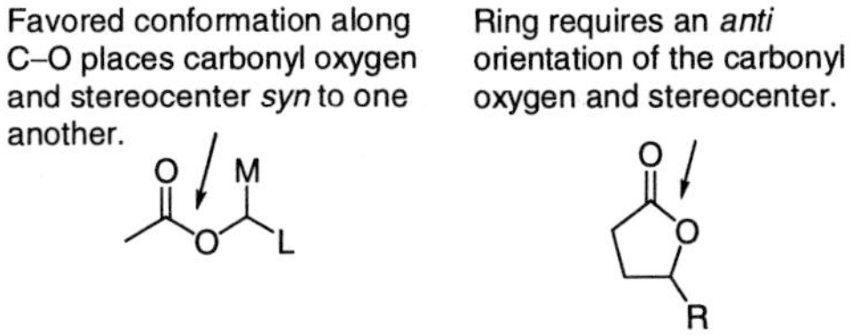

Favored conformation along
C–O places carbonyl oxygen
and stereocenter *syn* to one
another.

Ring requires an *anti*
orientation of the carbonyl
oxygen and stereocenter.

Fig. 89.   The secondary alcohol rule cannot be used for lactones because the stereocenter lies in a different position. Acyclic esters adopt a *syn* conformation along the carbonyl C–alcohol–O-bond. The crystal structure of transition state analogs bound to lipases suggest that this conformation persists in the active site. On the other hand, the lactone ring forces an *anti* conformation along the carbonyl C–alcohol–O-bond which places the stereocenter in a different part of the enzyme. In particular, the lactone stereocenter appears to lie entirely within the L-pocket of the alcohol binding crevice. Indeed, many of the lactone examples in this section do not follow the secondary alcohol rule.

CAL-B, E >100
PCL or PFL, E = 10-74
alcoholysis w/ benzyl alcohol
Adam *et al.* (1997a)

PCL, E = 8
methanolysis
Xu *et al.* (1996)

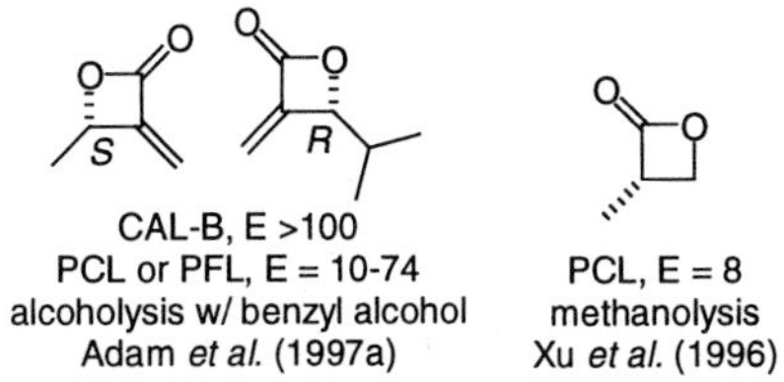

Fig. 90.   Seven- and four-membered ring lactones resolved by lipase-catalyzed alcoholysis.

A more common route to enantiomerically pure lactones is to prepare a precursor (usually an efficiently resolved secondary alcohol) and convert it using lipase to the desired lactone. Selected examples are shown in Fig. 91 (see also Sugai et al., 1990b). Another special case are lactones with additional alcohol or acid functional groups. Researchers resolved several such lactones without affecting the lactone ring; more examples are included in the surveys above. The advantages of these indirect methods are higher enantioselectivity and faster reaction times.

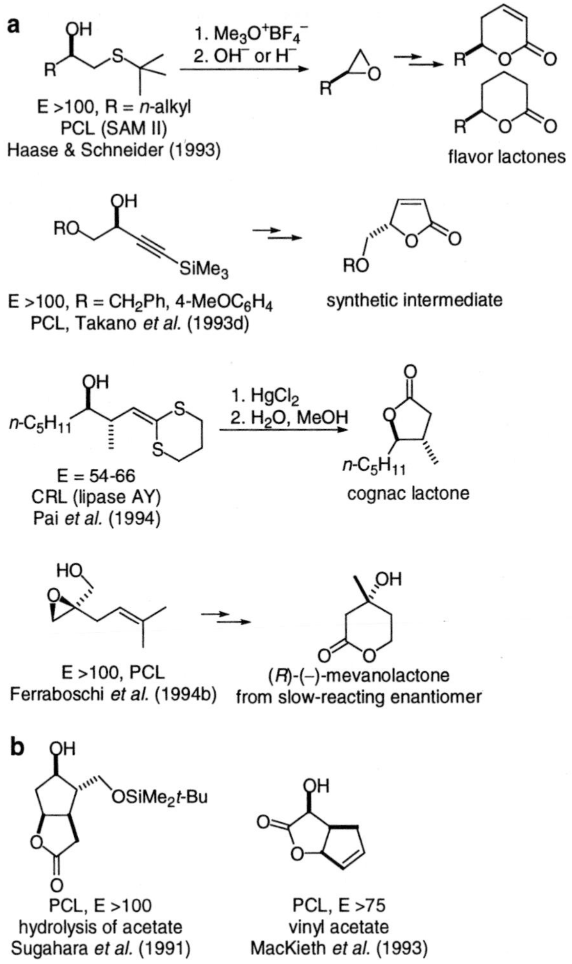
a
1. Me3O+BF4−
2. OH− or H−
E >100, R = n-alkyl
PCL (SAM II)
Haase & Schneider (1993)
flavor lactones
SiMe3
E >100, R = CH2Ph, 4-MeOC6H4
PCL, Takano et al. (1993d)
synthetic intermediate
1. HgCl2
2. H2O, MeOH
E = 54-66
CRL (lipase AY)
Pai et al. (1994)
cognac lactone
E >100, PCL
Ferraboschi et al. (1994b)
(R)-(−)-mevanolactone
from slow-reacting enantiomer
b
OSiMe2t-Bu
PCL, E >100
hydrolysis of acetate
Sugahara et al. (1991)
PCL, E >75
vinyl acetate
MacKieth et al. (1993)

Fig. 91. Indirect resolutions of lactones with lipases. **a** Lipase-catalyzed resolution of lactone precursors and their conversion to lactones. **b** Functionalized lactones can be resolved by reaction at the secondary hydroxyl group without affecting the lactone ring.

Lipases also catalyze the efficient macrolactonization of hydroxy acids as well as macrolactonization of diacids with diols (Fig. 92). Such macrolactonizations are difficult to perform chemically and require high dilution to minimize the competing oligomerization. Below 45 °C, oligomers are also the main products of the lipase catalyzed reactions, but at higher temperatures macrolactonization dominates even without high dilution conditions. Many different lipases catalyze such macrolactonizations. Guo and Sih (1988) reported that the free acids give higher yield of macrolactone than the esters for another group of hydroxy acids. Lobell and Schneider (1993) reported that only the vinyl esters lactonize efficiently.

Although noone knows why lipases favor the formation of macrolactones over oligomers, one possibility is that the hydrophobic binding pocket of lipases favors folded
conformations of the hydroxy acid. These folded conformations place the alcohol and
acid closer to one another and thus favor intramolecular cyclization.

The ring-opening oligomerization of lactones is discussed in Sect. 6.3. The macrolactonization reaction was enantioselective favoring the (R)-enantiomer for lactones (Lobell
and Schneider, 1993) and dilactones (Guo and Sih, 1988) (Fig. 93).

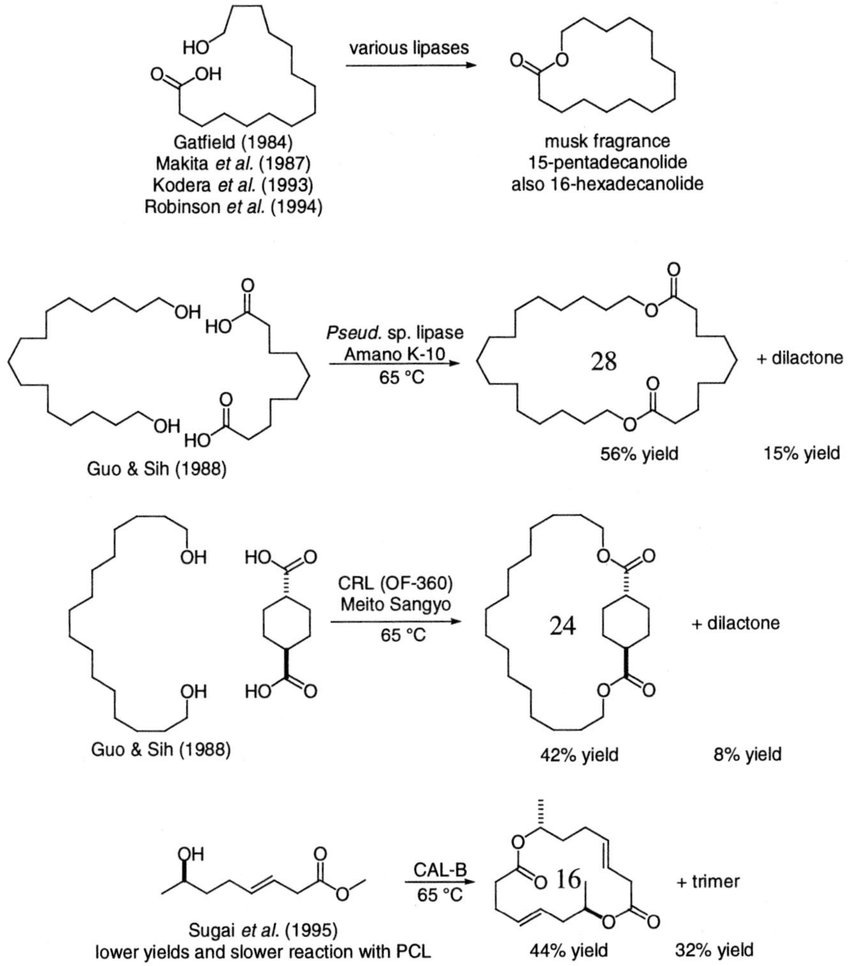

Gatfield (1984)
Makita *et al.* (1987)
Kodera *et al.* (1993)
Robinson *et al.* (1994)

musk fragrance
15-pentadecanolide
also 16-hexadecanolide

Guo & Sih (1988)

Pseud. sp. lipase
Amano K-10
65 °C

28   + dilactone

56% yield   15% yield

Guo & Sih (1988)

CRL (OF-360)
Meito Sangyo
65 °C

24   + dilactone

42% yield   8% yield

Sugai *et al.* (1995)
lower yields and slower reaction with PCL

CAL-B
65 °C

16   + trimer

44% yield   32% yield

Fig. 92.   Lipase-catalyzed formation of macrolactones and macrodiolides (cyclic dimers). Examples not shown include cyclization of 16-hydroxyhexadecanoic acid to a 34-membered diolide
(Guo et al., 1988; Zaidi et al., 1995).

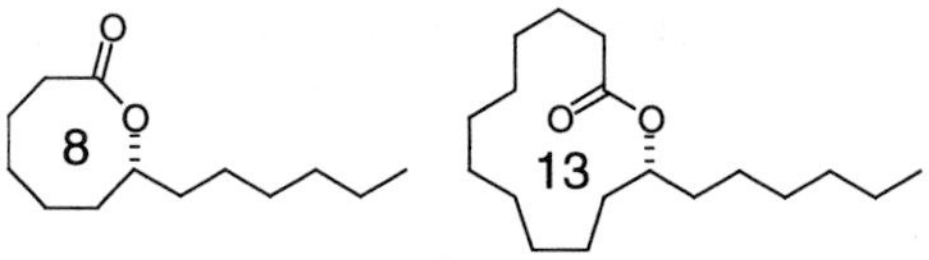

PCL, lactonization of vinyl ester,
dilactones also formed
Lobell & Schneider, 1993

Fig. 93.   Enantioselective macrolactonization.

5.4   Dynamic Kinetic Resolutions

Kinetic resolution limits the yield of the pure enantiomer to 50 %. However, if the substrate racemizes quickly in the reaction mixture, then the yield can be 100 %. This resolution with *in situ* racemization is called dynamic kinetic resolution or second order asymmetric transformation (for reviews see Ward, 1995; Stecher and Faber, 1997). The requirements for a dynamic kinetic resolution are: (1), the substrate must racemize faster than the subsequent enzymatic reaction, (2), the product must not racemize, and (3), as in any asymmetric synthesis, the enzymic reaction must be highly stereoselective. Equations for asymmetric syntheses (Sect. 3.1.3) also apply to dynamic kinetic resolutions. Normal alcohols and carboxylate esters racemize only with difficulty, so this method requires either substrates that racemize spontaneously or addition of a racemization catalyst.

The earliest dynamic kinetic resolutions involved 5-arylsubstituted hydantoins, which racemize spontaneously at pH > 8 via an enolate (Tsugawa et al., 1966; Takahashi et al., 1979; Olivieri et al., 1981). Hydantoinases catalyze the highly enantioselective hydrolysis of 5-monosubstituted hydantoins to *N*-carbamoyl α-amino acids (Fig. 94).

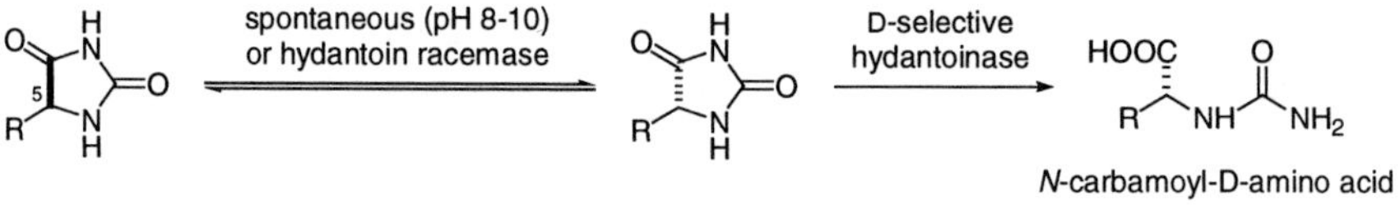

N-carbamoyl-D-amino acid

Fig. 94.   Hydantoinases catalyze the hydrolysis of 5-monosubstituted hydantoins to *N*-carbamoyl α-amino acids. 5-Substituted hydantoins, especially 5-aryl hydantoins, racemize readily either enzymatically or chemically at pH 8–10 via the enolate. The equation shows a D-selective hydanoinase; L-selective hydanoinases are rare.

Both Kanegafuchi Industries (Japan) and DEBI Recordati (Italy) produce unnatural D-amino acids using D-selective hydantoinases. The most important amino acids are phenylglycine and 4-hydroxyphenylglycine for production of the semisynthetic penicillins, ampicillin and amoxicillin, respectively.

The structurally similar 4-substituted-2-phenyloxazolin-5-ones also racemize readily, but finding enantioselective hydrolases has been more difficult (Eq. 16; Tab. 12). For R = Me, Bn, and several others, Bevinakatti et al. (1990, 1992) used an RML-catalyzed alcoholysis in organic solvents, but the enantioselectivity was only E = 3–5. Sih's group screened a dozen lipases for hydrolysis of the phenylalanine derivative (R = Bn) and found that PPL favored the natural (R)-enantiomer (E > 100), while ANL favored the unnatural (S)-enantiomer (E > 100) (Gu et al., 1992; Crich et al., 1993) (Tab. 12). However, these enzymes were less enantioselective toward other, similar derivatives. Several *Pseudomonas* lipases (PCL, Amano AK, Amano K-10) at 50°C in *t*-BuOMe catalyzed methanolysis of a variety of 4-substituted 2-phenyloxazolin-5-ones with enantioselectivities of 5–39, usually favoring the (S)-enantiomer.

$$(16)$$

Tab. 12.   Lipase-Catalyzed Ring-Opening of 2-Phenyloxazolin-5-ones.

Lipase	Reaction	R =	E	Reference
RML	alcoholysis	Bn, Me, *n*-Pr, CH$_2$*i*-Pr	3–5 (S)	Bevinakatti et al. (1990, 1992)
PPL	hydrolysis	Bn	> 100 (S)	Gu et al. (1992)
ANL	hydrolysis	Bn	> 100 (R)	Gu et al. (1992)
ANL, PPL	hydrolysis	Ph, 4-OMePh, CH$_2$CH$_2$Ph, several CH$_2$Ar, CH$_2$*i*-Pr, CH$_2$CH$_2$SMe	2–12	Gu et al. (1992); Crich et al. (1993)
PXL[a]	alcoholysis	13 different examples	5–39	Crich et al. (1993)
RML	alcoholysis	*t*-Bu	> 100 (S)	Turner et al. (1995)

[a] One of several *Pseudomonas* lipases: PCL, Amano AK, or Amano K-10. Most reactions favored the (S)-enantiomer, but in some cases the enantiopreference was either (R) or (S) depending on the amount of added water.

In several cases, the enantioselectivity reversed depending on whether the reaction mixture contained added water or not. The lipase usually hydrolyzed substrates with larger R groups (e.g., Ph, CH$_2$*i*-Pr) more selectively than small ones (e.g., Me). For preparative use, Crich et al. (1993) further resolved the enantiomerically-enriched methyl esters of N-benzoyl amino acids by protease-catalyzed cleavage of the ester. Turner et al. (1995) found that RML-catalyzed alcoholysis of the *t*-butyl derivative was highly enantioselective (99.5 %ee, 94 % yield), but only when the reaction mixture contained a catalytic amount of triethylamine. The authors suggested that the triethylamine inhibits a less enantioselective isozyme.

Simple esters of chiral carboxylic acids also undergo base-catalyzed racemization if the acid contains electron-withdrawing substituents. Fülling and Sih (1987) reported the first enzyme-catalyzed example using a *Streptomyces* protease (Eq. 17).

85% ee, 92% yield

Tan et al. (1995) resolved 2-(phenylthio)propanoic acid by PCL-catalyzed hydrolysis of the thioester in the presence of trioctylamine (Eq. 18). Both the thioester and the trioctylamine promote racemization via an enolate mechanism. Similarly, Um and Drueckhammer (1998) resolved thioesters of 2-aryl and 2-aryloxypropanoic acids using subtilisin ($E = 7–11$) combined with *in situ* racemization promoted by trialkylamines. Chosing the type of thioester (e.g., 2,2,2-trifluoroethyl thiol ester) was the key to rapid racemization. Vörde et al. (1996) suggested that even simple esters may racemize in the presence of both CAL-B and α-phenylethylamine (Eq. 19). They did not detect racemization in the presence of only one of these.

For chiral alcohols, Inagaki et al. (1991, 1992) racemized cyanohydrins by the reversible base-catalyzed addition of HCN to aldehydes. Enantioselective acetylation of the (*S*)-cyanohydrin catalyzed by PCL yielded the acetate in good to moderate yields and enantiomeric purity. In general, PCL showed higher enantioselectivity toward cyanohydrins derived from aromatic aldehydes than from aliphatic aldehydes. PCL did not catalyze acylation of the HCN donor, acetone cyanohydrin, a tertiary alcohol, presumably because it is too hindered (Fig. 95).

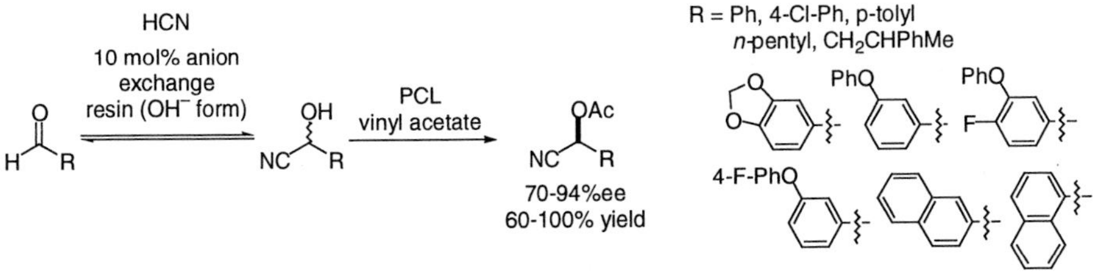

Fig. 95.   Dynamic kinetic resolution of cyanohydrins.

Two similar examples are the resolution of hemithioacetals where a thiol adds reversibly to an aldehyde (Brand et al., 1995) and of a cyclic hemiacetal which reversibly opens (van den Heuvel et al., 1997) (Fig. 96). Acyloin derivatives also racemize quickly in the presence of a catalytic amount of triethylamine (Taniguchi et al., 1997; Taniguchi and Ogasawara, 1997), while butenolides racemize readily at room temperature and pyrrolinones racemize at 69 °C (Fig. 96) (van der Deen et al., 1996, Thuring et al., 1996a). In each case, lipases catalyzed selective acetylation of one enantiomer in excellent yield.

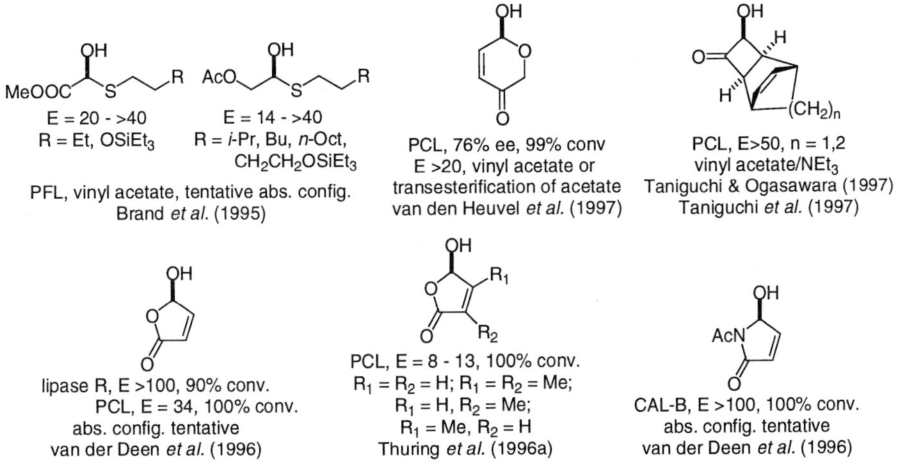

Fig. 96.   Dynamic kinetic resolution of alcohols that racemize rapidly: hemithioacetals, a hemiacetal, acyloins, butenolides and pyrrolinones. In each case, the product isolated was the corresponding acetate.

A potentially more general reaction is the racemization of simple secondary alcohols by temporary oxidation followed by reduction using hydrogen transfer catalysts (Eq. 20) (Dinh et al., 1996; Larsson et al., 1997). Similarly, Reetz and Schimossek (1996) catalyzed the racemization of amines with palladium during a resolution. Koh et al. (1998) reported new racemization catalysts for secondary alcohols. These may be useful in future dynamic kinetic resolutions.

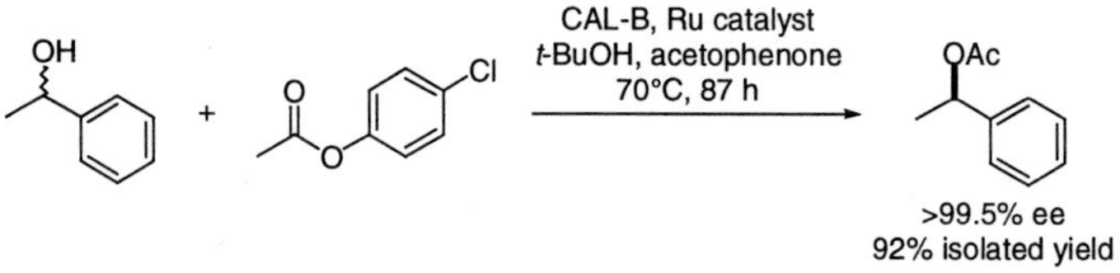

$$\text{(diagram of reaction 20)}$$

CAL-B, Ru catalyst
t-BuOH, acetophenone
70°C, 87 h

>99.5% ee
92% isolated yield

(20)

Acetates of simple secondary alcohols do not readily racemize. The only example is a palladium catalyzed *in situ* racemization of allylic acetates, such as the 1-acetoxy-3-phenyl-2-cyclohexene (Fig. 97). Although slow racemization limited the rate of the reaction, both the yield and enantioselectivity were good.

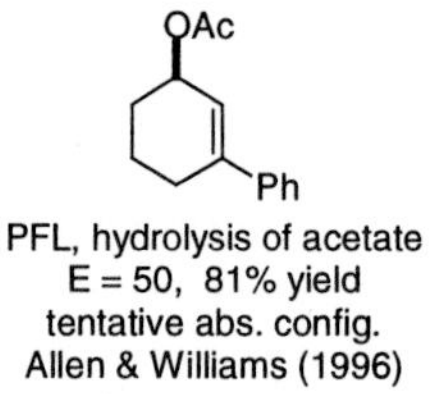

$$\text{(structure with OAc and Ph)}$$

PFL, hydrolysis of acetate
E = 50, 81% yield
tentative abs. config.
Allen & Williams (1996)

Fig. 97.   Dynamic kinetic resolution of an allylic acetate. Palladium catalyzed the racemization of the allylic acetate. The product was the allylic alcohol.

A related strategy, although it is not a dynamic kinetic resolution, is to invert the configuration of one enantiomer (Schneider and Goergens, 1992). Vänttinen and Kanerva (1995) resolved α-phenylethanol by PCL-catalyzed acetylation with vinyl acetate yielding a mixture of the (*R*)-acetate and the (*S*)-alcohol. Treating the mixture as shown in Eq. 21 converted the alcohol to the acetate while inverting the configuration. The net reaction was converting a racemic alcohol to the (*R*)-acetate.

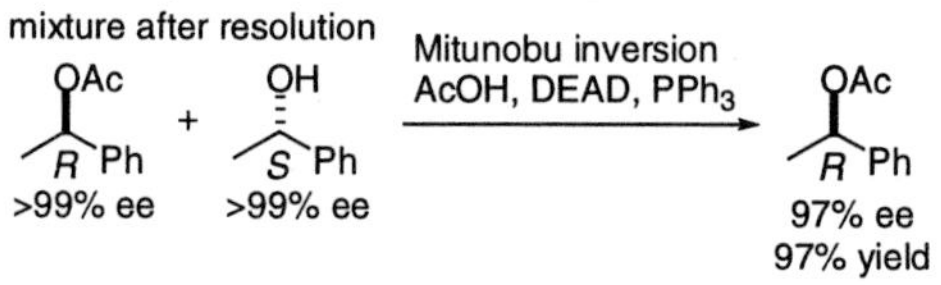

mixture after resolution

$$\text{(structures of R-OAc and S-OH)}$$

>99% ee   >99% ee

Mitunobu inversion
AcOH, DEAD, PPh$_3$

$$\text{(structure of R-OAc)}$$

97% ee
97% yield

(21)

5.5   Commercial Enantioselective Reactions

5.5.1   Enantiomerically-Pure Chemical Intermediates

DSM-Andeno (Netherlands) produce (*R*)-glycidol butyrate using a PPL-catalyzed resolution, but they did not reveal details of the process (Ladner and Whitesides, 1984; Kloosterman et al., 1988). Several groups have since studied this reaction and its scale-up in more detail (Walts and Fox, 1990; Wu et al., 1993; van Tol et al., 1995a, b) (Eq. 22).

$$\text{(22)}$$

| | | |
| (±)-glycidyl butyrate | (*R*)-glycidyl butyrate | (*R*)-glycidol |

BASF produces enantiomerically-pure amines using a *Pseudomonas* lipase-catalyzed acylation (Balkenhohl et al., 1997). A key part of the commercialization of this process was the discovery that methoxyacetate esters reacted much fast than simple esters. Activated esters are not suitable due to a competing uncatalyzed acylation (Eq. 23).

$$\text{(23)}$$

racemate
R = H, 4-Me, 3-OMe

Chiroscience (Cambridge, UK) has scaled-up the dynamic kinetic resolution of (*S*)-*tert*-leucine, an intermediate for the synthesis of conformationally-restricted peptides and chiral auxiliaries (see Sect. 5.4) (Turner et al., 1995; McCague and Taylor, 1997).

5.5.2   Enantiomerically-Pure Pharmaceutical Intermediates

Both DSM-Andeno (Netherlands) and Tanabe Pharmaceutical (Osaka, Japan) in collaboration with Sepracor (Marlborough, MA) have commercialized lipase-catalyzed resolutions of (+)-(2*S*,3*R*)-MPGM, a key precursor to diltiazem (Hulshof and Roskam, 1989; Matsumae et al., 1993, 1994; Furui et al., 1996). The DSM-Andeno process uses RML, while the Tanabe process uses a lipase secreted by *Serratia marcescens* Sr41 8000. In both cases the lipase catalyzed hydrolysis of the unwanted enantiomer with high enantioselectivity (E > 100). The resulting acid spontaneously decomposed to an aldehyde (Fig. 98).

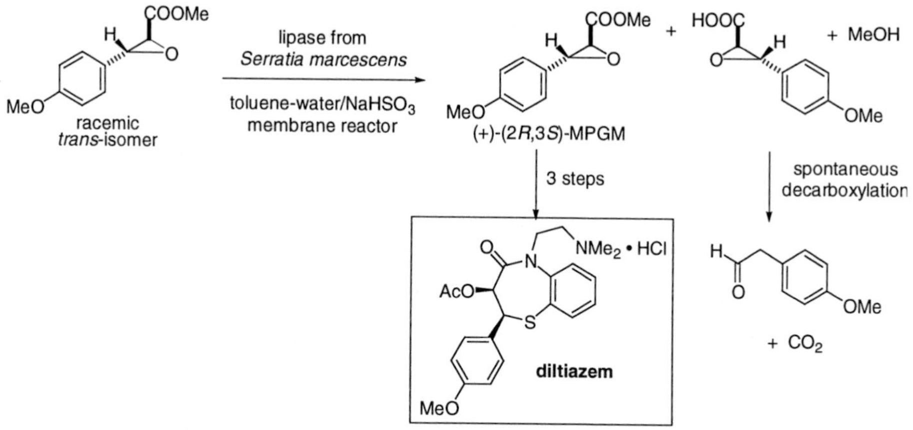

Fig. 98.   Commercial synthesis of diltiazem by Tanabe Pharmaceutical uses a kinetic resolution catalyzed by lipase from *Serratia marcescens*.

In the Tanabe process, a membrane reactor and crystallizer combine hydrolysis, separation, and crystallization of (+)-(2R,3S)-MPGM. Toluene dissolves the racemic substrate in the cystallizer and carries it to the membrane containing immobilized lipase. The lipase catalyzes hydrolysis of the unwanted (−)-MPGM to the acid, which then passes through the membrane into an aqueous phase. Spontaneous decarboxylation of the acid yields an aldehyde which reacts with the bisulfite in the aqueous phase. In the absence of bisulfite, this aldehyde deactivates the lipase. The desired (+)-MPGM remains in the toluene phase and circulates back to the cystallizer where it crystallizes. Lipase activity drops significantly after eight runs and the membrane must be recharged with additional lipase. Although the researchers detected no lipase-catalyzed hydrolysis of (−)-MPGM, chemical hydrolysis lowered the apparent enantioselectivity to E = 135 under typical reaction conditions. The yield of crystalline (+)-(2R,3S)-MPGM is > 43 % with 100 % chemical and enantiomeric purity.

Glaxo resolves (1S,2S)-*trans*-2-methoxycyclohexanol, a secondary alcohol, on a ton scale for the synthesis of a tricyclic β-lactam antibiotic (Stead et al., 1996). The slow reacting enantiomer needed for synthesis is recovered in 99 %ee from an acetylation of the racemate with vinyl acetate in cyclohexane. Immobilized CAL-B and PFL (Biocatalysts, Ltd.) both showed high enantioselectivity, but CAL-B was more stable over multiple use cycles. Other workers had resolved this alcohol by hydrolysis of its esters with PCL, CRL, or pig liver acetone powder (Laumen et al., 1989; Hönig and Seufer-Wasserthal, 1990; Basavaiah and Krishna, 1994), but Glaxo chose resolution by acylation of the alcohol because it yields the required slow-reacting alcohol directly (Fig. 99).

CAL-B, 37 g/L
vinyl acetate, 1.7 M
triethylamine, 0.16 M
cyclohexane, 6-8 h
racemic
>99% ee
36% yield
antibiotic

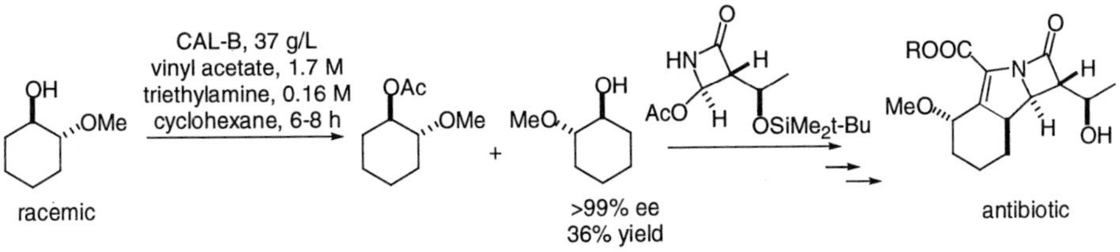

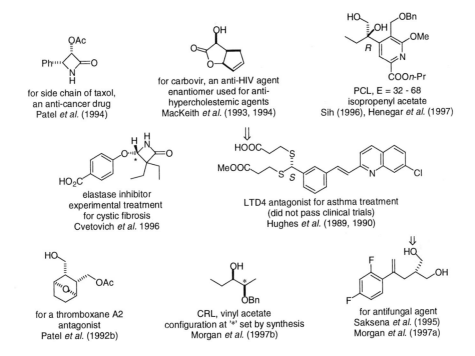

Fig. 99. Glaxo resolves a building block for antibiotic synthesis.

Researchers reported a number of other kilogram scale routes to pharmaceutical precursors that involve lipases. Selected examples are summarized in Fig. 100.

for side chain of taxol,
an anti-cancer drug
Patel et al. (1994)
for carbovir, an anti-HIV agent
enantiomer used for anti-
hypercholestemic agents
MacKeith et al. (1993, 1994)
PCL, E = 32 - 68
isopropenyl acetate
Sih (1996), Henegar et al. (1997)
elastase inhibitor
experimental treatment
for cystic fibrosis
Cvetovich et al. 1996
LTD4 antagonist for asthma treatment
(did not pass clinical trials)
Hughes et al. (1989, 1990)
for a thromboxane A2
antagonist
Patel et al. (1992b)
CRL, vinyl acetate
configuration at '*' set by synthesis
Morgan et al. (1997b)
for antifungal agent
Saksena et al. (1995)
Morgan et al. (1997a)

Fig. 100. Kilogram-scale routes to pharmaceutical precursors involving lipases.

6 Chemo- and Regioselective Lipase-Catalyzed Reactions

6.1 Protection and Deprotection Reactions

6.1.1 Hydroxyl Groups

The most difficult part of carbohydrate chemistry is the selective protection and deprotection of the various hydroxyl groups. The difficulty stems from their similar chemical
reactivity, so researchers have searched for enzymic methods to simplify this problem.
For example, Fink and Hay (1969) investigated the selective deprotection of peracylated
sugars thirty years ago, but only more recently have researchers found enzymes and reaction conditions sufficiently selective for synthetic use (for reviews see Waldmann and
Sebastian, 1994; Thiem, 1995; Wong, 1995; Bashir et al., 1995; Riva, 1996).

The simplest examples are sugars with a single ester group. For example, PPL-catalyzed hydrolysis of the esters in the glucopyranose and -furanose shown in Fig. 101
(Kloosterman et al., 1987). However, lipases provide little advantage over chemical
methods in these reactions.

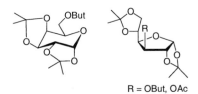

R = OBut, OAc

Fig. 101. PPL-catalyzed hydrolysis of the single ester group in protected sugars.

The most useful reactions are those which selectively protect or deprotect one hydroxyl
in the presence of several others. The selectivity of lipases usually parallels the chemical
reactivity of the hydroxyls, but with increased selectivity. Thus, in hydrolysis reactions of
peracylated sugars, the ester at the anomeric carbon (a secondary hydroxyl) reacts first,
followed by the ester at the primary hydroxyl. The remaining esters at the secondary
hydroxyls react next. In acylation reactions, the primary hydroxyl group reacts first, followed by the secondary hydroxyls. The relative reactivity among the secondary hydroxyls in either acylation or hydrolysis of the esters remains difficult to predict because it
varies with the lipase, reaction conditions and structure of the sugar. Not all reactions
follow generalizations. For example, lipases sometimes acylate a secondary hydroxyl
group in the presence of a primary hydroxyl.

6.1.1.1   Primary Hydroxyl Groups in Sugars

Hydrolysis of Esters of Primary Hydroxyl Groups. Sweers and Wong (1986) found that CRL selectively hydrolyzed the ester of the primary alcohol of methyl-2,3,4,6-tetra-*O*-pentanoyl-D-glycopyranosides of glucose, galactose and mannose yielding the corresponding tri-*O*-pentanoates. A later paper included methyl-2-acetamido-2-deoxy-3,4,6-tri-*O*-pentanoyl-D-mannoside (Hennen et al., 1988). The solvent was water containing 9 % acetone and the isolated yields were good for the glucoside, but moderate for the galactoside and the two mannosides. The corresponding acetyl esters did not react under these conditions and the octanoyl esters formed emulsions which made isolation difficult (Fig. 102).

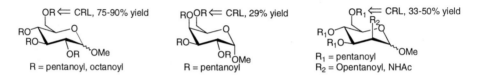

Fig. 102. Selective hydrolysis of esters at the primary position.

Hennen et al. (1988) extended this work to the furanosides shown in Fig. 103 using 10 % DMF in buffer. In most cases the ester at the primary hydroxyl reacted selectively. In methyl α-D-2-deoxyriboside, both the primary ester at C-5 and the secondary ester at C-3 reacted at similar rates, while in methyl-β-D-xyloside, the secondary ester at C-3 reacted faster than the primary ester at C-5. Kloosterman et al. (1987) used PPL to selectively hydrolyze the ester from the primary alcohol in the protected D-riboside (Fig. 103).

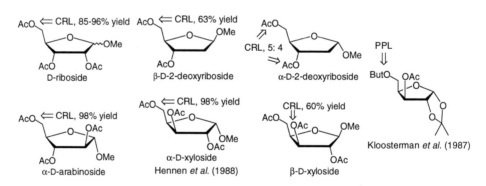

Fig. 103. Hydrolysis usually favors the primary position.

Digestive lipases do not catalyze hydrolysis of Olestra® (sucrose esterified with six to eight fatty acids), even though it contains three esters of a primary alcohol (Fig. 104). This inability to digest Olestra® permits food manufacturers to make foods (e.g., potato chips) with lower calories, while retaining the desired flavor of a fat.

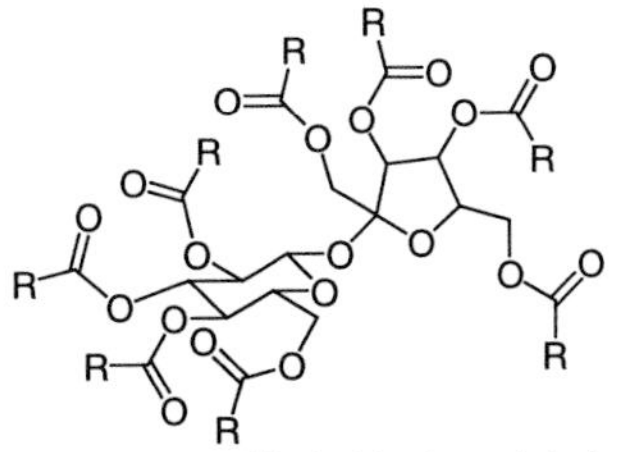

Olestra®, sucrose esterified with six to eight fatty acids

Fig. 104. Lipases do not catalyze hydrolysis of Olestra®.

Acylation of Primary Alcohols in Unmodified Sugars. For the reverse reaction, acylation, the biggest problem is finding an organic solvent that dissolves the polar sugar, but does not inactivate the lipase. Therisod and Klibanov (1986) were the first to find that warm pyridine dissolved sugars, yet did not denature crude PPL. They used PPL to selectively acylate glucose at C-6 with 2,2,2-trichloroethyl laurate giving 40 % conversion after two days with 95 % regioselectivity (Eq. 24). Similarly, PPL selectively acylated the primary alcohol in mannose and galactose, but in fructose, which has primary alcohols at C-1 and C-6, both reacted at similar rates.

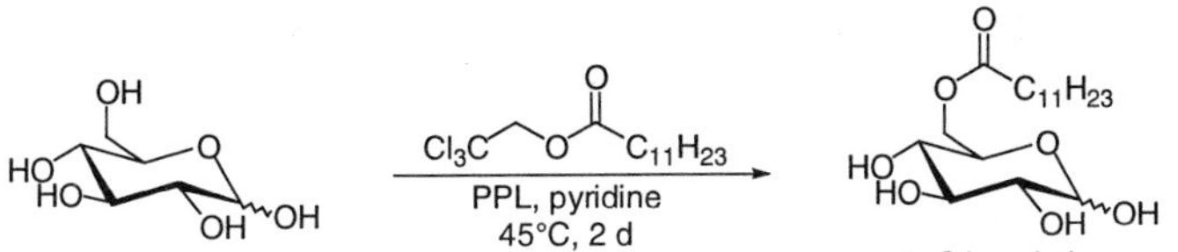

6-*O*-lauryl glucose
40% conversion
95% regioselectivity

(24)

In 2:1 benzene/pyridine, Wang et al. (1988) found that CRL also retained activity (Tab. 13). They acylated mannose and *N*-acetylmannosamine with the more active acyl donor, vinyl acetate. Using oxime esters as acyl donors, Gotor and Pulido (1991) found that PCL was active in pyridine or 3-methyl-3-pentanol and acylated glucose, L-arabinose, galactose, mannose and sorbose. All acylations favored the primary hydroxyl groups, but the oxime esters were more selective since Gotor and Pulido detected no diacylation. With the thermostable CAL-B, Pulido and Gotor (1993) raised the temperature to 60 °C and used the more convenient solvent dioxane and alkoxycarbonyl oximes as acyl donors. This reagent introduced the carbobenzyloxy (Cbz) protective group among others (Eq. 25). More active acyl donors, such as acid anhydrides and even vinyl esters in pyridine, gave nonselective background reactions. None of the conditions above are suitable for the acylation of disaccharides, presumably because they are too insoluble.

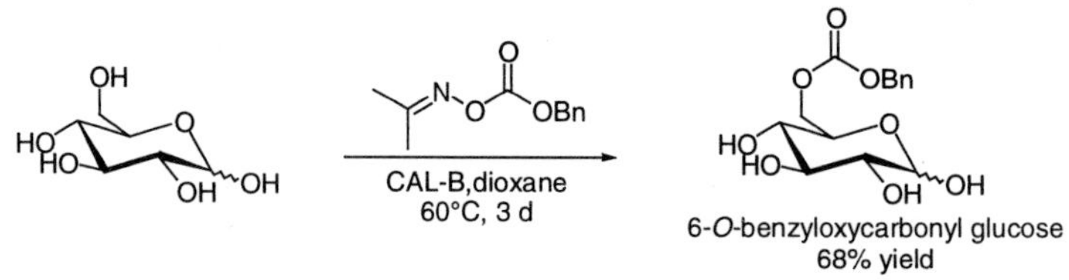

(25)

6-O-benzyloxycarbonyl glucose
68% yield

Another application of sugar esters is as nonionic surfactants for the food and cosmetic industries (Fig. 105). The advantage of an enzymic route over chemical processes, besides milder reaction conditions and fewer side reactions, would be the ability to label the surfactant 'natural' (Sarney and Vulfson, 1995). In most countries, products produced from natural starting materials using enzymic catalysts are still considered natural. The acylation reactions above all use expensive acylating agents, toxic solvents, and far too much lipase (sometimes four times the weight of sugar). Although they are convenient on a lab scale, they are not practical for surfactant production. For these applications, researchers directly esterified sugars with fatty acids and optimized the reactor configuration to increase yields and reaction rate (Tab. 13).

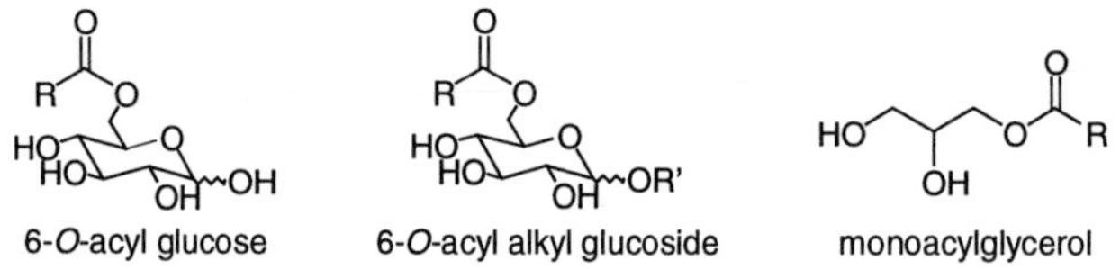

6-O-acyl glucose   6-O-acyl alkyl glucoside   monoacylglycerol

Fig. 105. Examples of surfactants prepared by lipase-mediated reactions. R = C_7–C_{17} chain.

This section reviews the synthesis of acylated carbohydrates, such as 6-O-acyl glucose and 6-O-acyl alkyl glucosides; Sect. 6.2.1.3 reviews the synthesis of monoglycerides (reviewed by Fiechter, 1992; Bornscheuer, 1995).

Although Sieno et al. (1984) reported esterification of sugars and fatty acids in aqueous solution, Janssen et al. (1990) found only small amounts of ester, which they extracted using a membrane reactor. Others used polar organic solvents such as 2-pyrrolidone and hindered tertiary alcohols and vacuum or drying agents to remove the water released in the esterification. This removal increased the reaction rate and the yield. Cao et al. (1996, 1997) crystallized the product ester to shift the equilibrium. A suspension of glucose, stearic acid, immobilized CAL-B, and molecular sieves in acetone yielded solid 6-O-stearoyl-D-glucose in 92 % conversion after 72 h at 60 °C. The acetone created a small catalytic phase, while allowing the product to precipitate (solubility: glucose in acetone = 0.04 mg/mL, glucose stearate = 3.3 mg/mL). No sugar esters formed in reverse micelles (Hayes and Gulari, 1992, 1994). Direct esterification usually works better with longer fatty acids (C_{14}–C_{18}) than with medium chain fatty acids (C_8–C_{12}). This reaction system could also be applied to the CAL-B-catalyzed synthesis of sugar esters based on arylaliphatic carboxylic acids (Fig. 106). CAL-B selectively esterified the primary hydroxyl at C-6 of the glucose moiety of D-(–)-salicin (2-(hydroxymethyl)-phenyl-β-D-glucopyranoside) (Otto et al., 1998a).

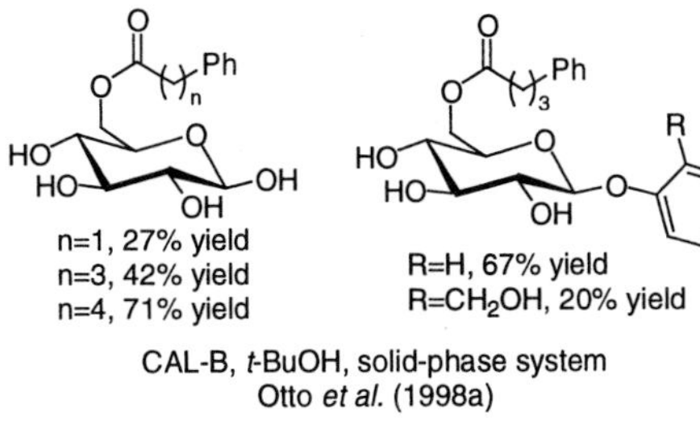

Fig. 106. CAL-B-catalyzed synthesis of sugar esters based on arylaliphatic carboxylic acids.

Acylation of Primary Alcohols in Alkyl Glycosides and Other Modified Sugars. Since the poor solubility of sugars in organic solvents is a major limitation of lipase-catalyzed acylations of sugars, many researchers modified the sugars to increase their solubility (Tab. 14). Holla (1989) used glycals (sugar precursors) which are more soluble because they lack two hydroxyl groups. Acetalisation of sugars with acetone increased the solubility so much that researchers eliminated the solvent and dissolved the sugar acetal in the cosubstrate fatty acid (Fregapane et al., 1991). Ikeda and Klibanov (1993) complexed glucose with phenylboronic acid (Fig. 107). The complex dissolved in *t*-butanol and PCL efficiently catalyzed the acylation of the primary hydroxyl group with vinyl or trifluoroethyl butyrate. Solubilization of fructose in hexane with phenylboronic acid allowed selective acylation of the C-1 primary hydroxyl, with no reaction at the C-6 primary hydroxyl (Schlotterbeck et al., 1993; Scheckermann et al., 1995). However, the reaction was one hundred times slower than the glucose reaction reported by Ikeda and Klibanov. Complexation with boronic acids or acetalisation also allow acylation of disaccharides (Oguntimein et al., 1993; Sarney et al., 1994).

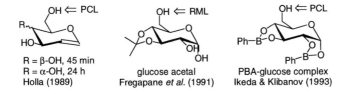

Fig. 107. Increased solubility of modified sugars in organic solvents simplifies acylation reactions.

Tab. 13.   Lipase-Catalyzed Acylation of Unmodified Sugars.

Sugar	Acyl donor	Solvent	Lipase	Rate[a]	Reference
Glucose, mannose, galactose, fructose[b]	trichloro ethylester	Pyridine	PPL	0.02	Therisod and Klibanov (1986)
Mannose	vinyl acetate	benzene/pyridine	CRL	0.004	Wang et al. (1988)
Glucose, L-arabinose, galactose, mannose, sorbose	oxime ester	pyridine, 3-methyl-3-pentanol	PCL	0.02–0.03	Gotor and Pulido (1991); Pulido et al. (1992)
Glucose, galactose, mannose, ribose, arabinose	O-(alkoxylcarbonyl) oxime	Dioxane	CAL-B	0.08	Pulido and Gotor (1993)
Glucose	fatty acid	water, EMR[c]	CRL	0.02[d]	Janssen et al. (1990)
Glucose	fatty acid	water	CRL	0.04	Sieno et al. (1984)[e]
Sorbitol	fatty acid	2-pyrrolidone, EMR[c]	CVL	1.4[d]	Janssen et al. (1991a)
Glucose	fatty acid	reverse micelles	ROL, CRL	no rxn	Hayes and Gulari (1992)
Glucose, fructose	fatty acid	t-butanol	CAL-A & B	0.3	Oguntimein et al. (1993)
Glucose	fatty acid	acetonitrile	CAL-B	0.04	Ljunger et al. (1994)
Sorbitol	fatty acid	2-methyl-2-butanol	CAL-B	0.2	Ducret et al. (1995)
Gructose	fatty acid	2-methyl-2-butanol	RML, CAL-B	0.05	Scheckermann et al. (1995)
Glucose	fatty acid	acetone/crystallization	CAL-B	0.2–0.4	Cao et al. (1996, 1997)
Fructose	vinyl laurate	THF	PFL	0.05	Sin et al., (1996)
Ribose, 2-deoxyribose	propanoic anhydride	THF	CAL-B	3.3	Prasad et al. (1995)

[a] mmol sugar ester produced per gram enzyme and hour calculated from literature data. [b] Both primary alcohol groups in fructose reacted at similar rates. [c] Initial rate [d] Initial rate [e] Later workers found little ester formation under these conditions.

Alkyl glucosides are more soluble in organic solvents, hence, lipase-catalyzed acylations of these sugar derivatives is simpler than unmodified sugars. In addition, the rate of
acylation increases as the size of the acyl group increases. The Wong group acylated
methyl-β-D-glucoside using CRL and vinyl acetate in a mixture of benzene and pyridine
and several methylfuranosides (D-ribose, D-arabinose or D-xylose) using PPL and 2,2,2-
trifluoroethyl acetate in THF (Hennen et al., 1988; Wang et al., 1988). Theil and Schick
(1991) significantly improved the rate of acylation using ethyl glycosides, crude PPL and
vinyl acetate in a mixture of THF and triethylamine. In all cases, acylation was selective
for the primary alcohol group.

For surfactant applications, the Novo group catalyzed the direct esterification of alkyl
glucosides in molten fatty acid using immobilized CAL-B (Björkling et al., 1989; Adelhorst et al., 1990). Typical reactions showed excellent yield and good regioselectivity
(Eq. 26). Upon scale-up, Unichema International (a subsidiary of Unilever) encountered
difficulties with the viscous and heterogenous reaction mixture. Adding 25 vol % t-buta-
nol reduced the viscosity and adding 5 mol % product (e.g., 6-O-laurylglucopyranoside)
emulsified the reactants. In a packed bed reactor with a separate pervaporation compartment to remove water, they achieved 90 % conversion in 25 h for 40 batch reactions
(Macrae, personal communication, 1996).

Unichema International also produces esters such as decyl oleate, octyl palmitate, isopropyl myristate, isopropyl palmitate, and PEG400 monostearate for skin care products
using CAL-B catalyzed esterification (Bosley, 1997). Water is removed by vacuum.
Unichema calls these 'bioesters' because products made by lipase-catalyzed process
starting from natural materials retain their 'natural' designation. Other companies may
produce flavor esters such as isoamyl acetate or geranyl acetate by lipases.

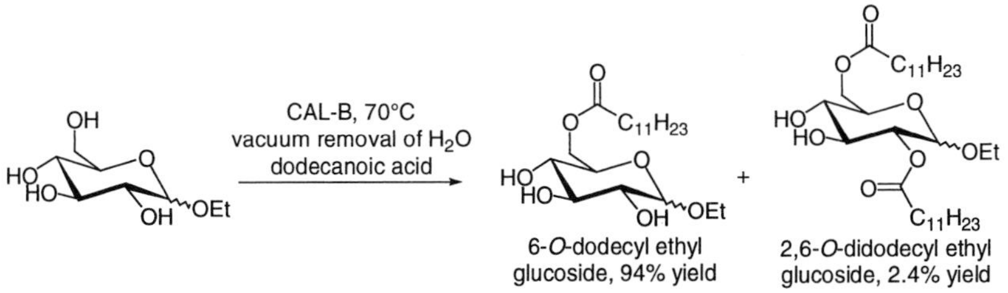

(26)

6-O-dodecyl ethyl
glucoside, 94% yield

2,6-O-didodecyl ethyl
glucoside, 2.4% yield

RML was less regioselective (14 % of the diester), but adding hexane improved the
regioselectivity (Fabre et al., 1993a). Pelenc et al. (1993) further combined this process
with an α-transglucosidase-catalyzed synthesis of α-butyl glucoside from maltose and
butanol (Eq. 27).

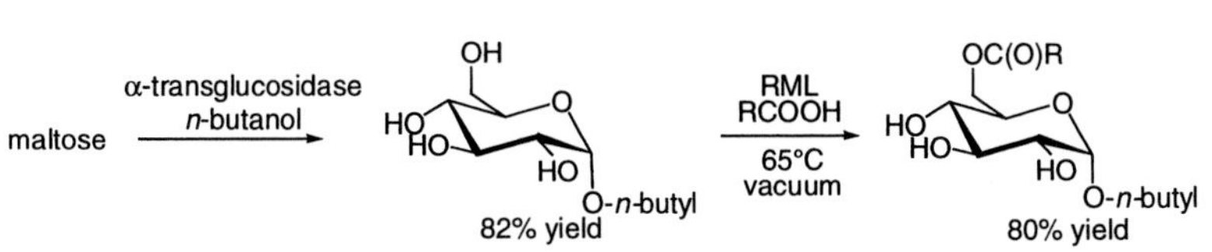

(27)

82% yield

80% yield

RML also catalyzed the 6-*O*-selective acylation of α-butyl glucoside with more complex acids, for example, the protected amino acid BocNH(CH$_2$)$_3$COOCH$_2$CCl$_3$ (Fabre et al., 1994) and bis(2,2,2-trichloroethyl)adipate (Fabre et al., 1993b).

CAL-B selectively acylates the primary alcohol in a wide variety of nucleosides. Gotor and Morís (1992) found that oxime esters of simple acids or protected amino acids selectively acylated the primary hydroxyl, while oxime carbonates gave the 5'-*O*-carbonates. Interestingly, PCL selectively acylated the secondary 3'-hydroxyl even when the primary alcohol was unprotected (Fig. 108).

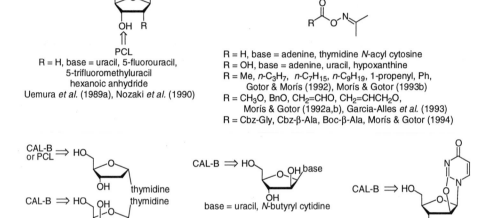

<table>
<tr><td>

CAL-B ⇒ HO. base

⇑
PCL
R = H, base = uracil, 5-fluorouracil,
5-trifluoromethyluracil
hexanoic anhydride
Uemura *et al.* (1989a), Nozaki *et al.* (1990)

</td><td>

acylation with oxime esters

R = H, base = adenine, thymidine *N*-acyl cytosine
R = OH, base = adenine, uracil, hypoxanthine
R = Me, *n*-C$_3$H$_7$, *n*-C$_7$H$_{15}$, *n*-C$_9$H$_{19}$, 1-propenyl, Ph,
    Gotor & Morís (1992), Morís & Gotor (1993b)
R = CH$_3$O, BnO, CH$_2$=CHO, CH$_2$=CHCH$_2$O,
    Morís & Gotor (1992a,b), Garcia-Alles *et al.* (1993)
R = Cbz-Gly, Cbz-β-Ala, Boc-β-Ala, Morís & Gotor (1994)

</td></tr>
</table>

CAL-B or PCL ⇒ HO

OH   thymidine

CAL-B ⇒ HO OH   thymidine

CAL-B ⇒ HO   OH, base

OH
base = uracil, *N*-butyryl cytidine

CAL-B ⇒ HO

oxime butyrate or butyric anhydride Morís & Gotor (1993a)

Fig. 108. CAL-B selectively acylates the primary position, while PCL favors the secondary position.

Subtilisin, a protease, also catalyzes the selective acylation of carbohydrates (Riva et al., 1988; Carrea et al., 1989; Riva, 1996). Subtilisin is better suited than lipases for the acylation of disaccharides, and often shows complementary selectivity to lipases (Sect. 8.1.1.2) (Kazlauskas and Weissfloch, 1997).

Danieli et al. (1995) selectively acylated one of the two primary hydroxyl groups in the triterpene oligoglycoside ginsenoside Rg$_1$ using CAL-B and vinyl acetate or bis(2,2,2-trichloroethyl)malonate (Fig. 109).

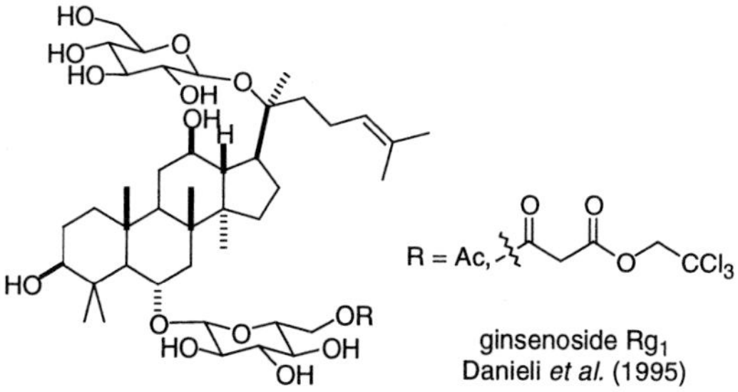

ginsenoside Rg$_1$
Danieli *et al.* (1995)

Fig. 109. CAL-B selectively acylates one of the two primary hydroxyls.

Eberling et al. (1996) used lipase from wheat germ to hydrolyze all the acetates from the sugar portion of *O*-glycosyl amino acid (methoxyethoxy)ethyl (MEE) esters. The MEE esters were removed later using RJL (see Sect. 6.1.3). For example, a glucosyl serine derivative is shown in Fig. 110; Eberling et al. (1996) also deprotected a number of similar compounds (galactose, galactosamine and xylose) for the sugar portion and threonine for the amino acid portion.

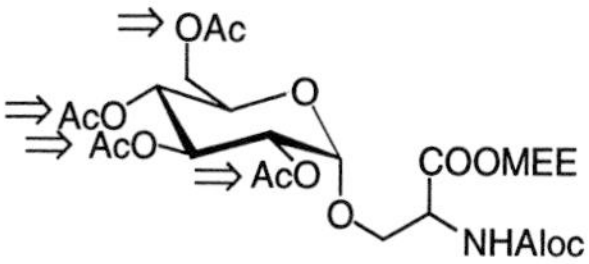

73% yield, wheat germ lipase
Eberling, *et al.* (1996)

Fig. 110. Wheat germ lipase hydrolyzed the acetyl groups at both primary and secondary positions.

Riva et al. (1996) selectively acylated the only primary hydroxyl group in the flavonoid glycosides isoquercitin and naringin using CAL-B (Fig. 111).

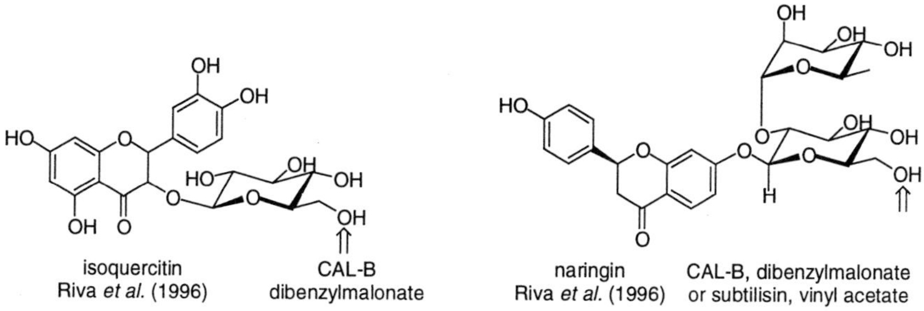

isoquercitin   CAL-B
Riva *et al.* (1996)   dibenzylmalonate

naringin   CAL-B, dibenzylmalonate
Riva *et al.* (1996)   or subtilisin, vinyl acetate

Fig. 111. CAL-B selectively acylates the primary position.

Achromobacter sp. lipase regioselectively acylated the hydroxyl group on *sn*-1 carbon among the two primary hydroxyl groups of 3-*O*-β-D-galactopyranosyl-*sn*-glycerol (Fig. 112).

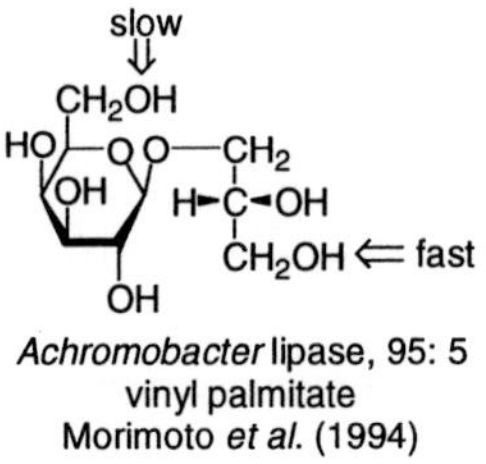

slow
⇓
CH₂OH

HO⎯O O⎯CH₂
‌OH   H⎼C⎼OH
         CH₂OH ⇐ fast
OH

Achromobacter lipase, 95: 5
vinyl palmitate
Morimoto *et al.* (1994)

Fig. 112. *Achromobacter* sp. lipase regioselectively acylated one of two primary hydroxyl groups.

Tab. 14. Lipase-Catalyzed Reactions of Modified Sugars[a].

Sugar-derivative	Acyl donor	Solvent	Lipase	Rate[b]	Reference
IPG-sugar	fatty acid	none	RML	4.540	Fregapane et al. (1991, 1994)
PBA-α-D- & β-D-glucose and others[c]	vinyl ester	t-BuOH	PCL	4.166	Ikeda and Klibanov (1993)
PBA-fructose	fatty acid	hexane	RML CAL-B	0.068	Scheckermann et al. (1995); Schlotterbeck et al. (1993)
Methyl xylose	TFEA	THF	PPL	0.007	Hennen et al. (1988)
Methyl glucose	vinyl ester	pyridine/benzene	CRL	0.006	Wang et al. (1988)
Ethyl-sugar	vinyl ester	THF/Et$_3$N	PPL	1.880	Theil and Schick (1991)
Alkyl-sugar	fatty acid	none	CAL-B	na[d]	Adelhorst et al. (1990); Björkling et al. (1989)
Alkyl-sugar	fatty acid	hexane	RML	na[d]	Fabre et al. (1993a)

[a] IPG-sugar: isopropylidene glucose, galactose, or xylose, TFEA: 2,2,2-trifluoroethyl acetate; PBA: phenylboronic acid complex; THF: tetrahydrofuran; DMF: dimethylformamide. [b] mmol sugar ester produced per gram enzyme and hour calculated from literature data. [c] D-galactose, D-fructose, sucrose, lactose maltose, D-glucosamine, D-gluconic acid. [d] na = data not available.

6.1.1.2   Secondary Hydroxyl Groups

Hydrolysis of Acylated Secondary Hydroxyl Groups. In peracylated sugars, the most chemically reactive ester is the one at the anomeric position. Although it is more hindered than the primary alcohol ester, the anomeric hydroxyl is the best leaving group, because it is a hemiacetal. The ester at the anomeric position is also the most reactive in lipase-catalyzed deacylations. Hennen et al. (1988) found that PPL or ANL in buffer containing 9 % DMF catalyzed selective hydrolysis of the acetate at the anomeric position for pyranoses and furanoses (Fig. 113). In most cases, the yields were above 70 %. One exception was peracetyl β-D-glucopyranose, where CRL catalyzed selective hydrolysis of the esters at positions 4 and 6 leaving the triacetyl derivative.

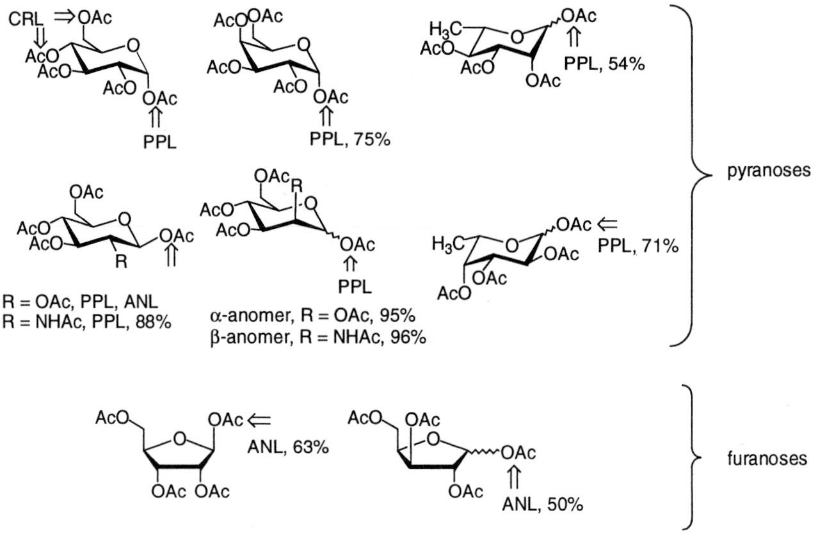

Fig. 113. Selective hydrolysis of the acetate at the anomeric position.

If the anomeric position lacks an ester group, then the most reactive ester is the one at the primary hydroxyl (see Sect. 6.1.1.1 for details). When the sugar lacks esters both at the anomeric and at the primary hydroxyls, it is not easy to predict which secondary hydroxyl ester will react most rapidly, even for the same lipase. In the series of anhydropyranoses in Fig. 114, CRL catalyzed hydrolysis of both the butyrates at the 2- and 4-positions in the glucose derivative, while several other lipases catalyzed hydrolysis of only the butyrate at the 4-position (Kloosterman et al., 1989). However in the 3-azido-3-deoxy-glucose derivative, CRL selectively hydrolyzed the acetate at the 4-position (Holla et al., 1992), while subtilisin favored the acetate at the 2-position. In the galactose derivative, CRL favored the butyrate at the 2-position (Ballesteros et al., 1989) (Fig. 114). Subtilisin and lipases often show an opposite regioselectivity in reactions involving secondary alcohols (Kazlauskas and Weissfloch, 1997).

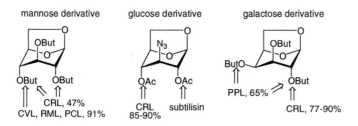

Fig. 114. Selectivity among secondary hydroxyl groups is hard to predict.

PCL selectively hydrolyzed the acetate at the 3-position or 4-position in the structures of Fig. 115 (Holla, 1989; López et al., 1994). The yields were 90 % in both cases. Note that this selectivity follows the secondary alcohol rule in Fig. 34.

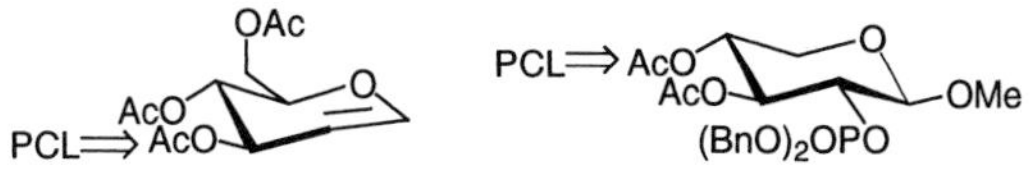

Fig. 115. Regioselectivity sometimes follows the secondary alcohol rule.

CRL and wheat germ lipase selective hydrolyzed an acetyl group from a secondary hydroxyl in octa-*O*-acetyl sucrose (Fig. 116). CRL selectively removed the acetyl group at the 4'-position, while wheat germ lipase selectively removed acetyl groups at the 4'- and 6'-positions. Kloosterman et al. (1989) attributed this surprising selectivity to selective hydrolysis of the acetyl group at the 6'-primary hydroxyl followed by acetyl migration from the 4'- to the 6'-position.

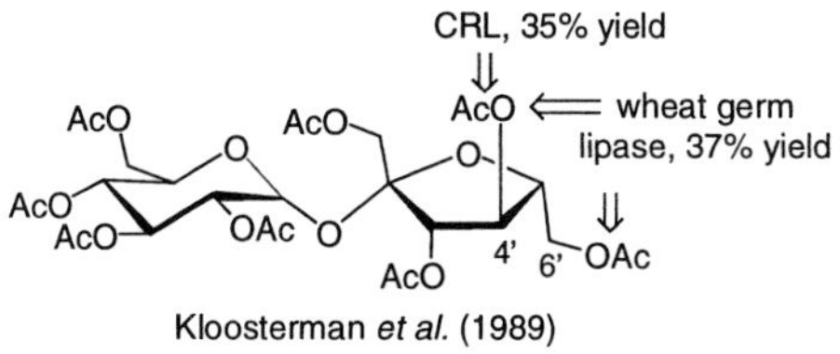

Kloosterman *et al.* (1989)

Fig. 116. Lipases selectively deacetylated octa-*O*-acetyl sucrose at a secondary position.

Several researchers reported selective reaction at the secondary alcohol position in the presence of a chemically more reactive primary alcohol position. For example, PCL selectively cleaved the hexanoate ester of secondary alcohol in several 2-deoxyribonucleosides (Uemura et al., 1989b). Subtilisin selectively cleaved the ester at the primary position. In the reverse reaction, PCL also catalyzed the selective acylation of this secondary hydroxyl (see above) (Uemura et al., 1989a; Garcia-Alles et al., 1993). Protecting the primary alcohol as a hindered ester also allowed selective hydrolysis of the ester at the secondary position in the protected arabinose in Fig. 117 (Kloosterman et al., 1987).

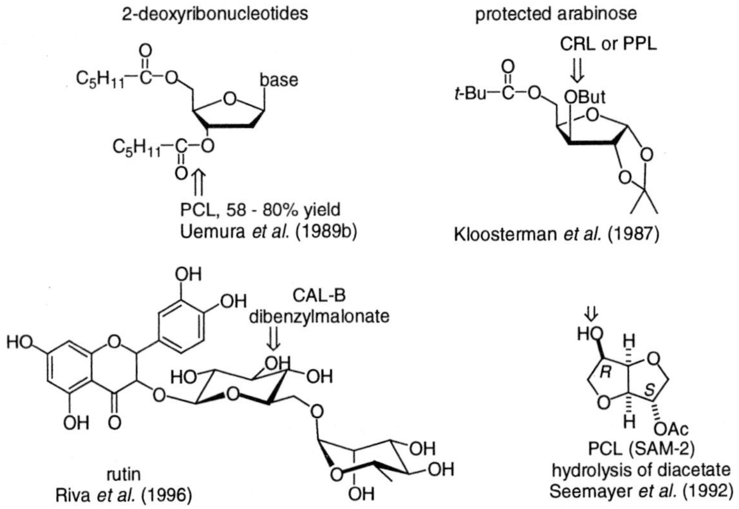

Fig. 117. Lipases sometimes favor hydrolysis of estes at secondary positions over primary positions. Lipases also show selectivity among secondary positions.

To form the monoacetate of 1,4:3,6-dianhydro-D-glucitol, Seemayer et al. (1992) started with the diacetate and selectively hydrolyzed the acetate at the (*R*)-stereocenter. This selectivity fits the secondary alcohol rule discussed in Sect. 5.1.1, but in this case the starting material was derived from a sugar and thus was already enantiomerically pure.

Acylation of Secondary Hydroxyl Groups. Therisod and Klibanov (1987) found that lipases catalyzed the regioselective acylation of the C-2 or C-3 hydroxyl group in C-6 protected glucose. The regioselectivity depended on the lipase. For example, CVL catalyzed the butyrylation of the C-3 hydroxyl of 6-*O*-butanoyl glucopyranose with trichloroethyl butyrate in THF, while PPL catalyzed butyrylation of the C-2 hydroxyl. Chemical or enzymatic methods removed the protecting groups at the 6-position leaving a C-2 or C-3 hydroxyl protected glucose. No lipase acylated at the C-4 hydroxyl. For 6-*O*-butyryl mannose and galactose the selectivity was low (5:1 at best, typically 2:1). Another example of lipase-dependent regioselectivity is the butyrylation of 1,4-anhydro-5-*O*-hexadecyl-D-arabinitol with trichloroethyl butyrate in benzene (Nicotra et al., 1989). HLL catalyzed butyrylation of the C-2 hydroxyl, while *R. japonicus* lipase favored the hydroxyl at C-3 (Fig. 118). In castanospermine, lipases PPL and CVL favored acylation of one hydroxyl group, while subtilisin favored another (Fig. 118) (Margolin et al., 1990).

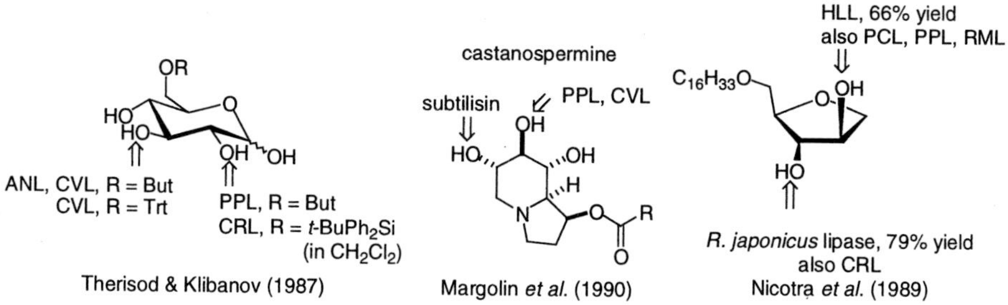

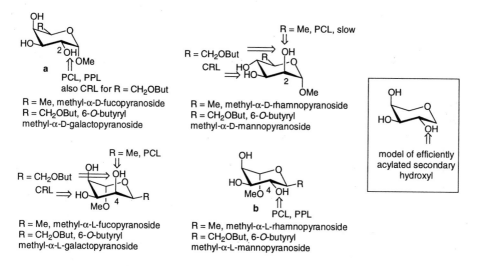

Fig. 118. Selectivity among secondary positions when the primary position is protected.

For methyl glycosides, several lipases (PCL, PPL, CRL) selectively acylated the C-2 hydroxyl in 6-O-butyryl methyl α-D-galactopyranoside, but the regioselectivity for the corresponding mannoside was still low (Fig. 119, right column). In a series of methyl pyranosides, Ciuffreda et al. (1990) found that PCL acylated the C-2 hydroxyl in the D-series of sugars, but the C-4 hydroxyl in the L-series. Acylation was much slower when the reacting hydroxyl group was axial (D-rhamnose and L-fucose derivatives). They suggested that efficient acylation requires an axial-equatorial-equatorial arrangement of hydroxyls with acylation occuring at the last equatorial hydroxyl.

Fig. 119. In methyl α-D- and α-L-glycopyranosides, PCL regioselectively acylated the C-2 hydroxyl group in the D-series (top two structures), but the C-4 hydroxyl group in the L-series (bottom two structures) using trifluoroethyl butyrate in THF (Ciuffreda et al., 1990). Only the sugars **a** and **b** reacted quickly, thus Ciuffreda et al. (1990) suggested that an efficiently acylated sugar contains an axial-equatorial-equatorial arrangement of hydroxyls as shown in the model.

The regioselectivity of the PCL- and PFL-catalyzed acylation of methyl 4,6-O-benzyl-idene glycopyranosides depended on the configuration at the anomeric carbons

(Fig. 120) (Chinn et al., 1992; Panza et al., 1993a, b; Iacazio and Roberts, 1993). The α-anomers yielded the C-2 monoester with typical reaction times of 7 h, while the β-anomers reacted in about 1 h and yielded the C-3 monoester. The Roberts group used vinyl acetate as the solvent and acylating agent, while Panza et al. (1993a, b) used a variety of vinyl esters and trifluoroethyl esters in THF. The galactose derivatives reacted significantly more slowly, possibly due to steric hindrance.

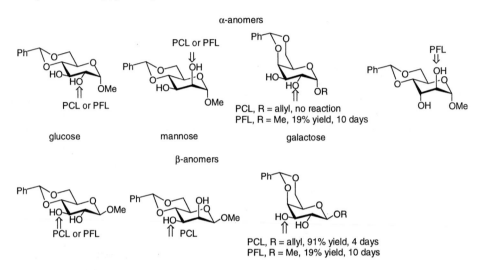

Fig. 120. The configuration at the anomeric carbon determines the regioselectivity of the acylation of methyl 4,6-O-benzylidene glycopyranosides. PCL and PFL acylate the C-2 hydroxyl in the α-anomers and the C-3 hydroxyl in the β-anomers, regardless of the orientation of the reacting hydroxyl.

PCL also catalyzed acylation of the C-3 hydroxyl in 6-O-acetyl D-glucal and 6-O-acetyl D-galactal with vinyl esters (Holla, 1989) (Fig. 121).

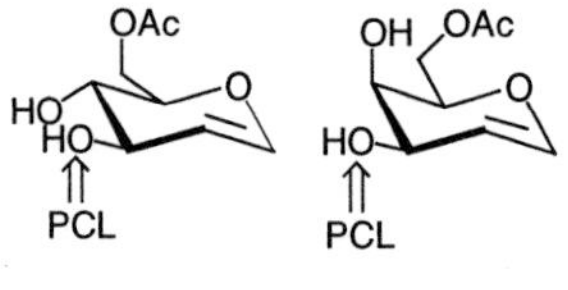

Fig. 121. Regioselectivity among secondary hydroxyl groups sometimes follows the secondary alcohol rule.

López et al. (1994) found that the regioselectivity also varies with the nature of the substituent at the anomeric position. PCL catalyzed formation of the 3,4-diacetate of methyl-β-D-xylopyranoside using vinyl acetate in acetonitrile, whereas the octyl derivative in acetonitrile or hexane gave a mixture of the 2,4- and 3,4-diacetates. At short reaction times, the 2-monoacetate predominated. The choice of solvent and reaction conditions is less critical than for sugars because these sugar derivatives are more soluble (Fig. 122).

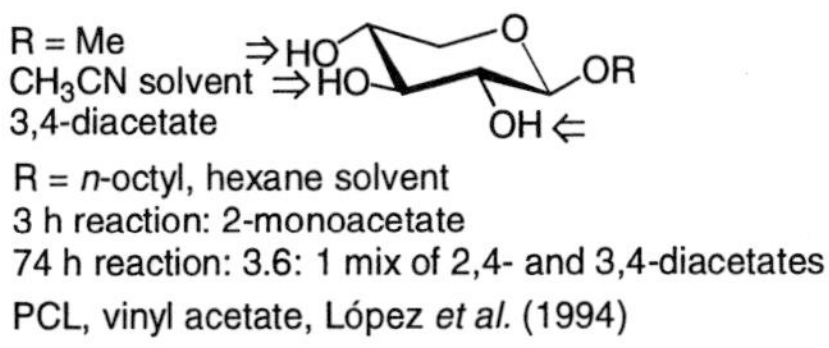

R = Me
CH$_3$CN solvent
3,4-diacetate

R = n-octyl, hexane solvent
3 h reaction: 2-monoacetate
74 h reaction: 3.6: 1 mix of 2,4- and 3,4-diacetates
PCL, vinyl acetate, López *et al.* (1994)

Fig. 122. Regioselectivity varies with the nature of the substituent at the anomeric position.

In summary, lipases can react selectively at the different secondary hydroxyls. The selectivity varies with lipase and substrate structure (anomeric orientation, anomeric substituent, orientation of hydroxyl) and one cannot make broad generalizations yet.

6.1.1.3   Hydroxyl Groups in Non-Sugars

Phenolic hydroxyls. Several lipases, especially PCL and PPL, catalyze the deacetylation of peracetylated polyphenols by transesterification with *n*-butanol in organic solvents (Figs. 123–125). Researchers deacetylated by transesterification instead of hydrolysis because the substrates do not dissolve in water. The regioselectivity of lipases toward phenolic hydroxyls usually paralleled their chemical reactivity – less hindered positions reacted more quickly. For flavone acetates and related compounds, a generalization in Fig. 124 summarizes some of the observed regioselectivity. In addition, PCL catalyzed the regioselective acetylation of polyphenols with vinyl acetate (Fig. 123 and 125). Both acetylation and deacetylation favor the less hindered positions, thus the two reactions yield complementary products. Nicolosi et al. (1993) used the deacetylation reaction in the synthesis of a rare *O*-methyl flavonoid, ombuin.

In a number of symmetrical acylated catechols, PPL selectively removed only one acyl group (Parmar et al., 1996, 1997). Lipases CRL and PPL also showed excellent chemoselectivity. They cleaved the phenolic ester, while leaving the benzoate ester intact.

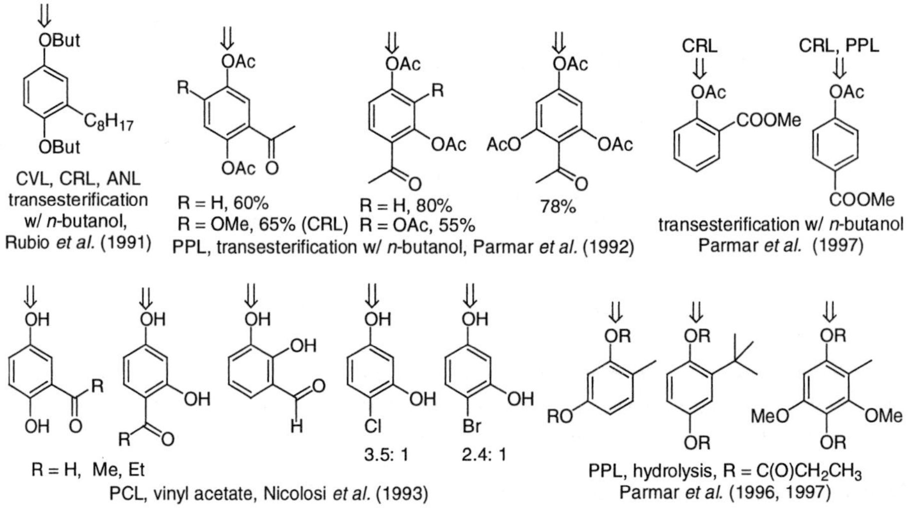

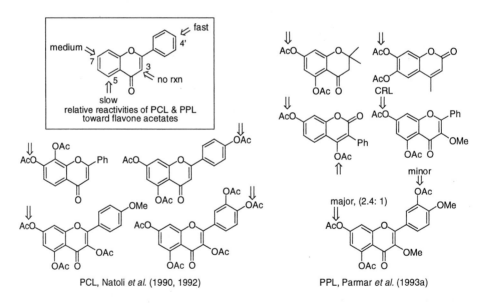

Fig. 123. Regioselectivity of lipases toward polyhydroxylated benzenes. Lipases favored the less hindered position in both deacetylation of peracetylated phenols by transesterification with *n*-butanol and in acetylation of phenols with vinyl acetate. Note that the first two examples of Parmar et al. (1996, 1997) show deacylation at the more hindered ester.

Fig. 124. Regioselectivity of lipases toward flavone acetates and related compounds. Lipases catalyzed the deacetylation by transesterification with *n*-butanol. Less hindered acetates react more quickly; a generalization for the observed regioselectivity is suggested above.

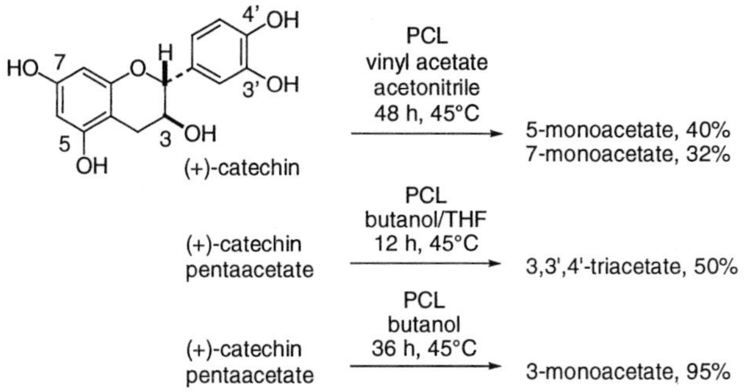

Fig. 125. Regioselective acetylation and deacetylation of catechin. Hydroxyls or acetates at positions 5 and 7 react most quickly, while those at position 3 do not react. Acetylation and deacetylation yield complementary acetates (Lambusta et al., 1993).

Aliphatic Hydroxyls. PPL in acetone selectively acylated the primary hydroxyl group in several diols using trifluoroethyl butyrate (Parmar et al., 1993b). Deacylation of the corresponding diesters showed apparent selectivity for the secondary hydroxyl, but later work showed that deacylation occurred at the primary position, followed by acyl migration to the secondary position (Bisht et al., 1996) Several lipases selectively hydrolyzed one primary hydroxyl in a diacetate (Itoh et al., 1996c) (Fig. 126).

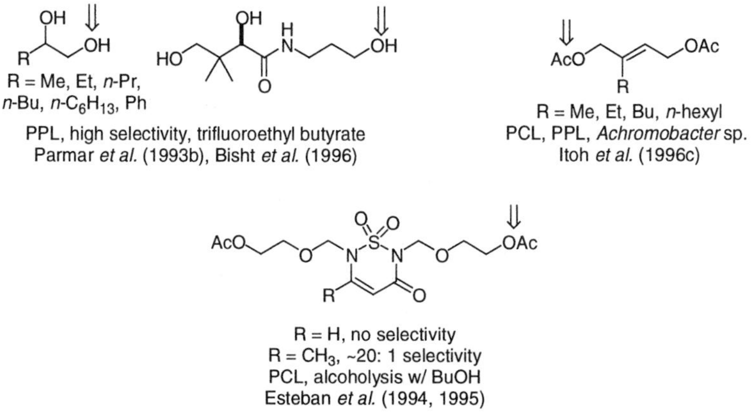

Fig. 126. Selective acylation and deacylation of primary alcohols.

In some cases, the configuration of nearby stereocenters changed the selectivity. For example, PCL showed a low selectivity for the less hindered primary hydroxyl in the (*R*)-enantiomer in Fig. 127, but a moderate selectivity for the more hindered primary hydroxyl in the (*S*)-enantiomer. In another case, CRL acetylated the hydroxyl at the (*S*)-stereocenter only in the (*S,S*)-stereoisomer, not in the (*S,R*)-stereoisomer (Fig. 127).

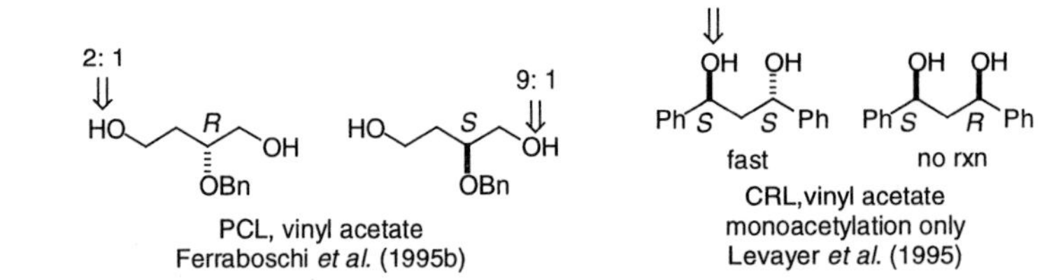

Fig. 127. Selectivity varies with the configuration of nearby stereocenters.

Sattler and Haufe (1995) selectively acylated the primary over the secondary alcohol in a mixture of diastereomers. This regioselective reaction is a more convenient way of separating the isomers than the orginal method of flash chromatography (Fig. 128).

Garcia-Granados et al. (1998) separated a mixture of three diastereomeric sesquiterpene lactones (formed by hydrogenation of a natural product) by selective acetylation. First, either PPL or RML selectively acetylated the diastereomer having the (R)-carbinol. After separation of the unreacted diastereomers, CAL-B selectively acetylated the trans diastereomer (Fig. 128).

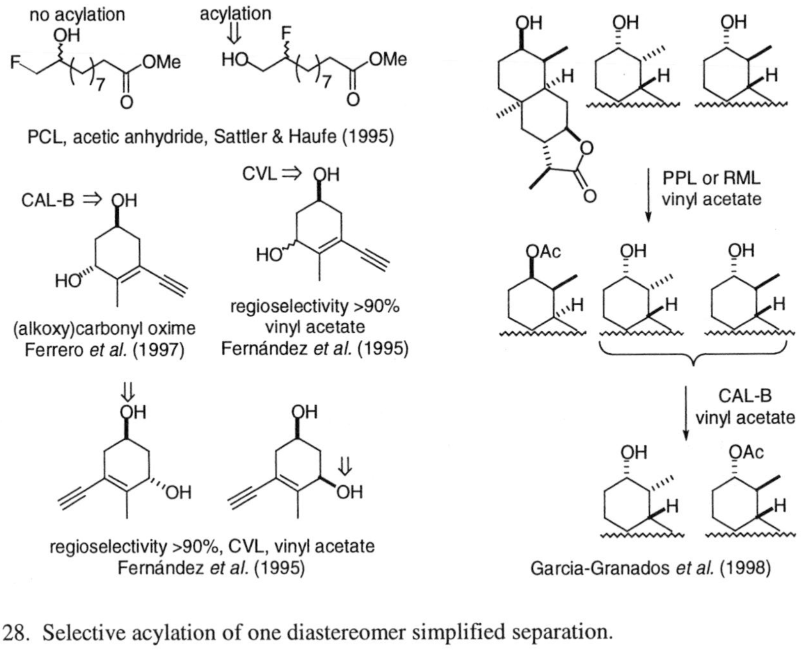

Fig. 128. Selective acylation of one diastereomer simplified separation.

Several lipases (PCL, ANL, CRL, GCL) also resolved hydrophobic amino acids (Ala, Leu, Met, Phe, Val) by hydrolysis of methyl ester in N-Cbz protected amino acids (Fig. 129). The enantioselectivity was usually moderate and favored the L-enantiomer (Chiou et al., 1992).

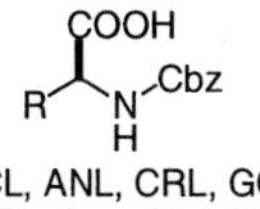

PCL, ANL, CRL, GCL
E >15 (Ala, Leu, Met, Phe, Val)
hydrolysis of Me ester
Chiou *et al.* (1992)

Fig. 129. Resolution of *N*-Cbz protected amino acids by lipase-catalyzed hydrolysis.

   Lipases catalyze the chemoselective acylation of 2-mercaptoethanol and similar compounds. Acylation occured on the oxygen yielding *O*-acyl esters, not *S*-acyl thioesters (Baldessari et al., 1994; Iglesias et al., 1996). This chemoselectivity may be in part due to the choice of an unactivated acyl donor – simple carboxylic ethyl esters – which lack the thermodynamic driving force to form a thioester. In steroids, lipases favor the less-hindered and equatorially oriented hydroxyls (Fig. 130).

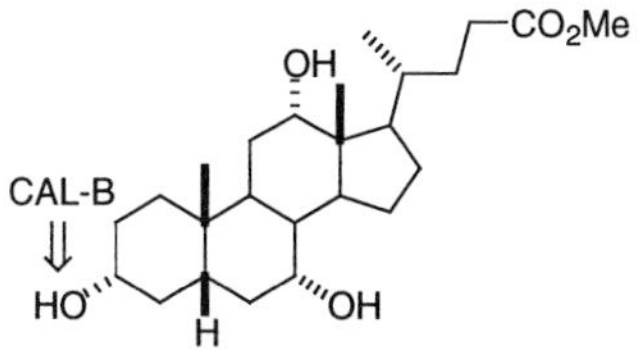

acylation wth methyl alkanoate
Sugai *et al.* (1996)

Fig. 130. CAL-B favored the less-hindered and equatorially oriented hydroxyls.

6.1.2   Amino Groups

Amines react spontaneously with most acylating agents so few lipase-catalyzed reactions have been reported. Gardossi et al. (1991) used dilute solutions and a large amount of lipase to selectively acetylate the ε-amino group in L-Phe-α-L-Lys-O-*t*-Bu and L-Ala-α-L-Lys-O-*t*-Bu with trifluoroethyl acetate. Pozo et al. (1992) used the less reactive vinyl carbonate and CAL-B to form a carbamate, one of the more common amino protective groups (Eq. 28).

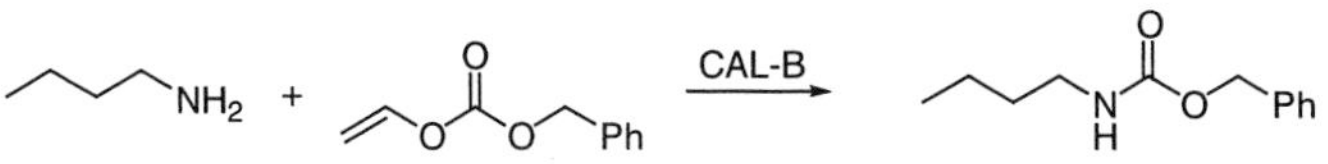

(28)

58%

   Adamczyk and Grote (1996) protected amines by PCL-catalyzed acylation using benzyl esters.

Lipases are not used to deprotect amines because lipases rarely cleave amides or carbamates, the most common amino protective groups. Proteases such as penicillin G acylase are normally used for deprotection (reviewed by Waldmann and Sebastian, 1994). However, Waldmann and Naegele (1995) reported an indirect removal of carbamate protective group with an esterase. Upon cleavage of the acetyl group from a *p*-acetoxybenzyloxycarbonyl-protected peptide with acetylesterase, the carbamate link cleaved spontaneously. Lipases should also catalyze this reaction.

PCL catalyzed the hydrazidolysis of α,β-unsaturated esters such as methyl acrylate (Eq. 29) (Gotor et al., 1990; Astorga et al., 1991, 1993). These reactions occur at room temperature with simple esters, while chemical methods require higher temperatures, activated esters or acid chlorides, and suffer from competing Michael additions.

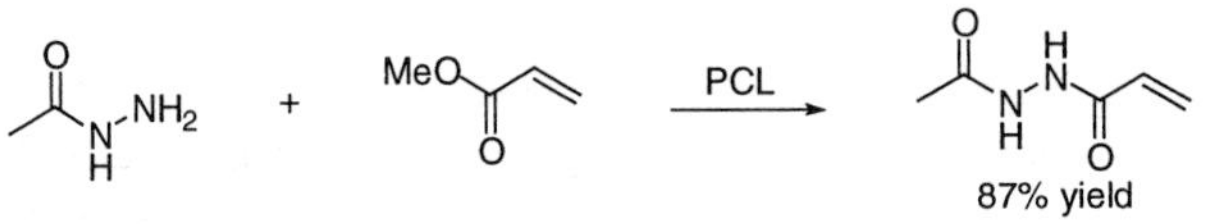

$$(29)$$

87% yield

ANL was effectively used for the mild removal of an acetyl protective group in a precursor of *N*-glycolylneuraminic acid – a member of the sialic acid family – to afford *N*-glycosylmannosamine (Fig. 131) (Kuboki et al., 1997). Khmelnitsky et al. (1997) regioselectively acylated paclitaxel to introduce water-solubilizing groups. The same group also used regioselective acylation of a flavonoid to generate a combinatorial library of derivatives (Fig. 131) (Mozhaev et al. (1998).

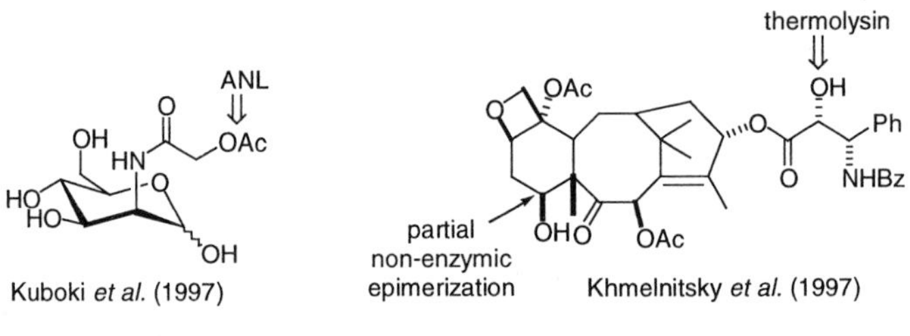

Kuboki *et al.* (1997)   Khmelnitsky *et al.* (1997)

Fig. 131. ANL catalyzed mild removal of an acetyl protective group. Thermolysin regioselectively acylated only one hydroxyl in paclitaxel.

Penicillin G acylase (PGA, penicillin amidase) is highly selective for phenylacetyl groups. Researchers exploited this chemoselectivity to selectively remove *N*-protective groups from peptides (Waldmann et al., 1991). In addition, Pohl and Waldmann (1996) made a carbamate protective group that can be removed with PGA.

6.1.3   Carboxyl Groups

Although many chemical methods exist to protect and deprotect carboxyl groups in amino acids for peptide synthesis, many of these are incompatible with sensitive functional groups such as thioesters, phosphate esters, and polyenes (farnesyl groups). The mild reaction conditions for enzymic reactions makes them ideal for reactions involving sensitive substrates. Chemists have developed a variety of methods, most of which involve proteases, esterases or lipases (Waldmann and Sebastian, 1994). Only the lipase examples are reviewed below.

The advantages of lipases over proteases are that they tolerate water-insoluble substrates, do not cleave peptide bonds (a potential side reaction in protease-catalyzed reactions of peptides), and tolerate both L- and D-amino acids.

tert-Butyl esters are usually inert to lipases and proteases, but thermitase (see Sects. 2.2.2.2 and 8.2.1) is an exception. Thermitase deprotected the carboxyl group in several peptides and glycopeptides by hydrolysis of the *t*-butyl esters (Schultz et al., 1992). Thermitase-catalyzed hydrolysis of ester links is much faster than of peptide links, so the peptide remained intact.

Many groups have used lipases to deprotect carboxyl groups in peptides. Braun et al. (1990, 1991) cleaved the heptyl ester carboxyl protective group from a wide range of dipeptides using ROL   (Amano N). This lipase did not cleave the amino protective groups Cbz, Boc, Alloc, or Fmoc and the heptyl protective group survived conditions used to remove these amino protective groups (hydrogenation, HCl/ether, or Pd(0)/C-nucleophile). Although the crude lipase (Amano N) also hydrolyzed the peptide link, pretreatment with PMSF, a serine protease inhibitor, eliminated this side reaction. Hydrolysis of the heptyl group slowed and sometimes did not proceed when the peptide was hindered and/or hydrophobic. In some of these cases, replacing the heptyl ester with the more reactive 2-bromoethyl ester or the more water-soluble 2-(*N*-morpholino)ethyl ester or MEE ester increased reaction rate (Waldmann et al., 1991; Braun et al., 1993; Kunz et al., 1994; Eberling et al., 1996) (Fig. 132).

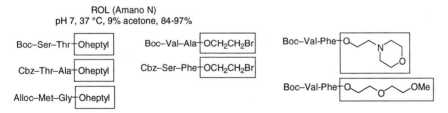

Fig. 132. Deprotection of carboxyl groups in more hydrophobic peptides requires more reactive or more soluble esters.

In other cases where hydrolysis catalyzed by ROL (Amano N) was slow, researchers substituted another lipase. During a glycopeptide synthesis researchers used RJL to cleave the heptyl protective group (Braun et al., 1992, 1993). In a similar glycopeptide, RJL did not cleave the heptyl ester, so Eberling et al. (1996) used the more reactive MEE ester. To cleave *C*-terminal proline MEE esters, researchers used HLL (Kunz et al., 1994) (Fig. 133).

Sutherland and Willis (1997) used CRL to deprotect several α-keto acids by hydrolysis of their methyl esters (Fig. 133). The mild conditions prevented decarboxylation and allowed the researchers to perform the next step, a dehydrogenase-catalyzed reduction, in the same pot.

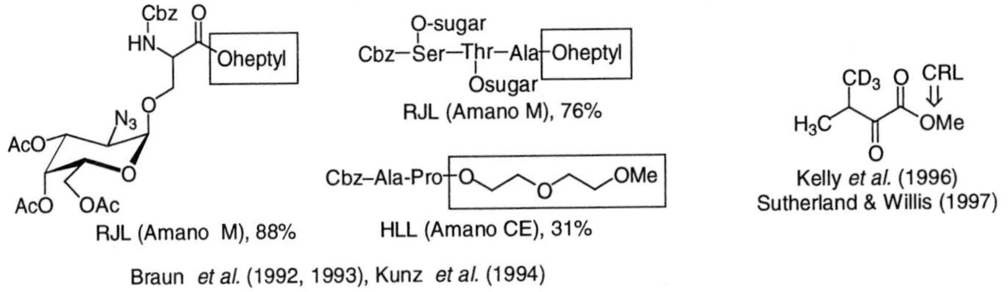

Fig. 133. Deprotection of carboxyl groups using other lipases.

Lipases can also selectively deprotect the two carboxylate groups in glutamic acid. For the uncommon enantiomer glutamic acid diesters (D-), both CRL and CAL-B favored reaction at the less hindered ester. CRL catalyzed the selective hydrolysis of the dicyclopentylester (Wu et al., 1991), while CAL-B catalyzed the selective amidation of the diethyl ester with pentylamine (Chamorro et al., 1995). On the other hand, in the more common enantiomer L-glutamate diethylester, CAL-B selectively amidates the more hindered ester (Chamorro et al., 1995) (Fig. 134). Proteases subtilisin and α-CT showed an opposite regioselectivity toward the cyclohexyl diester of L-aspartic acid. While α-CT favored the 'normal' carboxylate at the α-position, subtilisin favored the less hindered carboxylate at the β-position (Fig. 134).

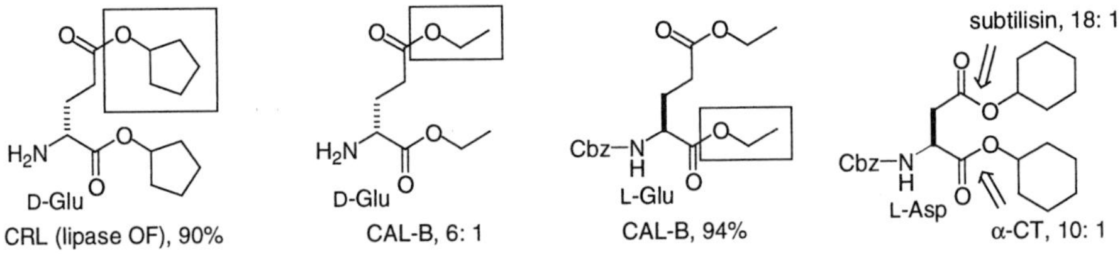

Fig. 134. Regioselective deprotection of glutamic and aspartic acid esters.

To develop immunoassays for drugs, researchers must immunize animals with the drug linked to a protein. In several cases, lipases have simplified this linking process (Adamczyk et al., 1994, 1995). PCL selectively hydrolyzed one of the ester groups in the diacid linker molecule for both the rapamycin and the digoxigenin derivatives. For the digoxigenin derivatives, ester groups on shorter linkers did not react (Fig. 135).

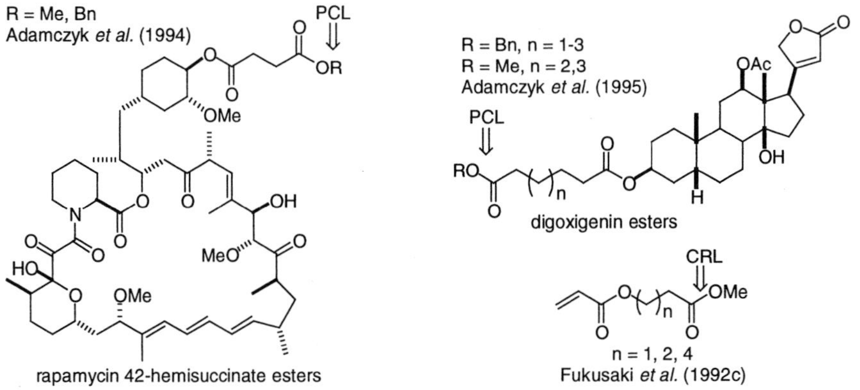

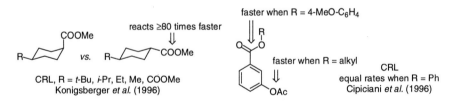

Fig. 135. Regioselective deprotection of carboxyl groups.

Sharma et al. (1995) used PPL to selectively monoesterify aliphatic dicarboxylic acids with butanol. Acid groups containing a carbon-carbon double bond at the α,β-position reacted more slowly than acid groups adjacent to saturated chains. Some CRL-catalyzed chemoselective reactions are summarized in Fig. 136.

Fig. 136. Chemoselective reactions catalyzed by CRL.

6.2   Lipid Modifications

Of the 60 million metric tons of fats and oils produced each year world-wide, most are used directly in food, but about 2 million tons undergo chemical processing such as hydrolysis, glycerolysis and alcoholysis. Current chemical processes require high temperature and pressure which degrade the fats and introduces impurities. For example, sodium methoxide-catalyzed interesterification of triglycerides also catalyzes Claisen condensations and imparts a brown color. Lipases require milder conditions – lower temperatures, near neutral pH – and are also regioselective for the primary vs. secondary positions in glycerol and chemoselective for different fatty acids. Researchers have developed several lipase-catalyzed processes for specialty fats. These processes exploit either the regioselectivity of lipases, the fatty acid selectivity or the mild reaction conditions. However, lipases are more expensive, so they will not replace chemical catalysts for processing of low-value fats. For recent reviews on lipase-catalyzed modification of lipids see Nielsen (1985); Bühler and Wandrey (1987a, b); Yamane (1987); Baumann et al. (1988); Mukherjee (1990); Björkling et al. (1991); Casey and Macrae (1992); Prazeres and Cabral (1994); Adlercreutz (1994); Vulfson (1994); Haas and Joerger (1995); Marangoni and Rousseau (1995); Bornscheuer (1995); Villeneuve and Foglia (1997) and Gunstone (1998).

6.2.1   1,3-Regioselective Reactions of Glycerides

Many lipases, called 1,3-selective lipases, catalyze reactions at the primary alcohol positions of glycerol and glycerol derivatives, while other lipases, called nonselective lipases, react at all three positions (Tab. 15) (see Sect. 5.1.2.4 for the glycerol nomenclature). The *Rhizomucor* and *Penicillium* lipases are all 1,3-selective, but the *Candida* lipases include both 1,3-selective and nonselective lipases. A single microorganism, *Candida antarctica*, produces two lipases: A, which is nonselective, and B, which is 1,3-selective. For lipase from *Pseudomonas fluorescens*, some researchers reported 1,3-selectivity, while others reported no selectivity. This may be due to the *P. cepacia* vs. *P. fluorescens* confusion mentioned in Sect. 2.1.1.

The 1,3-selective lipases differ in the degree of selectivity as indicated qualitatively in Tab. 15. Quantitative measure of selectivity is difficult in water due to facile acyl migration (Sect. 6.2.1.2), but Schneider's group measured the selectivity for several lipases in nonpolar organic solvents where acyl migration is slow. The highly 1,3-selective ROL (Amano D) reacted at the primary position 76 times faster than at the secondary position, while the moderately selective CVL and RML reacted 26 and 11 times faster, respectively. The nonselective PFL reacted only 1.4 times faster (Berger and Schneider, 1991b; Berger et al., 1992). Note also that lipases with 1,3-selectivity toward triglycerides still hydrolyze esters of secondary alcohols. For examples see Sect. 5.1.1.6.

Rogalska et al. (1993) reported that CAL-A selects for the secondary alcohol position (*sn*-2) in monolayer films. In other cases, apparent 2-selectivity of a lipase was later attributed to nonselective hydrolysis combined with acyl migration from the 2 position to the 1,3 positions (for an example see Briand et al., 1995). Rogalska et al. (1993) found no acyl migration under their reaction conditions, but the 2-selectivity of this lipase awaits independent verification.

6.2.1.1 Modified Triglycerides

Structured triglycerides (ST's) are triglycerides modified in either the type of fatty acid or the position of the fatty acids. The synthesis of ST's such as cocoa butter equivalents and MLM lipids (triglycerides with medium-chain fatty acids at *sn*-1 and *sn*-3 and a long chain fatty acid at *sn*-2) relies on the 1,3-selectivity of lipases. The synthesis of ST's enriched in polyunsaturated fatty acids exploits the fatty acid selectivity of lipase and will be discussed in Sect. 6.2.2.2. Christophe (1998) provides an excellent overview about nutritional properties and enzymatic synthesis of structured triglycerides.

Cocoa Butter Equivalents. Cocoa butter is predominantly 1,3-disaturated-2-oleyl-glyceride, where palmitic, stearic and oleic acids account for more than 95 % of the total fatty acids. Cocoa butter is crystalline and melts between 25 and 35 °C imparting the desirable 'mouth feel'. Unilever (Coleman and Macrae, 1977) and Fuji Oil (Matsuo et al., 1981) filed the first patents for the lipase-catalyzed synthesis of cocoa butter equivalents. Both companies currently manufacture cocoa butter equivalents on a multi-ton scale using 1,3-selective lipases to replace palmitic acid with stearic acid at the *sn*-1 and *sn*-3 positions (for reviews see Macrae, 1983; Macrae and Hammond, 1985; Quinlan and Moore, 1993). The Unilever subsidiary Quest-Loders Croklaan (Holland) uses RML and palm oil mid fraction or high oleate sunflower oil as starting materials (Eq. 30). Other suitable starting oils are rape seed (Adlercreutz, 1994) or olive oils (Chang et al., 1990).

$$2 \begin{array}{l} \text{—OC}_{16:0} \\ \text{—OC}_{18:1} \\ \text{—OC}_{16:0} \end{array} + 3\,C_{18:0} \xrightarrow{\text{1,3-selective lipase}} \begin{array}{l} \text{—OC}_{16:0} \\ \text{—OC}_{18:1} \\ \text{—OC}_{18:0} \end{array} + \begin{array}{l} \text{—OC}_{18:0} \\ \text{—OC}_{18:1} \\ \text{—OC}_{18:0} \end{array} + 3\,C_{16:0} \qquad (30)$$

palm oil fraction   cocoa butter equivalent

Recent work on cocoa butter equivalents focused on optimizing this process. Mohamed et al. (1993) used CAL-B, while others complexed lipases from *Pseudomonas* sp. (Isono et al., 1995) or RJL (Basheer et al., 1995a, b) with surfactants to either increase reaction rates or selectivity. Cho and Rhee (1993) and Cho et al. (1994) investigated continuous packed bed reactors.

Another structured triglyceride is Betapol, a formula additive for premature infants. Human milk contains the more-easily digested OPO but vegetable oils contain the POO isomer (OPO = 1,3-dioleyl-2-palmitoyl glyceride, POO = mixture of 1,2-dioleyl-3-palmitoyl glyceride and its enantiomer). Interesterification of tripalmitin with oleic acid using RML at low water activity yields OPO. Evaporation removes excess fatty acids and crystallization removes remaining PPP.

Tab. 15. Regioselectivity of Some Lipases Toward Hydrolysis or Transesterification of Triglycerides.

Lipase[a]	Regioselectivity	Reference
ANL	moderately 1,3-selective	Macrae and Hammond (1985); Amano Pharmaceutical
CAL-B	1,3-selective	Novo Nordisk
CAL-A	nonselective or 2-selective	Novo Nordisk; Rogalska et al. (1993)
CRL	nonselective	Amano Pharmaceutical
CLL	moderately 1,3-selective	Amano Pharmaceutical
CVL	moderately 1,3-selective	Berger et al. (1992); Macrae and Hammond (1985)
GCL	nonselective	Amano Pharmaceutical
HLL	slightly 1,3-selective	Macrae and Hammond (1985); Amano Pharmaceutical
RJL	slightly 1,3-selective	Macrae and Hammond (1985); Amano Pharmaceutical
PPL	1,3-selective	Macrae and Hammond (1985)
PcamL	highly 1,3-selective	Amano Pharmaceutical
ProqL	moderately 1,3-selective	Amano Pharmaceutical
PCL	nonselective	Amano Pharmaceutical
PFL	nonselective or 1,3-selective	Berger et al. (1992); Berger and Schneider (1991b); Macrae and Hammond (1985); Amano Pharmaceutical
ROL (Amano D)	highly 1,3-selective	Berger et al. (1992); Berger and Schneider (1991b); Amano Pharmaceutical
ROL (Amano N)	highly 1,3-selective	Amano Pharmaceutical
RML	moderately 1,3-selective	Berger et al. (1992); Berger and Schneider (1991b), Novo Nordisk

[a]See Tab. 2 for lipase abbreviations. Additional data is given in Ader et al. (1997).

Synthesis of MLM's. MLM's are triglycerides containing medium chain fatty acids (usually $C_{8:0}$ or $C_{10:0}$) in the *sn*-1 and *sn*-3 positions of the glycerol backbone and a long chain fatty acid in the *sn*-2 position (reviewed by Akoh, 1995). Several chemically synthesized MLM's are commercially available (Tab. 16) but these have a random distribution of fatty acids.

Tab. 16.   Commercially-Available Chemically Synthesized MLM's.

Product	Composition	Company
Captex	$C_{8:0}$, $C_{10:0}$, $C_{18:2}$	ABITEC, Columbus, OH
Neobee	$C_{8:0}$, $C_{10:0}$, LCFA[a]	Stepan Co., Maywood, NJ
Caprenin	$C_{6:0}$, $C_{8:0}$, $C_{22:0}$	Procter & Gamble, Cincinnati, OH
Salatrim	$C_{3:0}$, $C_{4:0}$, $C_{18:0}$	Nabisco Foods, East Hanover, NJ

[a]LCFA = long-chain fatty acids

MLM's are an efficient food source for persons with pancreatic insufficiency and other forms of malabsorption (Babayan, 1987; Megremis, 1991; Babayan and Rosenau, 1991). Pancreatic esterases catalyze hydrolysis of medium chain triglycerides faster than long chain triglycerides and the resulting 2-monoglycerides are absorbed efficiently (Jandacek et al., 1987). Further, the 2-monoglycerides transfer directly from the bloodstream to the cells without forming chylomicrons (Kennedy, 1991). Triglycerides with three medium chain fatty acids (MMM's) are also easily digested, but lack essential fatty acids such as linoleic or linolenic acid, which can be included in MLM's.

Another application of MLM's is as low-calorie fats. Shorter chain fatty acids have fewer calories per unit weight than long chain fatty acids. Stearic acid is poorly absorbed so stearic acid-containing fats impart less energy. Thus, a triglyceride containing stearic acid and short chain fatty acids contains fewer calories. Current syntheses of these materials (e.g., Salatrim (Smith et al., 1994), Caprenin) use chemical catalysts yielding a random distribution of fatty acids, but several researchers made these materials using lipases (Eq. 31) (Shieh et al., 1995; McNeill and Sonnet, 1995; Akoh, 1995; Soumanou et al., 1997, 1998a, b).

$$
\begin{bmatrix} OC_{18:1} \\ OC_{18:1} \\ OC_{18:1} \end{bmatrix} + 2\,C_{10:0} \; \underset{}{\overset{RML}{\rightleftharpoons}} \; \begin{bmatrix} OC_{10:0} \\ OC_{18:1} \\ OC_{10:0} \end{bmatrix} + 2\,C_{18:1} \tag{31}
$$
$$\text{triolein} \qquad\qquad\qquad\qquad \text{an MLM}$$

For instance, the interesterification of trisun 90 (an oil containing 90 % triolein) with capric acid using RML yielded 69 % COC (also containing CCO and OCC) after 30 h reaction in hexane (Shieh et al., 1995) A two-step process gave significantly higher yields and purer products (Soumanou et al., 1997, 1998a). In the first step, alcoholysis of pure triglycerides or natural fats with 1,3-regiospecific lipases (e.g., ROL) yielded *sn*-2 monoglycerides (2-MG) in up to 72 % yield after crystallization. In the second step, the same lipases catalyzed esterification of these 2-MG with caprylic acid. The final product contained more than 90 % caprylic acid in *sn*-1 and *sn*-3 positions, whereas the *sn*-2 position was composed of 98.5 % unsaturated long chain fatty acids. This two-step process was found to be also very useful for the synthesis of Betapol (see above). Alcoholysis of tripalmitin catalyzed by RDL or RML gave 2-monopalmitin (72 % yield, > 95 % purity), which was then esterified with oleic acid using the same lipases to yield up to 72 % OPO containing > 94 % palmitic acid in *sn*-2 position (Bornscheuer et al., 1998b; Schmid et al., 1998). In addition, the alcoholysis can also be performed efficiently in acetone (unpublished), whilst esterification can be performed in the absence of solvent, thus allowing the product to be used in food applications.

Triglycerides Containing Polyunsaturated Fatty Acids. Polyunsaturated fatty acids (PUFA's) for example, eicosapentaenoic acid (EPA, 20:5n-3) and docosahexaenoic acid (DHA, 22:6n-3), are essential fatty acids and may be beneficial in cardiovascular and inflammatory diseases (Kosugi and Azuma, 1994; Pedersen and Holmer, 1995; Akoh, 1995). PUFA's are most efficiently absorbed as triglycerides. Elevated temperatures and extreme pH's promote side reactions, oxidation, *cis-trans* isomerization or double-bond migrations in PUFA's (Haraldsson et al., 1993), so lipase are ideal for syntheses of PUFA-containing triglycerides.

EPA is a better substrate than DHA for all commercial lipases (Haraldsson et al., 1995). The reason is probably that the *cis* double bond is four carbons from the reaction site (carbonyl group) in EPA, but only three carbons away in DHA (Fig. 137).

EPA (*cis*-5,8,11,14,17-eicosapentaenoic acid)   DHA (*cis*-4,7,10,13,16,19-docosahexaenoic acid)

Fig. 137. EPA is a better substrate for lipases than DHA.

Researchers incorporated EPA or DHA into soybean oil to a final content of 10–35 % (Huang and Akoh, 1994), into sardine oil up to 70 % in the presence of ethylene glycol as water mimic (Hosokawa et al., 1995b), into trilinolein up to 70 and 81 % using RML and EPA or DHA ethyl esters (Akoh et al., 1995), and completely into glyceryl ether lipids (Haraldsson and Thorarensen, 1994). Shimada et al. (1995) enriched arachidonic acid (up to 50 % of the remaining glyceride fraction) present in single cell oil from *Mortierella alpina* by hydrolytic removal of other fatty acids with CRL. Pedersen and Holmer (1995) reported that RML favored monounsaturated fatty acids (20:1n-9, 22:1n-9) over EPA or DHA. For this reason, incorporation of PUFA's into triglycerides containing monounsaturated acids was slow.

Direct esterification of PUFA's (or their ethyl esters) and glycerol yielded monoglycerides, and some di- and triglycerides (Li and Ward, 1993; Haraldsson et al., 1993; Kosugi and Azuma, 1994). Esterification of isopropylidene glycerol with a mixture of EPA and DHA using RML yielded protected monoglycerides with a maximum yield of 80 % (Li and Ward, 1994).

Other Triglycerides. Another potential application of lipases is in the synthesis of 'zero-*trans*' margarines (Marangoni and Rousseau, 1995). Partial hydrogenation of oils to increase the melting point also introduces *trans*-isomers of unsaturated fatty acids. Natural oils contain only *cis*-isomers and the *trans*-isomers may contribute to heart disease. Interesterifications similar to those for the synthesis of cocoa butter equivalent also raise the melting points of oils without introducing *trans*-isomers, but the cost is much higher than hydrogenation.

An alternative to triglyceride modification is genetic engineering of the metabolic pathways in plants that produce the oils, so they produce more of the desired oils. For example, high laurate canola has the melting behaviour of a saturated fat, but retains the unsaturated oleate chain at *sn*-2 (Kinney, 1996).

6.2.1.2   Diglycerides

Diglycerides (DG's) occur as 1,3-DG's and the racemic mixture of 1,2-DG and 2,3-DG which we abbreviate as 1,2(2,3)-DG's. Mixtures of isomeric DG's are used with MG's as emulsifiers, but pure regioisomers of DG's are most useful as chemical intermediates (Berger and Schneider, 1993). Researchers used DG's for the synthesis of phospho- (van Deenen and de Haas, 1963) and glycolipids (Wehrli and Pomeranz, 1969), and conjugated DG's to drugs to create more lipid-soluble prodrugs (Garzon-Aburbeh et al., 1983, 1986; Saraiva-Conçalves et al., 1989).

Of the many possible routes to 1,3-DGs (Fig. 138) the best one is the esterification of glycerol with a free fatty acid (or derivatives like ethyl- or vinyl esters) using a 1,3-selective lipase (Tab. 17). To combine the immiscible glycerol and fatty acid, Berger et al. (1992) adsorbed glycerol onto silica gel. This adsorption creates a large surface area and also prevents the glycerol from coating the lipase (Castillo et al., 1997). RML-catalyzed esterification with fatty acid methyl esters or ROL-catalyzed esterification with fatty acid vinyl esters, both yielded 1,3-DG. Vinyl ester reacted faster than simple esters, but are more expensive. Both the yields (> 70 %) and the regioisomeric purity (> 98 %) were high. Although ROL is more regioselective, they preferred RML because it had a higher activity. To increase yields, Weiss (1990) and Yamane et al. (1994) carried out reactions in the solid phase, where crystallization of the diglyceride minimized further reactions and shifted the equilibrium toward 1,3-DG's.

The other regioisomer – 1,2(2,3)-DG – is more difficult to prepare. Hydrolysis or alcoholysis of triglycerides initially yields 1,2(2,3)-DG, but cleavage continues to 2-MG. In addition, the water promotes acyl migration so that 1,2(2,3)-DG isomerises to 1,3-DG. Millqvist-Fureby et al. (1997) formed 1,2(2,3)-DG in 75 % yield by alcoholysis of trilaurin using ProqL. Reaction condition control minimized overhydrolysis, while minimizing water activity avoided acyl migration.

Tab. 17. Lipase-Catalyzed Syntheses of Diglycerides (DG's).

Acyl acceptor	Acyl donor	Reaction system	Yield (%)[a]	Lipase[b]	Reference
Glycerol	palmitic acid	solid-phase	80 (1,3)	RJL	Weiss (1990)
Glycerol	capric acid	free evaporation	73 (1,3)	RML	Kim and Rhee (1991)
Glycerol	vinyl caprylate	t-BuOMe	80 (1,3)	ROL (Amano D)	Berger et al. (1992)
Glycerol	lauric acid	t-BuOMe	80 (1,3)	RML	Berger et al. (1992)
Glycerol	oleic acid	2-butanone	28 (1,3)	CVL	Janssen et al. (1993b)
Glycerol	beef tallow	solid-phase	90 (mix)	PCL	Yamane et al. (1994)
Glycerol	palmitic acid	hexane	63 (1,3)	RML	Kwon et al. (1995)
Glycerol	ethyl caproate	free evaporation/precipitate	88 (1,3)	ROL[c]	Millqvist-Fureby et al. (1996a)
Water	triolein	phosphate buffer	43 (mix)	PPL	Plou et al. (1996)
Ethanol	trilaurin	i-Pr$_2$O	75 (1,2)	ProqL	Millqvist-Fureby et al. (1997)

[a] 1,3: 1,3-DG; 1,2: 1,2 (2,3)-DG; mix: mixture of 1,3-DG and 1,2(2,3)-DG. [b] See Tab. 2 for lipase abbreviations. [c] Authors used lipase from *Rhizopus arrhizus* which has been reclassified as *Rhizopus oryzae*.

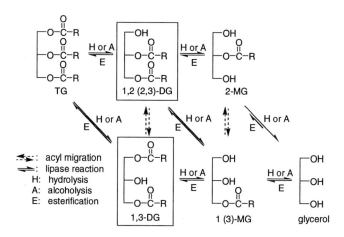

Fig. 138. Reactions during lipase-catalyzed synthesis of DG (after Millqvist-Fureby et al., 1997, modified).

Acyl Migration in Mono- and Diglycerides. Acyl migration is the intramolecular transfer of an acyl group to an adjacent hydroxyl group. This reaction isomerizes the regioisomers of mono- and diglycerides (Eq. 32 and 33) thus degrading the regioselectivity of lipases. The equilibrium favors acylation of the less hindered primary position in both mono- and diglycerides, although the equilibrium constant in diglycerides is only 1.5 (Serdarevich, 1967).

$$
\begin{array}{c}
\text{2-MG} \quad \xrightarrow{\;K_{eq}\sim 9\;} \quad \text{1(3)-MG}
\end{array}
\tag{32}
$$

$$
\begin{array}{c}
\text{1,2(2,3)-DG} \quad \xrightarrow{\;K_{eq}\sim 1.5\;} \quad \text{1,3-DG}
\end{array}
\tag{33}
$$

Acid or base catalyze acyl migration, but acyl migration remains fast even at neutral pH in polar solvents. To suppress acyl migration, researchers use aprotic solvents such as ketones, ether, toluene or alkanes (Sjursnes and Anthonsen, 1994; Millqvist-Fureby et al., 1996b). Surprisingly, acyl migration is faster in nonpolar solvents like hexane than in polar aprotic solvents like acetone (Sjursnes and Anthonsen, 1994; Millqvist-Fureby et al., 1996b). The rate is still hundred of times slower than in protic solvents.

To further reduce acyl migration, one should minimize the water content (Heisler et al., 1991; Millqvist-Fureby et al., 1996b) and use polypropylene supports for lipases instead of ion exchange resins (Millqvist-Fureby et al., 1996b). Hydrogen phosphate salts, used to control water activity, promoted acyl migration, but sulfate salts did not (Sjursnes et al., 1995). Dorset, 1987) even observed an intermolecular acyl migration in crystalline 1,2(2,3)-DG's.

Since the equilibrium constant favors the 1(3)-MG's over 2-MG's by 9:1, researchers often either ignore or even encourage acyl migration in this case. A 9:1 ratio of regioisomers is sufficient for emulsifier applications. For synthetic application a crystallization can give pure 1(3)-MG in many cases.

6.2.1.3   Monoglycerides

Monoglycerides are the predominant emulsifiers in food, pharmaceuticals and cosmetics (Sonntag, 1982; Baumann et al., 1988). In addition, MG's are also building blocks for syntheses of lipids, liquid crystals, and drug carriers (Berger and Schneider, 1993). Current MG manufacture involves continuous glycerolysis of fats and oils at 220–250 °C using inorganic alkaline catalysts. Manufacturers avoid unsaturated fats because they burn or polymerize causing a dark color, off-odor and burnt taste. This process yields technical MG of ~50 % purity which is suitable for many applications. Purification by molecular distillation yields pure MG: 90 % MG (an equilibrium mixture of regioisomers), 10 % DG (also mixed regioisomers) and < 1 % glycerol. Pure MG's have better emulsifying properties than mixed glycerides (Nagao and Kito, 1990). Yearly european production is 28 000 metric tons of technical MG and 42 000 metric tons of pure MG.

Researchers have developed three lipase-catalyzed routes to MG's: (1) hydrolysis or alcoholysis of triglycerides, (2) glycerolysis of triglycerides, and (3) esterification or transesterification of glycerol with fatty acids or esters (Tab. 18) (reviewed by Bornscheuer, 1995). The first method yields 2-MG's, while methods two and three usually yield an equilibrium mixture of MG's, from which the predominant 1(3)-MG's can be isolated in good yield

Hydrolysis or Alcoholysis of Triglycerides to 2-MG's. Hydrolysis or alcoholysis of triglycerides catalyzed by a 1,3-selective lipase yields 2-MG (Eq. 34; Tab. 18). Hydrolysis gave moderate yields of 2-MG's (78 %) (Holmberg and Österberg, 1988). Acyl migration probably limited the yield by forming 1(3)-MG's which underwent further hydrolysis. On the other hand, alcoholysis can be carried out in nonpolar solvents where acyl migration is slower. For this reason, alcoholysis of triglycerides gave higher yields (75–97 %) than hydrolysis. In addition alcoholysis reactions are often faster because there is no change in pH during the reaction and less inhibition of the lipase by free fatty acids. Addition of excess alcohol shifts the equilibrium toward MG formation (Zaks and Gross, 1990b; Li and Ward, 1993; Millqvist et al., 1994; Millqvist-Fureby et al., 1997).

$$
\begin{array}{l}
\text{—OC(O)R} \\
\text{—OC(O)R} \\
\text{—OC(O)R}
\end{array}
+ 2\ \text{R'OH}
\xrightarrow{\text{1,3-selective lipase}}
\begin{array}{l}
\text{—OH} \\
\text{—OC(O)R} \\
\text{—OH}
\end{array}
+ 2\ \text{RC(O)OR'}
\qquad (34)
$$

triglyceride   water or alcohol   2-MG   fatty acid or ester

Tab. 18.   Lipase-Catalyzed Syntheses of Monoglycerides (MG). [a]

Method[b]	Acyl acceptor	Acyl donor	Reaction system	Yield[c] (%)	Lipase[a]	Reference
1	water	palm oil	reverse micelles	(26) 78[d]	ROL	Holmberg and Österberg (1988)
1	ethanol	triolein	ethanol	(28) 84[e]	PFL	Zaks and Gross (1990b)
1	butanol	tricaprin	butanol/two phase	(29) 87[d]	HLL	Mazur et al. (1991)
1	isopropanol	cod liver oil	isopropanol	(25) 75[e]	PCL	Li and Ward (1993)
1	ethanol	tripalmitin	t-BuOMe	(32) 97[d]	ROL[f]	Millqvist et al. (1994)
1	ethanol	trilaurin	ethanol	(25) 75[d]	PCL	Millqvist-Fureby et al. (1997)
1	water	triolein	phosphate buffer	(22) 67[e]	PPL	Plou et al. (1996)
2	glycerol	palm oil	reverse micelles	(30) 30	ROL[f]	Holmberg et al. (1989b)
2	glycerol	beef tallow	precipitate/no solvent	(72) 72	PFL	McNeill et al. (1990)
2	glycerol	triolein	reverse micelles	(50) 50	CVL	Chang et al. (1991)
2	glycerol	triolein	precipitate/no solvent	(96) 96	CVL	Bornscheuer and Yamane (1994)
3	glycerol	palmitic acid	precipitate/no solvent	(31) 95	PcamL	Weiss (1990)
3	glycerol	oleic acid	molecular sieve/no solvent	(25) 74	PcamL	Yamaguchi and Mase (1991)
3	glycerol	lauric acid	reverse micelles	(18) 55	CRL	Hayes and Gulari (1991)
3	glycerol	lauric acid	reverse micelles	(20) 62	ROL	Hayes and Gulari (1994)
3	glycerol	oleic acid	molecular sieve/hexane	(24) 72	PcamL	Akoh et al. (1992)
3	IPG	oleic acid	no solvent	(26) 80	RML	Pecnik and Knez (1992)
3	glycerol	vinyl laurate	t-BuOMe	(30) 90	RML	Berger and Schneider (1992)
3	glycerol/PBA	17-OH stearic acid	hexane	(28) 84	RML	Steffen et al. (1992)
3	glycerol	oleic acid	reverse micelles	(14) 42	RML	Singh et al. (1994a)
3	glycerol	lauric acid	reverse micelles	(4) 11	PCL	Bornscheuer et al. (1994b)
3	glycerol	ethyl oleate	acetone	(22) 68	CAL-B	Pastor et al. (1995b)
3	glycerol	palmitic acid	hexane	(20) 61	ROL	Kwon et al. (1995)
3	IPG	vinyl palmitate	pentane	(31) 95	PCL	Bornscheuer and Yamane (1995)

[a] Abbreviations: IPG: isopropylidene glycerol; PBA: phenyl boronic acid; for lipase abbreviations see Tab. 2; [b] Method 1: hydrolysis or alcoholysis of triacylglycerides to 2-MG's, Method 2: glycerolysis of triacylglycerides yielding mixtures of 1(3)-MG's and 2-MG's, Method 3: esterification or transesterification of glycerol with fatty acids or esters yielding 1(3)-MG's; [c] Overall yields of MG relative to glycerolysis (per mol TG), hydrolysis (per mol TG) or esterification (per mol glycerol) are given in brackets; [d] 2-MG; [e] No clear data, probably mixture of 1(3) and 2-MG [f] Authors used lipase from *Rhizopus arrhizus* which has been reclassified as *Rhizopus oryzae*. *Rhizopus* species also include in *R. oryzae*.

Glycerolysis of Triglycerides to 1(3)-MG's. A disadvantage of the method above is the waste of the two fatty acids from the triglyceride. A more efficient approach is to use glycerol as the alcohol thereby using all three fatty acids (Eq. 35) (Yamane et al., 1986). One can use nonselective lipases for this reaction because even a 1,3-selective lipase yields one 2-MG and two 1(3)-MG's. In practice, reaction conditions usually promote acyl migration so that an equilibrium mixture of 9:1 1(3)-MG and 2-MG is formed.

$$
\begin{array}{l}
\left[\begin{array}{l}\text{OC(O)R}\\ \text{OC(O)R}\\ \text{OC(O)R}\end{array}\right]
+ 2 \left[\begin{array}{l}\text{OH}\\ \text{OH}\\ \text{OH}\end{array}\right]
\;\xrightleftharpoons{\text{lipase}}\;
3 \left[\begin{array}{l}\text{OC(O)R}\\ \text{OH}\\ \text{OH}\end{array}\right]
+ \text{2-MG}
\end{array}
\qquad (35)
$$

triglyceride glycerol   1(3)-MG

Glycerolysis of triglycerides in the liquid phase typically yields only 30–50 % MG, due to an insufficiently-favorable equilibrium. To shift the equilibrium, McNeill et al. (1990) crystallized the MG. First, the glycerolysis was carried out in a liquid-liquid emulsion of glycerol and triglyceride, then cooled to crystallize 1(3)-MG. Yields with this method increased to 70–99 % using different reactions (McNeill et al., 1990; McNeill and Yamane, 1991; Ferreira-Dias and Fonseca, 1993, 1995; Bornscheuer and Yamane, 1994; Myrnes et al., 1995). This crystallization also increases the relative amount of 1(3)-MG over 2-MG. Reaction temperature is critical and must be kept just below the melting temperature of the monoglycerides, e.g., for beef tallow, optimum temperature was 42 °C yielding 72 % 1(3)-MG using PFL or CVL (McNeill et al., 1990). In continuous processes, researchers used membranes or off-line extraction to remove the MG in place of crystallization (Koizumi et al., 1987; Gancet, 1990; Chang et al., 1991; Stevenson et al., 1993).

Esterification of Glycerol with Fatty Acids or Fatty Acid Esters Yielding 1(3)-MG's. Esterification of glycerol with a fatty acid or a fatty acid ester also yields MG's without wasting fatty acids (Eq. 36). Reaction mixtures contain polar protic reactants that promote acyl migration, so esterification yields an equilibrium mixture of MG's. To shift the reaction toward MG formation, researchers removed water or alcohol using vacuum or molecular sieves (Miller et al., 1988; Ergan et al., 1990; Gancet, 1990; Kim and Rhee, 1991; Yamaguchi and Mase, 1991; Millqvist-Fureby et al., 1996a).

$$
\left[\begin{array}{l}\text{OH}\\ \text{OH}\\ \text{OH}\end{array}\right]
+ \text{RC(O)OR'}
\;\xrightleftharpoons{\text{lipase}}\;
\left[\begin{array}{l}\text{OC(O)R}\\ \text{OH}\\ \text{OH}\end{array}\right]
+ \text{2-MG} + \text{R'OH}
\qquad (36)
$$

glycerol   fatty acid   1(3)-MG   water or
            or ester   alcohol

The low solubility of the glycerol in nonpolar organic solvents slows the reaction and promote a side reaction – further acylation of the more soluble MG's to DG's. To minimize this problem researchers used different solvents (Akoh et al., 1992; Janssen et al., 1993a; Pastor et al., 1995a, b; Kwon et al., 1995), reverse micelles (Fletcher et al., 1987; Hayes and Gulari, 1991, 1992, 1994; Singh et al., 1994a, b; Bornscheuer et al., 1994b), solvent free reactions (Weiss, 1990; Yamaguchi and Mase, 1991), or hollow-fiber membrane reactors (Yamane et al., 1983; Hoq et al., 1985; Padt et al., 1992).

As expected, polar solvents dissolve more glycerol and favor formation of MG's, while nonpolar solvents dissolve less glycerol and favor formation of DG's. Berger and Schneider (1992, 1993) dispersed the glycerol on silica gel to increase the surface area and added a separate cold compartment where the 1(3)-MG crystallized. This reaction system gave excellent yields (> 75 %, often ~90 %) and high regioisomeric purity (~95 %). Weiss (1990) increased the surface area by suspending solid fatty acid or ester in glycerol, but this required large excesses of glycerol.

Another way to increase the solubility of glycerol and avoid subsequent acylation of the MG's is to protect the hydroxyls in glycerol as acetonide (Eq. 37) (Wang et al., 1988; Omar et al., 1989; Pecnik and Knez, 1992; Akoh, 1993; Bornscheuer and Yamane, 1995; Hess et al., 1995). Both yields and isomeric purities were close to 100 %, but this method required extra chemical steps. In principle this route can also yield enantiomerically pure MG's, but the enantioselectivity of lipases is too low (see Sect. 5.1.2.1).

$$
\begin{array}{ccc}
\text{(±)-1,2-}O\text{-isopropyl-} & + \quad \text{RC(O)OR'} & \xrightarrow{\text{lipase}} \quad \quad + \quad \text{R'OH} \\
\text{idene glycerol} & \text{fatty acid} & \text{water or} \\
& \text{or ester} & \text{alcohol}
\end{array}
\tag{37}
$$

In a similar approach, researchers used phenylboronic acid (PBA) to dissolve and protect glycerol in an esterification with uncommon fatty acids such as (S)-17-hydroxystearic acid (Steffen et al., 1992, 1995; Multzsch et al., 1994).

6.2.2 Fatty Acid Selectivity

6.2.2.1 Saturated Fatty Acids

Most lipases show little preference for different saturated fatty acids. They catalyze hydrolysis or esterification of fatty acids with chain lengths between C_2 and C_{18} with similar rates (Tab. 19). Although each lipase shows some fatty acids selectivity, this selectivity is small (usually less than a factor of 4). Of the lipases that do show significant selectivity, ProqL favors small to medium chain length fatty acids, while ROL favors medium to long chain fatty acids.

Like other selectivities, fatty acid selectivity is best measured not by separate rate measurements of pure substrates, but by competition experiments where the lipase chooses among different substrates (Rangheard et al., 1989; Berger and Schneider, 1991a). Fatty acid selectivities change slightly with reaction conditions (Adlercreutz, 1994), most likely due to changes in the solvation energies of the fatty acids (Janssen and Halling, 1994). Using site-directed mutagenesis, Klein et al. (1997) created a ROL mutant with high selectivity and activity toward tributyrin over tricaprylin (80-fold) and triolein (> 80-fold). Several other mutants, also predicted to have increased selectivity toward short chain fatty acids, showed only modest changes in selectivity and sometimes significant decreases in activity (Joerger and Haas, 1994; Atomi et al., 1996).

Tab. 19.   Fatty Acid Selectivities of Lipases.

Lipase	Selectivity	References
CAL-B, CRL, HLL, PCL, PPL, RML	little (usually < 4)[a]	Amano Enzyme Company; Rangheard et al. (1989); Berger and Schneider (1991a)
ProqL	S, M>>L fatty acids	Amano Enzyme Company
ROL	M to L fatty acids	Amano Enzyme Company
GCL	discriminates *against* erucic acid (*cis*Δ13 C$_{22}$)	Sonnet et al. (1993)
RML	discriminates *against* polyunsaturated fatty acids	Pedersen and Holmer (1995)
lipase from *Brassica napus* (rapeseed)	discriminates *against* fatty acids containg a *cis*Δ4 or *cis*Δ6 bond	Hills et al. (1990b)
GCL	selective for *cis*Δ9 double bonds	Jensen (1974); Baillargeon and McCarthy (1991); Charton and Macrae (1993)

[a] Within a factor of four for C$_4$, C$_6$, C$_8$, C$_{10}$, C$_{12}$, C$_{14}$, C$_{16}$, C$_{18}$. CRL is selective for C$_4$ by a factor of 4.7 over C$_6$. RML did not react with C$_2$.

6.2.2.2   Unsaturated Fatty Acids

Lipases show much higher selectivity toward unsaturated fatty acids. For example, GCL favors fatty acids having a *cis*Δ9 bond by a factor of 20 or more, 100 to 1 for oleic vs. stearic acids (Jensen, 1974; Baillargeon and McCarthy, 1991; Charton and Macrae, 1993). Nippon Oils and Fats produces ultrapure unsaturated fatty acids by a GCL-catalyzed process.

In another example, lipase from the seeds of *Brassica napu* (rapeseed or canola plant) discriminates *against* fatty acids containing a *cis*Δ4 or *cis*Δ6 bond by a factor of 14 to 50 (Hills et al., 1990b). Researchers used this discrimination to enrich seed oils in γ-linoleic acid (Hills et al., 1989, 1990a).

CRL discriminates against erucic acid (*cis*Δ13C$_{22}$) and γ-linoleic acid (*cis*Δ9, *cis*Δ12C$_{18}$) (Ergan et al., 1991; Trani et al., 1992). In esterification with lauryl alcohol, RDL discriminates against DHA, but in esterfication with small alcohols (e.g., ethanol), RDL showed little discrimination (Shimada et al., 1997).

Several lipases moderately favored polyunsaturated fatty acids (PUFA's) over saturated fatty acids, but no quantitative data are available. The usefulness, biotransformations, and biotechnological applications of PUFÁ have been reviewed recently (Gill and Valivety, 1997a, b). Fish oils contain up to 30 % PUFA's mainly in form of triglycerides. Researchers have used lipases to enrich oils in PUFA's. For example, Yadwad et al. (1991) used a ROL-catalyzed glycerolysis of cod liver oil (9.6 % PUFA) to yield 1(3)-mono-glycerides with 29 % PUFA (see also Zaks and Gross, 1990a). Hydrolysis of fish oils with CRL or GCL enriched the fatty acid product in PUFA's to 30–45 %. Unilever has tested this process at the pilot scale (Moore and McNeill, 1996; McNeill et al., 1996). Pedersen and Holmer (1995) reported that RML favored monounsaturated fatty acids (20:1n-9, 22:1n-9) over EPA or DHA. The incorporation of PUFA's into triglycerides is mentioned in Sect. 6.2.1.1.

Chen and Sih (1998) separated two isomeric octadecadienoic acids (conjugated lino-leic acids) using the selectivity of GCL for a 9-*cis* double bond in fatty acids. GCL selec-tively esterified the isomer containing the 9-*cis* double bond with a selectivity of 12. The less reactive isomer contained a 12-*cis* double bond. Another lipase, ANL, showed a higher selectivity (25), again favoring the isomer with the 9-*cis* double bond.

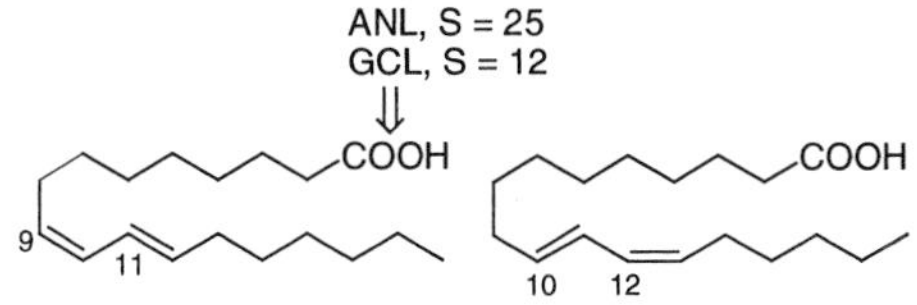

Fig. 139. In a mixture of two conjugated linoleic acids, both ANL and GCL favored esterification of the isomer with the 9-*cis* double bond over the isomer with the 12-*cis* double bond.

Lipases CAL-B, RML, and PPL catalyze hydrolysis and transesterification reactions of the C_{18} furanoid fatty acid in Fig. 140 (Lie-Ken-Jie and Syed-Rahmatullah, 1995).

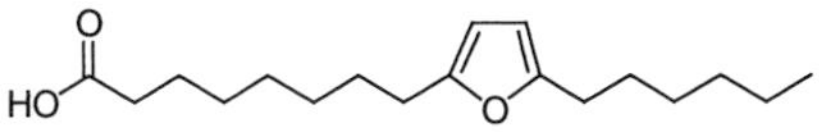

Fig. 140. Unusual fatty acid accepted by several lipases.

6.3   Oligomerization and Polymerizations

Lipases can catalyze polymerizations by formation of ester links (for reviews see Gutman, 1990; Kobayashi et al., 1994; Linko and Seppälä, 1996). The three main ap-proaches are (1) simple condensation of diacids with diols or hydroxy acids with them-selves, (2) transesterification of either hydroxy esters or diesters with diols and (3) ring-opening polymerization of lactones. The second two approaches are more common. The most useful lipases are CRL, PCL, PPL, and RML. The potential advantages of lipase-catalyzed polymerizations are their stereoselectivity and narrow range of molecular weights. Until now most polymerizations yielded polymers of too low molecular weight (usually 1 000–7 000), but the most recent reports include examples of high molecular weight (40 000).

Only a few groups reported simple condensation reactions. Either PCL (Ajima et al., 1985) or CRL (O'Hagan and Zaidi (1993, 1994a) catalyzed the condensation of 10-hydroxydecanoic acid to the corresponding polyester. At 55 °C in the presence of 3 Å molecular sieves, O'Hagan and Zaidi (1994a) obtained a polyester with a molecular weight of ~9 000 (50 repeat units) (Eq. 38).

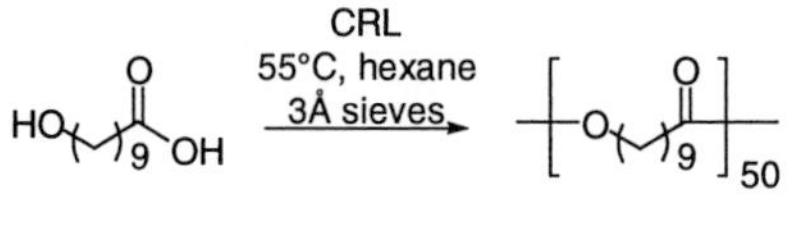

(38)

In the simple condensation of a diacid with a diol, Okumura et al. (1984) reported ANL-catalyzed formation of pentamers and heptamers. Binns et al. (1993) used a two-stage reaction to condense adipic acid with butane-1,4-diol. The first condensation formed oligomers, after isolation these oligomers were coupled in the second condensation to form a polyester with an average of 20 repeat units.

Most researchers used transesterification reactions to form the ester link. Early reports showed only oligomers such as pentamers or hexamers (Gutman et al., 1987; Margolin et al., 1987; Geresh and Gilboa, 1990, 1991; Knani and Kohn, 1993; Park et al., 1994; Chaudhary et al., 1995) even when using activated esters such as 2,2,2-trichloroethyl. However, Wallace and Morrow (1989a, b) obtained polyesters with degrees of polymerization up to 25 by using highly purified monomers. In addition, they used exactly two moles of diester for each mole of diol because only one enantiomer of the diester reacted (Fig. 141).

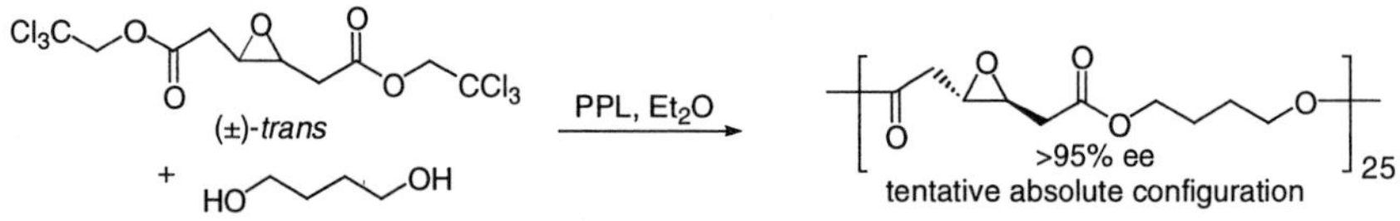

Fig. 141. PPL-catalyzed polymerization of a diester and a diol.

In later work Morrow suggested that the release of the alcohol, even a poorly nucleophilic alcohol like 2,2,2-trichloroethanol, limits the molecular weight of the polymer. Release of alcohol may also promote desorption of water from the enzyme which also limits the molecular weight by permitting hydrolysis of either the starting diester or the product polyester (Brazwell et al., 1995). To minimize this problem, researchers carried out polymerizations under vacuum in high boiling solvent to remove the released alcohol (Linko et al., 1995a, b; Brazwell et al., 1995). For example, a PPL-catalyzed transesterification of bis(2,2,2-trifluoroethyl)glutarate with 1,4-butanediol reached a molecular weight of 39 000 (> 200 repeat units) under vacuum, while only ~2 900 without vacuum (Fig. 142).

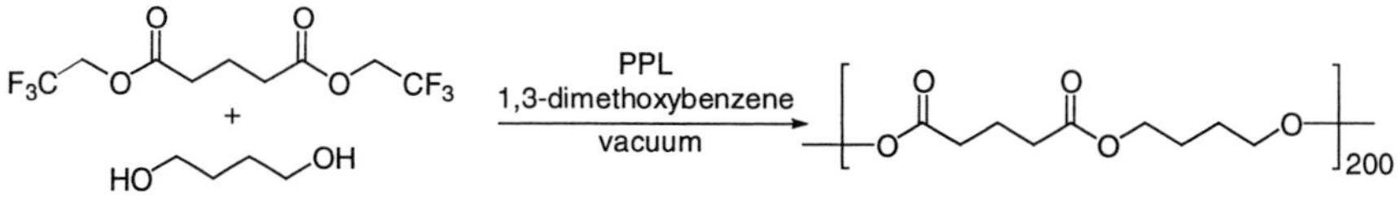

Fig. 142. PPL-catalyzed polymerization under vacuum gives high molecular weight polyester.

Polymerizations starting with vinyl esters (Uyama and Kobayashi, 1994) or oxime esters (Athawale and Gaonkar, 1994) gave molecular weights up to 7 000 (~35 repeat units). Chaudhary et al. (1995) lowered the molecular weight of polyesters from 2 600 to 800 in supercritical fluoroform by changing the pressure to decrease the solubility of the polymer (see Sect. 4.3.2).

Ring-opening polymerization is a special case of transesterification polymerization which does not release a molecule of alcohol. Lipase-catalyzed polymerization of ε-caprolactone with either PCL or PPL yields a polyester with a molecular weight up to 7 700 (67 repeat units) (Uyama and Kobayashi, 1993; Uyama et al., 1993; Knani et al., 1993; MacDonald et al., 1995). Researchers added a small amount of alcohol such as butanol to initiate the polymerization. MacDonald et al. (1995) suggested that water bound to the enzyme limits the molecular weight of the polymer by reacting with the oligomers (Fig. 143).

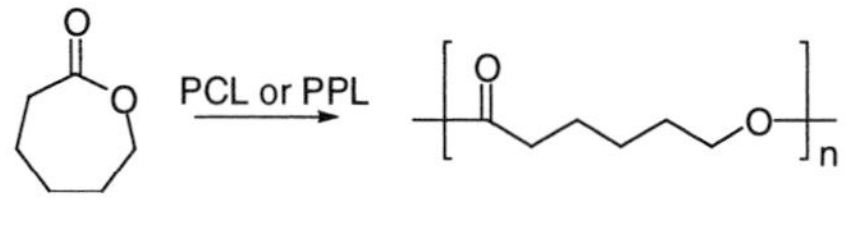

Fig. 143. Ring-opening polymerization of ε-caprolactone.

Similar polymers form upon ring-opening polymerization of the 12-membered 11-undecanolide and the 16-membered 15-pentadecanolide (Uyama et al., 1995) and also the four-membered β-propiolactones, including substituted β-propiolactones (Nobes et al., 1996; Svirkin et al., 1996). Ring-opening polymerization of succinic acid anhydride with diol gave polymers with degrees of polymerization up to 14 (Kobayashi and Uyama, 1993).

Lipases also catalyze the degradation of polyesters (for examples see Tokiwa et al., 1979; Koyama and Doi, 1996; Nagata, 1996).

6.4   Other Lipase-Catalyzed Reactions

In addition to various hydrolysis and transesterification reactions, CAL-B also catalyzed the "esterification" of carboxylic acids and hydrogen peroxide to peroxycarboxylic acids (Björkling et al., 1992; Kirk et al., 1994; Cuperus et al., 1994). Peroxycarboxylic acids are more reactive than hydrogen peroxide and reacted *in situ* with olefins to give epoxides (Eq. 39). Similarly, added ketones underwent Baeyer-Villiger-oxidation (Lemoult et al., 1995).

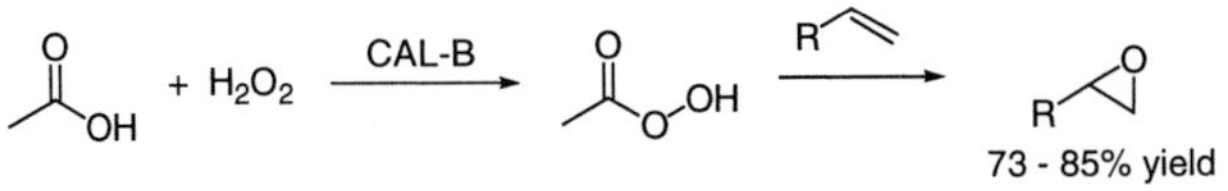

$$\text{(39)}$$

73 - 85% yield

This method can also be used for the 'self-epoxidation of unsaturated fatty acids (Fig. 144) such as oleic or linoleic acid and plant oils in high yields (e.g., rapeseed oil 91 %, sunflower oil 88 %) (Rüsch gen. Klaas and Warwel, 1996). The same group also described a three-step-one-pot synthesis of epoxyalkanolacylates, in which perhydrolysis, epoxidation and interesterification are integrated in a single process (Rüsch gen. Klaas and Warwel, 1998). For instance, 9-octadecen-1-ol was converted to the corresponding epoxyalkanol butyrate in 89 % yield after 22 h reaction time.

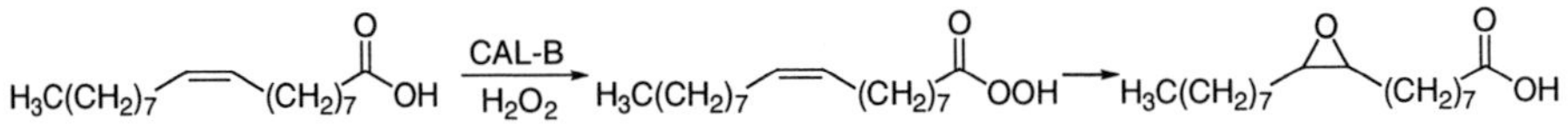

Fig. 144. CAL-B-catalyzed access to epoxidized fatty acids via peroxycarboxylic acids.

7   Phospholipases

Phospholipids are amphiphilic molecules that are ubiquitous in nature, where they are basic components of natural membranes and cell walls. Phospholipases catalyze the hydrolysis of these phospholipids. Four types of enzymes with different regioselectivities have been identified and their cleavage sites are given in Fig. 145 for a phosphatidyl choline. The physiological role of phospholipases is believed to be the degradation of phospholipid components of cell membranes and the digestion of phospholipid-containing fats in food.

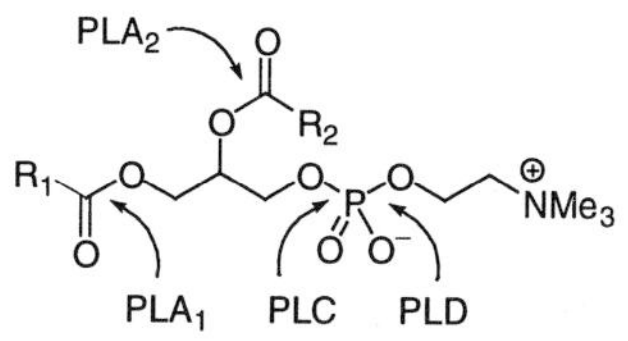

Fig. 145. Regioselectivity of different phospholipases.

The interest in new phospholipids and phospholipid analogs stems from their potential use as biodegradable surfactants, carriers of drugs or genes or as biologically active compounds in medicine and agriculture (New, 1993). However, synthesis of these substances is difficult by chemical means since control of regio- and stereoselectivity must be ensured.

The survey below focussed on the application of phospholipases in organic synthesis. Further applications and examples can be found in reviews (D'Arrigo et al., 1996) including detailed procedures (Kötting and Eibl, 1994).

7.1   Phospholipase A_1

PLA_1 (EC 3.1.1.32) is the only phospholipase, which is rarely used. This is mainly because the resulting phospholipid is unstable due to facile acyl migration of residues at the sn-2 position to the sn-1 position. In addition the enzyme is not commercially available and similar reactions are catalyzed by a number of lipases, e.g., from *Rhizopus* sp.

7.2   Phospholipase A$_2$

The best characterized enzyme is PLA$_2$ (EC 3.1.1.4) from cobra venom. Other sources are bovine or porcine pancreas or microorganisms (mainly *Streptomyces* sp.). Structural analysis of PLA$_2$ revealed that catalysis occurs at the interface of aggregated substrates (Scott et al., 1990). PLA$_2$ cleaves fatty acids from lecithin or cephalin forming lysolecithin or lysocephalin resulting in disruption of the cell wall structure and haemolysis of erythrocytes. Arachidonic acid (C$_{20:4}$) released from phospholipids by the action of PLA$_2$ is an important precursor of prostaglandins (Kudo et al., 1993).

PLA$_2$ usually shows higher efficiency in hydrolysis than in transesterification and the enzyme is specific for the natural absolute configuration at the *sn*-2 position. Thus, resolution of racemic phospholipids as well as the determination of absolute configuration are feasible. Sahai and Vishwkarma (1997) synthesized optically pure radiolabeled precursors of cell surface conjugates of the parasite *Leishmania donovani* by kinetic resolution of phospholipids using PLA$_2$ from the snake *Naja mocambique mocambique*. Hosokawa et al. (1995a) and Na et al. (1990) quantitatively incorporated polyunsaturated fatty acids into the *sn*-2 position of (lyso)phosphatidylcholine using PLA$_2$ from porcine pancreas.

7.3   Phospholipase C

Most microbial phospholipase C (EC 3.1.4.3) hydrolyse L-α-phosphatidylcholine to diglyceride and choline phosphate. Other microbial PLC are specific for phosphatidylinositol (Griffith et al., 1991; Iwasaki et al., 1994) or sphyngomyelin. Direct transphosphatidylation with PLC seems to be more difficult compared to PLD. Structural analogs of inositol phosphate diesters have been obtained by PLC-catalyzed hydrolysis of phosphatidylinositol, isolation of inositol-1,2-cyclic-phosphate and subsequent transesterification using the same enzyme to afford a broad range of *O*-alkyl inositol-1-phosphates (Fig. 146).

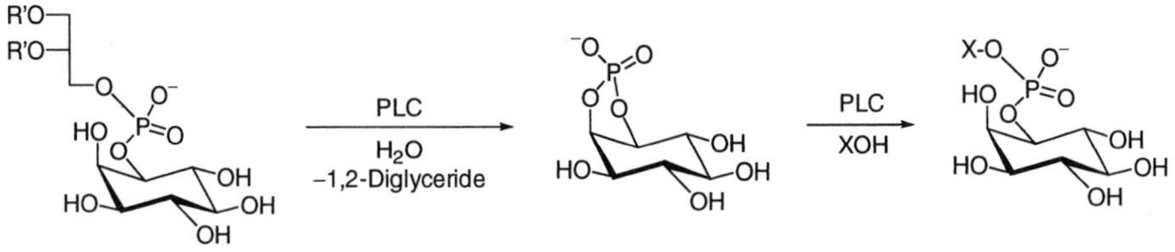

Fig. 146. Access to *O*-alkyl inositol-1-phosphates using PLC (Bruzik et al., 1996).

7.4   Phospholipase D

PLD (EC 3.1.4.4) is the most commonly used phospholipase in organic syntheses. Hydrolysis with PLD leads to phosphatidic acids, but in the presence of suitable alcohols, e.g., ethanolamine, glycerol, serine, PLD also catalyzes an efficient head group exchange. These reactions are usually performed in a biphasic mixture in order to ensure high enzymatic activity. A disadvantage of the biphasic system is the competing hydrolysis to phosphatidic acid. PLD from cabbage, peanut and several microbial sources, such as *Streptomyces* sp., are commercially available.

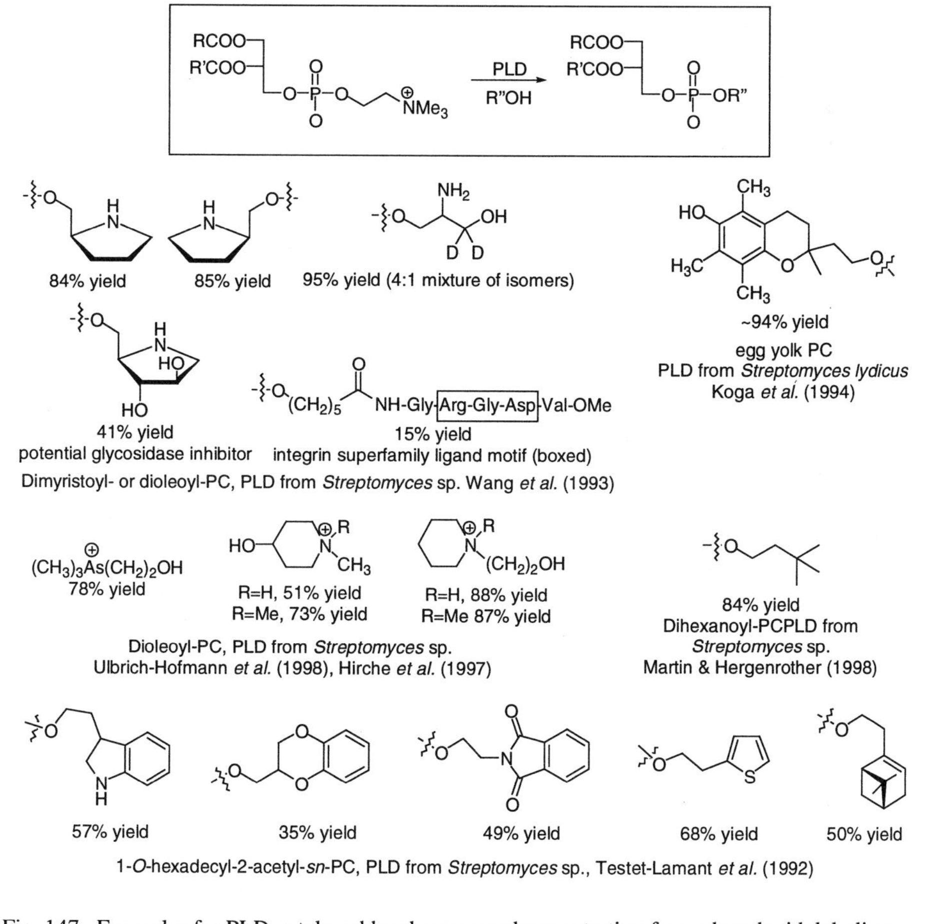

Fig. 147. Examples for PLD-catalyzed head group exchange starting from phosphatidylcholine.

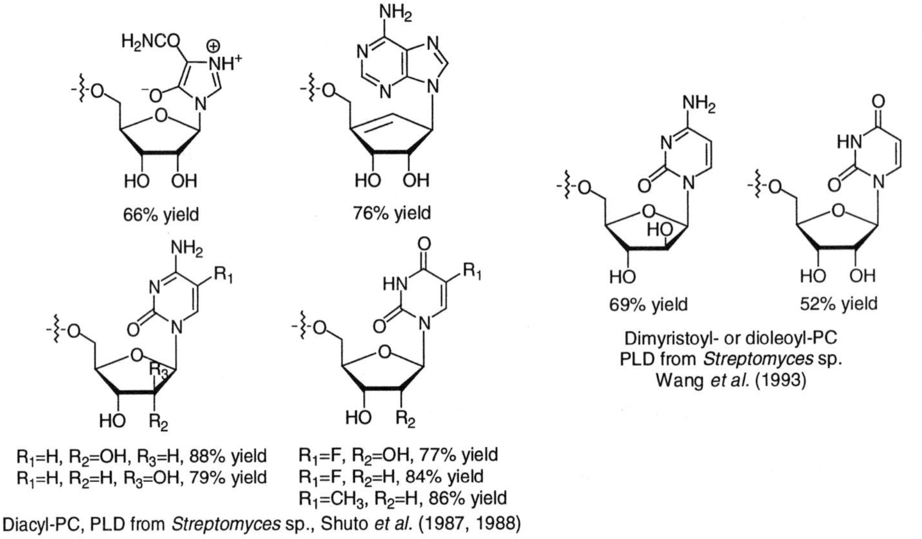

66% yield

76% yield

69% yield

52% yield

Dimyristoyl- or dioleoyl-PC
PLD from *Streptomyces* sp.
Wang *et al.* (1993)

R₁=H, R₂=OH, R₃=H, 88% yield
R₁=H, R₂=H, R₃=OH, 79% yield

R₁=F, R₂=OH, 77% yield
R₁=F, R₂=H, 84% yield
R₁=CH₃, R₂=H, 86% yield

Diacyl-PC, PLD from *Streptomyces* sp., Shuto *et al.* (1987, 1988)

Fig. 147. Examples for PLD-catalyzed head group exchange starting from phosphatidylcholine (continued).

Pisch et al. (1997) synthesized a range of phospholipids bearing acetylenic fatty acids (4- or 14-octadecynoic acid) by head group exchange from the corresponding phosphatidylcholine (PC) using a recombinant PLD from *Streptomyces antibioticus* (Iwasaki et al., 1995). Ethanolamine, glycerol or L-serine were quantitatively converted into the corresponding phospholipids within 1 h using a biphasic system (CHCl₃:buffer, 1:1.5) and isolated in 74–87 % yield. PLD also accepts a wide range of nucleophiles, e.g., N-heterocyclic and As-containing compounds (Hirche et al., 1997; Ulbrich-Hofmann et al., 1998), however amine or thiols do not react. Martin and Hergenrother (1998) synthesized a phospholipid bearing a *t*-butyl group by PLD-catalyzed head group exchange. The activity of PLC from *Bacillus cereus* – often used as a probe of the signal transduction pathway in mammalian systems – toward the resulting phospholipid was reduced by a factor > 1000 (Fig. 147).

Only a few examples can be found, where secondary alcohols or phenols were accepted by PLD (Takami et al., 1994; D'Arrigo et al., 1996). However, competing hydrolysis of PC to phosphatidic acid substantially affected transphosphatidylation (Fig. 148). Sugars containing solely secondary hydroxyls like L-Rhamnose and *myo*-inositol were not converted.

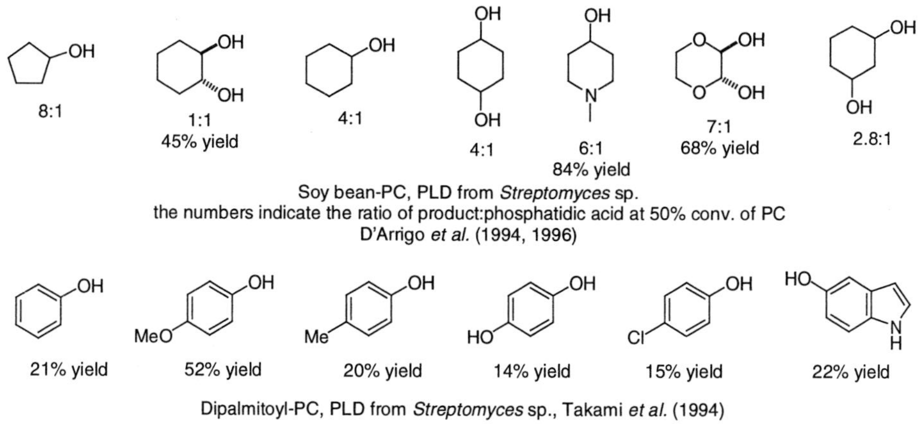

8:1

1:1
45% yield

4:1

4:1

6:1
84% yield

7:1
68% yield

2.8:1

Soy bean-PC, PLD from *Streptomyces* sp.
the numbers indicate the ratio of product:phosphatidic acid at 50% conv. of PC
D'Arrigo *et al.* (1994, 1996)

21% yield   52% yield   20% yield   14% yield   15% yield   22% yield

Dipalmitoyl-PC, PLD from *Streptomyces* sp., Takami *et al.* (1994)

Fig. 148. Secondary alcohols and phenols accepted by PLD from *Streptomyces* sp.

8 Survey of Enantioselective Protease- and Amidase-Catalyzed Reactions

8.1 Alcohols and Amines

Proteases and amidases usually show their highest selectivities toward the acyl part of an ester or amide, and lower selectivity toward the alcohol or amine part.

8.1.1 Secondary Alcohols and Primary Amines

8.1.1.1 Overview and Models

Like lipases, subtilisin is enantioselective toward secondary alcohols (HOCHRR') and the isosteric primary amines (H$_2$NCHRR'), but subtilisin favors the enantiomer opposite to the one favored by lipases (Fitzpatrick and Klibanov, 1991; Kazlauskas and Weissfloch, 1997). Empirical rules based on the size of the substituents predict the favored enantiomer (Fig. 149). In addition, lipases and subtilisin often show opposite regioselectivity (Sect. 6.1.1.2).

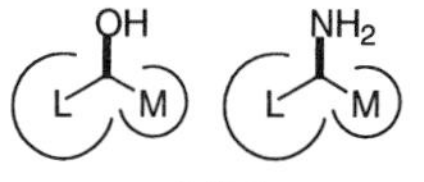

subtilisin

Fig. 149. Empirical rules predict the fast reacting enantiomer of secondary alcohols (HOCHRR') and isosteric primary amines (NH$_2$CHRR') with subtilisin. Like the rules for lipases, these rules are based on the size of the substituents, but the favored enantiomer is opposite for lipases and subtilisin. L represents a large substituent such as phenyl and M represents a medium substituent such as methyl.

Differences in protein structure are the likely origin of this opposite enantiopreference (Kazlauskas and Weissfloch, 1997). Like lipases, subtilisin contains a catalytic triad, an oxyanion hole and follows an acyl enzyme mechanism. However, the three-dimensional arrangement of the catalytic triad in subtilisin is the mirror image of that in lipases (Ollis et al., 1992). Folding the protein to assemble the catalytic machinery creates a pocket with restricted size, the 'M' or stereoselectivity pocket. The mirror image relationship places the catalytic histidine on opposite sides of this pocket in lipases and subtilisin. This difference likely accounts for the opposite enantiopreference. Note that an isolated catalytic triad cannot impart stereoselectivity toward alcohols or amines; the pockets created by the protein fold are essential. Chymotrypsin, which has a catalytic triad similar to that in subtilisin, but a different protein fold, shows little enantioselectivity toward alcohols (see Sect. 8.1.1.3).

8.1.1.2 Subtilisin

Ten of the twelve examples of secondary alcohols in Fig. 150 follow the empirical rule. The two exceptions are 3-quinuclidinol and one of the two reactive hydroxyls in the inositol derivative. A possible rationalization for the quinuclidinol exception is that solvation of the nitrogen increases its effective size. There is no obvious rationalization for the inositol exception. Note that the inositol was counted as two substrates because two of the hydroxyl groups reacted. The enantioselectivity of subtilisin toward secondary alcohols is usually only moderate.

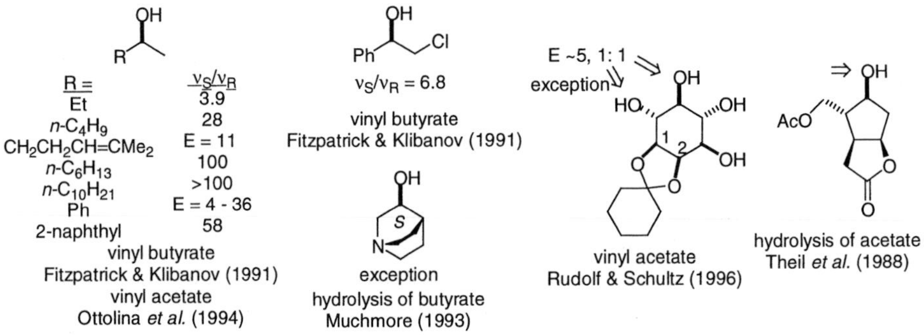

R =	v_S/v_R
Et	3.9
n-C$_4$H$_9$	28
CH$_2$CH$_2$CH=CMe$_2$	E = 11
n-C$_6$H$_{13}$	100
n-C$_{10}$H$_{21}$	>100
Ph	E = 4 - 36
2-naphthyl	58

vinyl butyrate
Fitzpatrick & Klibanov (1991)
vinyl acetate
Ottolina *et al.* (1994)

v_S/v_R = 6.8

vinyl butyrate
Fitzpatrick & Klibanov (1991)

exception
hydrolysis of butyrate
Muchmore (1993)

E ~5, 1 : 1
exception

vinyl acetate
Rudolf & Schultz (1996)

hydrolysis of acetate
Theil *et al.* (1988)

Fig. 150. Examples of secondary alcohols resolved by subtilisin. In some cases it was difficult to calculate the enantioselectivity from the reported data; for these, no enantioselectivities are given. In other cases, the authors estimated the enantioselectivity by separately measuring the initial reaction rates for the enantiomers (v_S/v_R). The exceptions to the empirical rule are marked.

Subtilisin also catalyzes the enantioselective acylation of various amines (Fig. 151). All examples follow the empirical rule above and the enantioselectivity is usually good to excellent. Cleavage of the resulting amides requires vigorous reaction conditions. For this reason, Orsat et al. (1996) introduced diallyl carbonate as a acylating agent (see Sect. 4.2.3) The allyl carbamate cleaves readily in the presence of palladium.

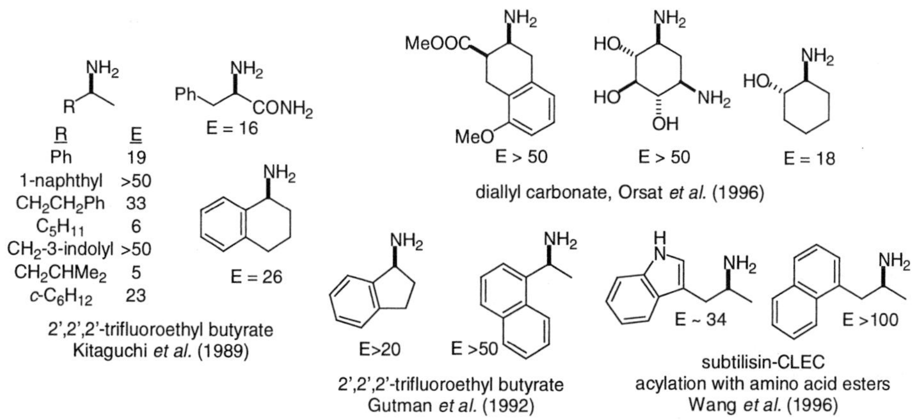

Fig. 151. Examples of primary amines of the type H_2NCHRR' resolved by subtilisin-catalyzed acylation.

8.1.1.3   Other Proteases and Amidases

Chymotrypsin shows low enantioselectivity toward secondary alcohols (E < 6 for esters of 2-butanol, 2-octanol or α-phenylethanol (Lin et al., 1974)). Chymotrypsin also showed low enantioselectivity (E < 3) in the acetylation of the prochiral 1,3-propanediol shown below in most organic solvents, but in diisopropyl ether the enantioselectivity was E = 16 (Fig. 152) (Ke and Klibanov, 1998).

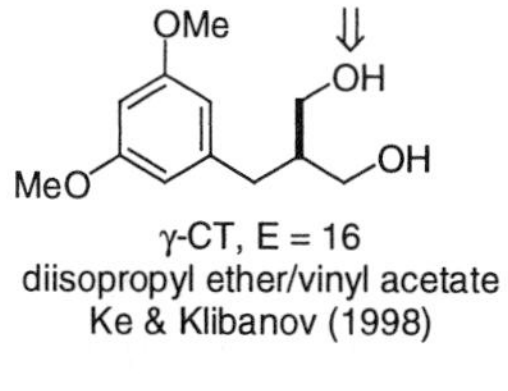

Fig. 152. Chymotrypsin shows moderate enantioselectivity in diisopropyl ether.

Penicillin amidase (penicillin G acylase, PGA) resolves alcohols and amines, usually by hydrolysis of their phenylacetyl derivatives (Fig. 153). The enantioselectivity of PGA toward alcohols is often lower than toward amines. To increase the enantioselectivity Pohl and Waldmann (1995) replaced the phenylacetyl ester with a 4-pyridylacetyl ester. For example, PGA catalyzed hydrolysis of the phenylacetyl ester of 1-phenyl propanol with an enantioselectivity of E = 31–36, but catalyzed hydrolysis of the 4-pyridylacetyl ester with an enantioselectivity of E > 100.

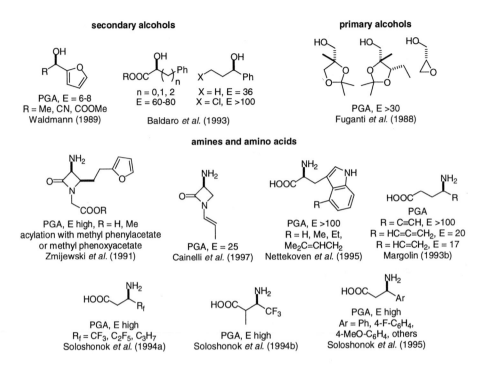

secondary alcohols

PGA, E = 6-8
R = Me, CN, COOMe
Waldmann (1989)

n = 0,1, 2
E = 60-80
Baldaro *et al.* (1993)

X = H, E = 36
X = Cl, E >100

primary alcohols

PGA, E >30
Fuganti *et al.* (1988)

amines and amino acids

PGA, E high, R = H, Me
acylation with methyl phenylacetate
or methyl phenoxyacetate
Zmijewski *et al.* (1991)

PGA, E = 25
Cainelli *et al.* (1997)

PGA, E >100
R = H, Me, Et,
Me₂C=CHCH₂
Nettekoven *et al.* (1995)

PGA
R = C≡CH, E >100
R = HC=C=CH₂, E = 20
R = HC=CH₂, E = 17
Margolin (1993b)

PGA, E high
R_f = CF₃, C₂F₅, C₃H₇
Soloshonok *et al.* (1994a)

PGA, E high
Soloshonok *et al.* (1994b)

PGA, E high
Ar = Ph, 4-F-C₆H₄,
4-MeO-C₆H₄, others
Soloshonok *et al.* (1995)

Fig. 153. Examples of alcohols and amines resolved by penicillin amidase (PGA). PGA resolves secondary and primary alcohols as well as amines. The amines include an α-amino acid, a γ-amino acid, and several β-amino acids. The enantioselectivity of PGA toward amines appears higher than toward alcohols. Unless otherwise noted, resolutions involve hydrolysis of the phenylacetyl derivative.

8.1.2   α-Amino Acids via Reactions at the Amino Group

8.1.2.1   Amino Acid Acylases

Amino acid acylases catalyze the hydrolysis of the *N*-acyl group from *N*-acyl-amino acids and shows high enantioselectivity for the natural L-enantiomer (Eq. 40) (for a review see Greenstein and Winitz, 1961).

$$\text{(40)}$$

Tanabe commercialized an acylase-catalyzed resolution of natural amino acids in 1954 (Chibata et al., 1992) and Degussa introduced a continuous process in 1981. Both acylase from porcine kidney and from *Aspergillus* sp. are available, but commercial processes (Bommarius et al., 1992; Chibata et al., 1992) use the *Aspergillus* enzyme because it is more stable. Beside natural amino acids, acylases accept non-proteinogenic amino acids (Chenault et al., 1989; Bommarius et al., 1997). A few examples are shown in Fig. 154. The carboxyl group must be free. The *N*-acetyl group may be replaced by chloroacetyl to increase the reaction rate. Chirotech (Cambridge, UK) sells many unnatural amino acids resolved by acylases.

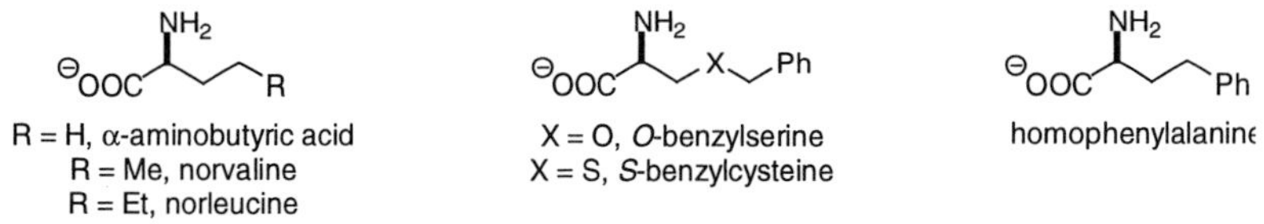

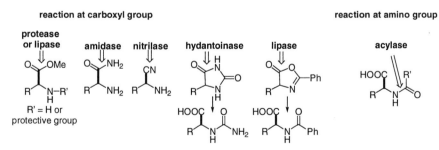

R = H, α-aminobutyric acid
R = Me, norvaline
R = Et, norleucine

X = O, *O*-benzylserine
X = S, *S*-benzylcysteine

homophenylalanine

Fig. 154. Examples of unnatural amino acids resolved by acylase.

8.2   Carboxylic Acids

8.2.1   α-Amino Acids via Reactions at the Carboxyl Group

Amino acids, especially unnatural α-amino acids (or nonproteinogenic α-amino acids), are the type of carboxylic acids most commonly resolved by proteases and amidases (for reviews see Williams, 1989; Kamphuis et al., 1992). Several derivatives are suitable starting materials (Fig. 155). Proteases or lipases can resolve esters of amino acids, where the amino group may be protected. Amidases can resolve amino acid amides, while nitrilases resolve α-amino nitriles. Both of these substrates come directly from a Strecker synthesis of amino acids. Hydantoinases resolve hydantoin derivatives of amino acids, while lipase also resolve 2-phenyloxazolin-5-one derivatives of amino acids.

reaction at carboxyl group   reaction at amino group

protease
or lipase   amidase   nitrilase   hydantoinase   lipase   acylase

R' = H or
protective group

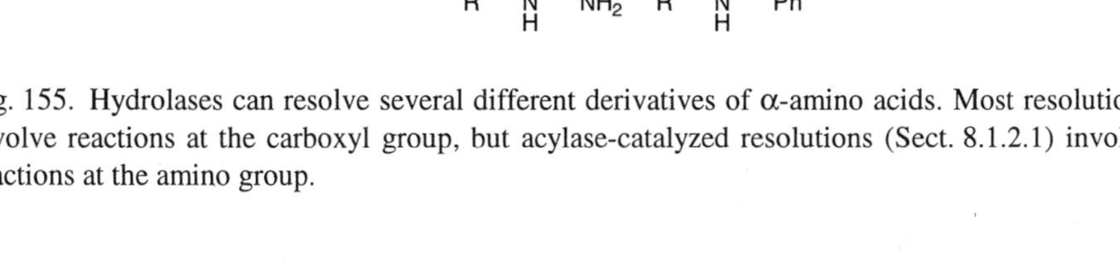

Fig. 155. Hydrolases can resolve several different derivatives of α-amino acids. Most resolutions involve reactions at the carboxyl group, but acylase-catalyzed resolutions (Sect. 8.1.2.1) involve reactions at the amino group.

All these resolutions involve reaction at the carboxyl group of the amino acid. Acylases resolve *N*-acyl amino acids by reaction at the amino group of the amino acid. None of these routes is clearly superior; which route is best depends on the cost of starting materials, ability to recycle the unwanted enantiomer, ease of isolation, as well as on other details for each amino acid.

The sections below cover resolutions involving proteases, amidases, and hydantoinases. The less common resolutions with nitrilases are covered in Sect. 11. Lipase-catalyzed resolutions are covered in Sect. 5.1 (hydrolysis of esters) and Sect 5.4 (hydrolysis of 2-phenyloxazolin-5-ones), while acylase-catalyzed resolutions are covered in Sect. 8.1.2.1.

Like proteases and lipases, carbonic anhydrase also catalyzes the hydrolysis of esters of amino acids. However, proteases and lipases favor the natural L-enantiomer, while carbonic anhydrases favors the unnatural D-enantiomer. The enantioselectivity is high for *N*-acetyl esters of Phe and *N*-acetyl diesters of Asp and Glu. For Glu, it is the δ-ester that reacts (Chênevert et al., 1993). Researchers also used other proteases to resolve natural and unnatural amino acids. For example, Lankiewicz et al. (1989) resolved several *N*-Boc protected analogs of phenylalanine by hydrolysis of their methyl esters catalyzed by thermitase.

8.2.1.1   Subtilisin

Subtilisin accepts a wide range of amino acids esters, but favors esters of hydrophobic amino acids. The approximate preference is Tyr, Phe > Leu, Met, Lys > His, Ala, Gln, Ser >> Glu, Gly (Estell et al., 1986; Wells et al., 1987). Subtilisin favors the natural L-enantiomer, but this preference is smaller than one might expect. For example, the enantiomeric ratio toward methyl esters of Tyr or Phe is only E ~ 100 and only E = 45 for Ala methyl ester (Fig. 156). Kawashiro et al. (1996) even reported a reversal of the enantioselectivity for the 2'2'2'-trifluoroethyl ester of *N*-trifluoracetyl phenylalanine. Transesterification of this ester in *tert*-amyl alcohol favored the unnatural D-enantiomer by ~2. In comparison, hydrolysis of racemic amino acid esters yielded the L-amino acid and the unreacted D-amino acid ester (Chen et al., 1986).

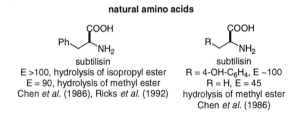

Fig. 156. Subtilisin resolves the hydrophobic natural amino acids with good to excellent enantioselectivity.

Subtilisin also accepts a wide range of unnatural amino acids (Fig. 157). Although chymotrypsin also resolves hydrophobic amino acids, often with higher enantioselectivity, most researchers favor subtilisin for large scale applications because of its low cost and stability. Cross-linked crystals of subtilisin are available from Altus Biologics (for a review see Wang et al., 1997b). The cross-linking simplifies recovery and also stabilizes subtilisin against autolysis.

unnatural amino acids

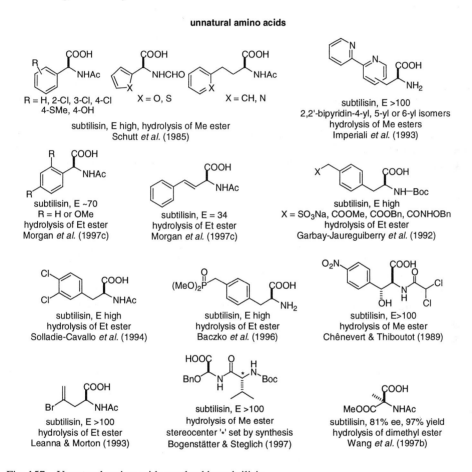

R = H, 2-Cl, 3-Cl, 4-Cl
4-SMe, 4-OH

subtilisin, E high, hydrolysis of Me ester
Schutt et al. (1985)

X = O, S

X = CH, N

subtilisin, E >100
2,2'-bipyridin-4-yl, 5-yl or 6-yl isomers
hydrolysis of Me esters
Imperiali et al. (1993)

subtilisin, E ~70
R = H or OMe
hydrolysis of Et ester
Morgan et al. (1997c)

subtilisin, E = 34
hydrolysis of Et ester
Morgan et al. (1997c)

subtilisin, E high
X = SO$_3$Na, COOMe, COOBn, CONHOBn
hydrolysis of Et ester
Garbay-Jaureguiberry et al. (1992)

subtilisin, E high
hydrolysis of Et ester
Solladie-Cavallo et al. (1994)

subtilisin, E high
hydrolysis of Et ester
Baczko et al. (1996)

subtilisin, E>100
hydrolysis of Me ester
Chênevert & Thiboutot (1989)

subtilisin, E >100
hydrolysis of Et ester
Leanna & Morton (1993)

subtilisin, E >100
hydrolysis of Me ester
stereocenter '*' set by synthesis
Bogenstätter & Steglich (1997)

subtilisin, 81% ee, 97% yield
hydrolysis of dimethyl ester
Wang et al. (1997b)

Fig. 157.   Unnatural amino acids resolved by subtilisin.

8.2.1.2   Chymotrypsin

Chymotrypsin (α-CT), like other proteases, shows high enantioselectivity for the natural enantiomer of amino acids. Esters and amides containing aromatic amino acids – Phe, Tyr, and Trp – as the acyl group react most readily. Esters hydrolyze more rapidly than amides, so most preparative reactions use esters. For example, α-chymotrypsin resolves the ethyl esters of N-acetyl of Phe, Trp, and Trp with very high enantioselectivity. One exception to the high enantioselectivity rule is the moderate enantioselectivity (E = 8) of α-CT toward the methyl ester of *N*-benzoyl alanine (Fig. 158). The lower enantioselectivity is likely due to the binding of the benzoyl group in the hydrophobic pocket in the 'wrong' enantiomer.

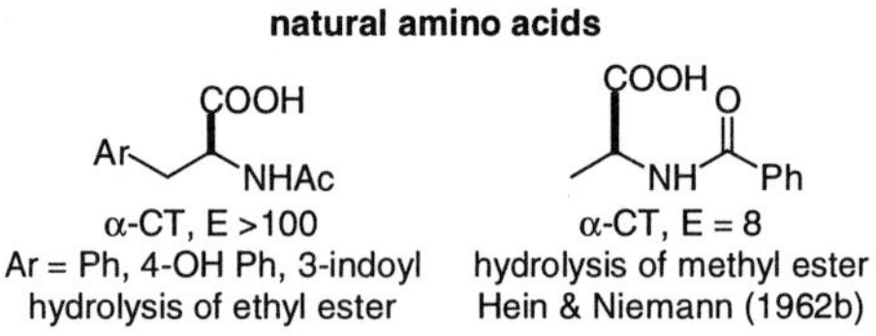

natural amino acids

α-CT, E >100
Ar = Ph, 4-OH Ph, 3-indoyl
hydrolysis of ethyl ester

α-CT, E = 8
hydrolysis of methyl ester
Hein & Niemann (1962b)

Fig. 158. α-Chymotrypsin shows high enantioselectivity toward *N*-acetyl amino acids, but low enantioselectivity toward the methyl ester of *N*-benzoyl alanine.

Chymotrypsin also resolves unnatural amino acids (Fig. 159). For example, α-CT resolves phenylglycine (Cohen et al., 1966) and ring substituted phenylalanine esters (3-OH, 3,4-OH$_2$, 4-Cl, 4-F, (Tong et al., 1971), 2-Me (Berger et al., 1973)). *O*-substitution of tyrosine by alkyl, aryl, or acyl groups completely abolishes activity (Kundu et al., 1972), presumably because the aromatic moiety is too large to fit in the hydrophobic pocket. Chymotrypsin also resolves analogs of tryptophan where the indoyl group is replaced by 2-naphthyl or 6-quinolyl (Berger et al., 1973) and unnatural amino acids containing an alkenyl side chain (Schricker et al., 1992). Esters of Glu and Asp, which contain a –COO– in the side chain, are not hydrolyzed, but upon esterification of the side chain, the diesters can be resolved (Cohen et al., 1963; Cohen and Crossley, 1964). Note that α-CT removes the ester group from the more hindered carbonyl, the one at the α-position.

unnatural amino acids

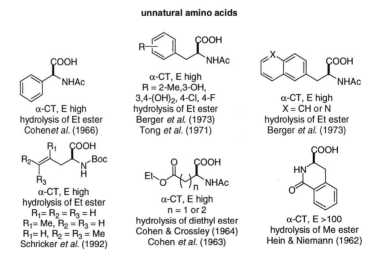

α-CT, E high
hydrolysis of Et ester
Cohen et al. (1966)

α-CT, E high
R = 2-Me,3-OH,
3,4-(OH)$_2$, 4-Cl, 4-F
hydrolysis of Et ester
Berger et al. (1973)
Tong et al. (1971)

α-CT, E high
X = CH or N
hydrolysis of Et ester
Berger et al. (1973)

α-CT, E high
hydrolysis of Et ester
R$_1$ = R$_2$ = R$_3$ = H
R$_1$= Me, R$_2$ = R$_3$ = H
R$_1$= H, R$_2$ = R$_3$ = Me
Schricker et al. (1992)

α-CT, E high
n = 1 or 2
hydrolysis of diethyl ester
Cohen & Crossley (1964)
Cohen et al. (1963)

α-CT, E >100
hydrolysis of Me ester
Hein & Niemann (1962)

Fig. 159. α-Chymotrypsin also resolves unnatural amino acids.

Chymotrypsin also resolves hindered α-methyl amino acids (Anantharamaiah and Roe-ske, 1982) and the related α-methyl α-nitro acids (Lalonde et al., 1988) (Fig. 160). The α-methyl α-nitro acids spontaneously decarboxylated, but the unreacted esters were recovered and reduced to α-methyl α-amino acid esters. Interestingly, in both cases chymotrypsin favors the D-enantiomer, suggesting that, for the favored enantiomer, the amino or nitro groups bind in the *h*-site. Consistent with this notion, addition of an *N*-acetyl group slows the reaction rate dramatically. Chymotrypsin also resolves cyclic, unnatural α-amino acids (Hein and Niemann, 1962a, b; Matta and Rohde, 1972; Dirlam et al., 1987).

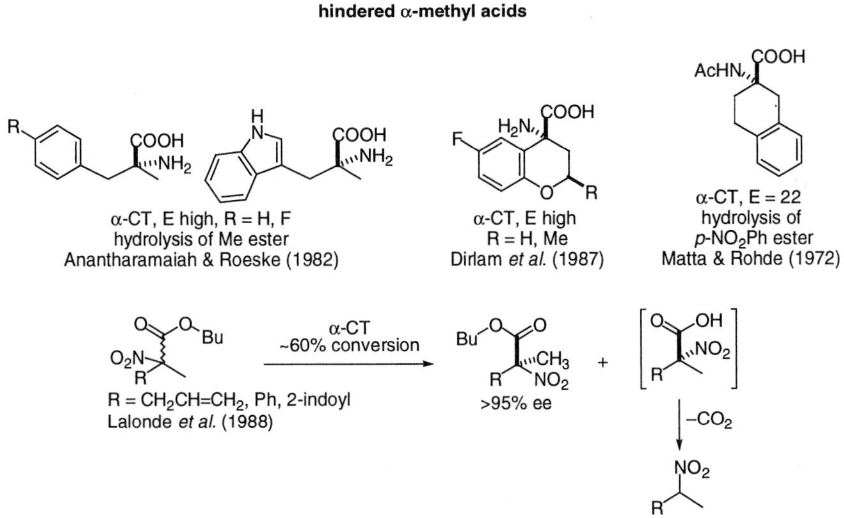

hindered α-methyl acids

Fig. 160. α-Chymotrypsin also resolves hindered α-methyl amino acids.

Chymotrypsin also catalyzes the hydrolysis of 2-phenyloxazolin-5-ones, but the enantioselectivity was low, e.g., E = 8 for the 2-phenyloxazolin-5-one derived from phenylalanine (Daffe and Fastrez, 1980). Lipases show much higher enantioselectivity toward these compounds (Sect. 5.4).

Researchers usually use Cohen's active site model (Cohen et al., 1969) to rationalize the selectivity of α-CT toward acids (Fig. 161). This model defines four sites that bind each of the four substituents at the stereocenter of a good substrate like *N*-acetyl-L-phenylalanine methyl ester. The strongest binding comes from the *ar*-site, a hydrophobic pocket, which binds aromatic or ether hydrophobic groups. When the substrate has the incorrect configuration, for example, *N*-acetyl-D-phenylalanine methyl ester, it still binds to the *ar*-site, but the remaining substituents cannot orient in a productive manner. One observes nonproductive binding. Jones and Beck (1976) reviewed the substrate specificity of α-CT in detail.

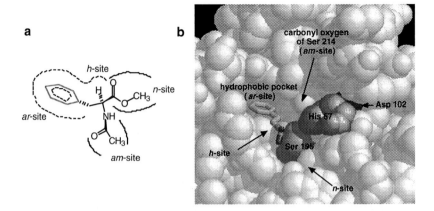

Fig. 161. The active site of α-chymotrypsin. **a** Cohen's model of α-CT (Cohen et al., 1969) containing a good substrate, *N*-acetyl-L-phenylalanine methyl ester. The *ar*-site (*aromatic*-site) is a hydrophobic pocket, the *am*-site (*amide*-site) is the amide carbonyl of Ser-124, which hydrogen bonds to the amide N-H of the substrate. An OH in the substrate can also form a hydrogen bond here. Other groups such as CH$_3$, Cl, AcO, also fit in the *am*-site, but do not form a hydrogen bond. The *am*-site is open to the solvent and can accommodate long groups. The *h*-site (*hydrogen*-site) is a small region in the active site. It is large enough to fit an H, Cl, or OH, but not a CH$_3$. The *n*-site (*nucleophile*-site) bind the leaving group, an ester or amide. Like the *am*-site, the *n*-site is open-ended, that is, it is open to the solvent and can accommodate large groups. In contrast, both the *ar*- and the *h*-sites are closed, that is, they point toward the center of the protein. **b** X-ray structure of α-chymotrypsin (spheres) contained transition state analog bound to the active site serine (Tulinsky and Blevins, 1987). The transition state analog is a 2-phenylethyl boronic acid (stick representation). The amino acid residues of the catalytic triad are labeled as are the regions that correspond to the four sites of Cohen's model.

Computer modeling can also help explain the enantioselectivity of chymotrypsin toward natural substrates (DeTar, 1981; Wipff et al., 1983), unnatural substrates (Norin et al., 1993) and explain more subtle phenomenon such as changes in enantioselectivity in different solvents (Ke and Klibanov, 1998).

8.2.1.3   Hydantoinases

Hydantoinases (for reviews see Ogawa and Shimizu, 1997; Syldatk et al., 1999) catalyze the hydrolysis of 5-monosubstituted hydantoins to *N*-carbamoyl α-amino acids (Fig. 162). Some hydantoinases are identical to dihydropyrimidase, an enzyme in pyrimidine degradation, but others are different enzymes. The most common hydantoinases favor the unnatural D-enantiomer, but L-selective hydantoinases are also known. 5-Substituted hydantoins, especially 5-aryl hydantoins racemize readily, thus permitting dynamic kinetic resolution (see Sect. 5.4). Synthetic applications use either isolated enzymes or whole microorganisms.

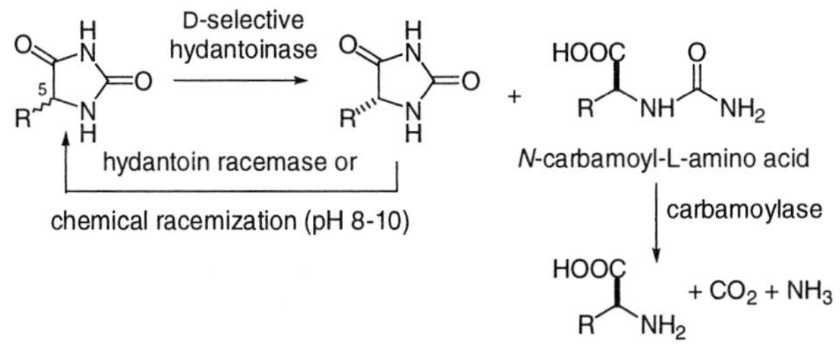

Fig. 162. Hydantoinases catalyze the hydrolysis of 5-monosubstituted hydantoins to *N*-carbamoyl α-amino acids. The equation shows a D-selective hydanoinase; L-selective hydanoinases are less common. 5-Substituted hydantoins racemize readily either enzymatically or chemically under basic reaction conditions (pH 8–10). An *in situ* racemization of the hydantoin permits a dynamic kinetic resolution (see Sect. 5.4). Many hydantoinase-catalyzed resolutions also include a carbamoylase that removes the *N*-carbamoyl group to give the free α-amino acid (Olivieri et al., 1979).

Both Kanegafuchi Industries (Japan) and DEBI Recordati (Italy) produce unnatural D-amino acids using D-selective hydantoinase. Most of the production focuses on D-phenylglycine and D-4-hydroxyphenyl glycine for the semi-synthetic penicillins ampicillin and amoxilicillin (for reviews see Syldatk et al., 1992a, b).

Hydantoinases tolerate a wide range of substituents at the 5-position, but 5,5-disubstituted react very slowly (Takahashi et al., 1979; Olivieri et al., 1981; Keil et al., 1995; Garcia and Azerad, 1997).

8.2.1.4   Amidases

DSM and Lonza resolve several unnatural amino acids using amidases (Kamphuis et al., 1992; Eichhorn et al., 1997) (Fig. 163). Both groups use whole cells of microorganisms that produce amidases. Unlike proteases, which always favor the natural L-enantiomer, amidases occasionally favor the D-enantiomer; for example, the *Burkholderia* amidase catalyzes hydrolysis of piperazine-2-carboxamide (Eichhorn et al., 1997).

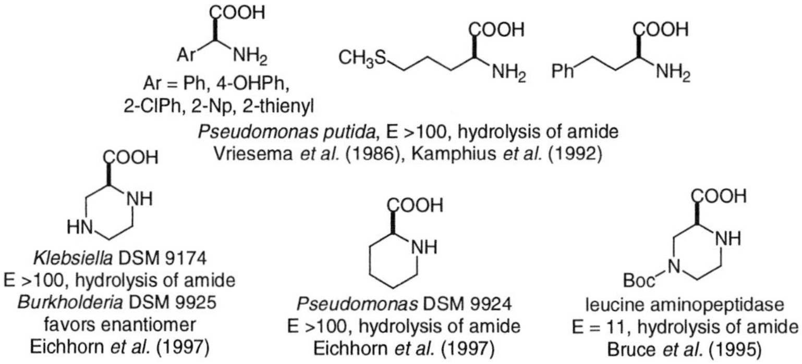

Fig. 163. Examples of unnatural α-amino acids resolved by amidases.

The DSM group also found an amidase that catalyzes hydrolysis of α,α-disubstituted α-amino acid amides (Kruizinga et al., 1988; Kamphuis et al., 1992; Kaptein et al., 1993) and proposed an active site model for its substrate specificity (Fig. 164).

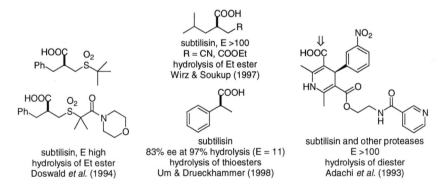

Fig. 164. Typical α,α-disubstituted α-amino acids resolved by amidase(s) from *Mycobacterium neoaurum*. The active site model summarizes the known substrate specificity. The P_{NH_2} site represents a polar site for the NH_2 group. The amidase does not tolerate substituents on the nitrogen. The H_S site is a small hydrophobic site that tolerates methyl, ethyl, and allyl groups. The H_L site is a large hydrophobic site that tolerated a range of groups.

8.2.2   Other Carboxylic Acids

8.2.2.1   Subtilisin

Subtilisin also catalyzes the enantioselective hydrolysis of esters of carboxylic acids that are not amino acids (Fig. 165). In the case of the 2-phenoxypropanoic acids chymotrypsin and subtilisin favored opposite enantiomers (Chênevert and D'Astous, 1988).

Fig. 165. Other carboxylic acids resolved by subtilisin.

8.2.2.2 Chymotrypsin

α-CT often shows lowered stereoselectivity upon replacing the amino group of α-amino acids with an α-hydroxy or α-*O*-acyl group. For example, α-CT showed no stereoselectivity toward mandelic acid, but high stereoselectivity toward the corresponding amino acid, phenylglycine (Fig. 165). On the other hand, α-CT efficiently resolves β-phenyllactate (analog of Phe) by hydrolysis of its ethyl ester (Cohen and Weinstein, 1964). However, hydrolysis of the same substrate in dichloromethane containing 0.2 % water showed no stereoselectivity (Ricca and Crout, 1993). Hydrolysis of *O*-benzoyl lactate ethyl ester favors the D-enantiomer (analog of the unnatural enantiomer of Ala). Binding of the benzoyl group in the *ar*-site presumably accounts for this reversal. Like the corresponding amino compound, the resolution of α-methyl-β-phenyllactate is efficient, but the stereoselectivity is reversed (Cohen and Lo, 1970). Presumably the OH group, but not the NH₂ group, favors the *h*-site. Replacement of the amino group in the hindered cyclic acid below with either hydrogen or *O*-acetyl abolished stereoselectivity (Matta and Rohde, 1972). α-CT efficiently resolved the dihydrobenzofuran derivative in Fig. 166 (Lawson, 1967). α-CT can also resolve acids without a polar substituent, for example the 3-cyclohexenyl acids in Fig. 166 (Cohen et al., 1969; Jones and Marr, 1973).

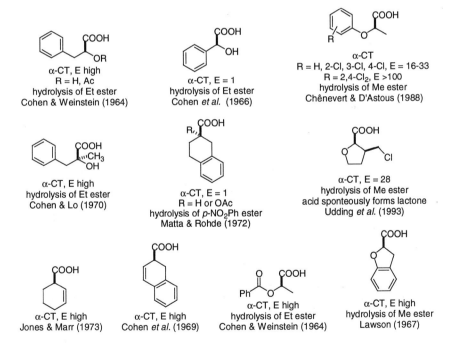

Fig. 166. Other carboxylic acids resolved by chymotrypsin.

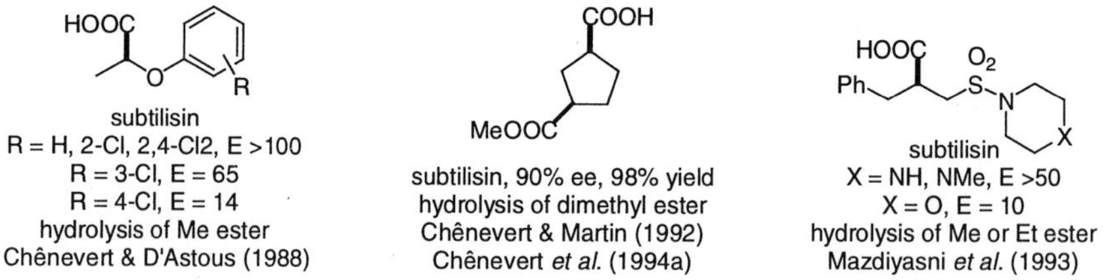

Fig. 166. Other carboxylic acids resolved by chymotrypsin (continued).

8.2.2.3   Other Proteases and Amidases

Other proteases and amidases can also resolve carboxylic acids that are not amino acids (Fig. 167). Penicillin acylase from *E. coli* strongly favors phenylacetyl esters or amides, but it also accepts close analogs. The 2-substituted phenylacetic acids are chiral and PGA showed high enantioselectivity toward several of these (Fig. 153).

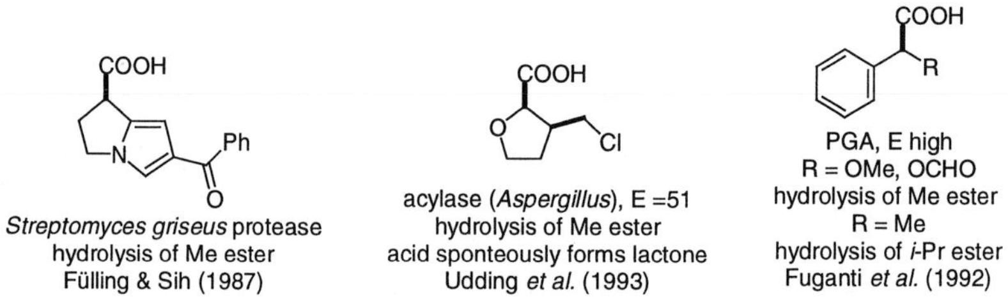

Fig. 167.   Non-amino acids resolved by proteases and amidases.

8.2.3   Commercial Enantioselective Reactions

8.2.3.1   Unnatural Amino Acids

Unnatural amino acids are produced commercially via acylase-, hydantoinase- and amidase-catalyzed routes (for reviews see Bommarius et al., 1992; Kamphuis et al., 1992). A few examples are shown in Fig. 168. Acylase-catalyzed routes are best for the L-amino acids because acylases favor the L-enantiomer. Racemization of unreacted N-acyl derivative of the D-amino acid avoids wasting half of the starting material. Hydantoinase-catalyzed routes are best for producing D-amino acids because the most common hydantoinases favor the D-enantiomer. Amidase-catalyzed resolution are best for unusual amino acids whose derivatives are not substrates for acylases or hydantoinases. For example, α-alkyl-α-amino acids are resolved by amidases as are the cyclic amino acids 2-piperidine carboxylic acid (pipecolic acid) and piperazine-2-carboxylic acid. Amidases usually favor the natural L-enantiomer.

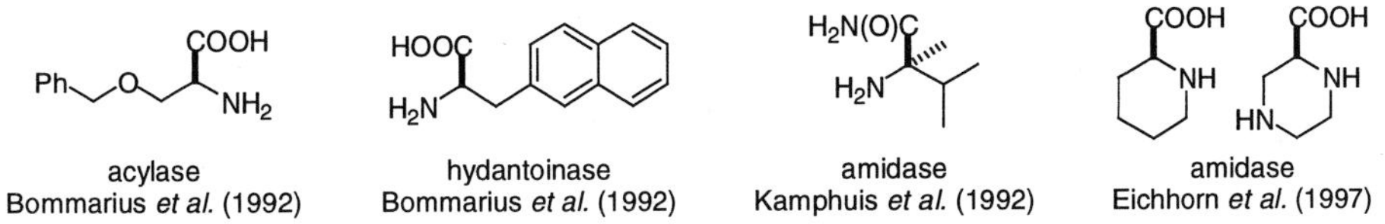

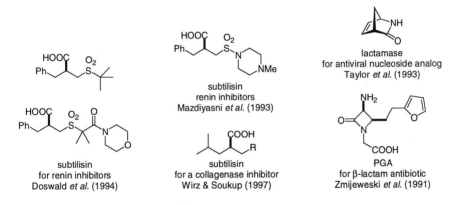

acylase
Bommarius *et al.* (1992)

hydantoinase
Bommarius *et al.* (1992)

amidase
Kamphuis *et al.* (1992)

amidase
Eichhorn *et al.* (1997)

Fig. 168. Examples of unnatural amino acids produced via acylase, hydantoinase or amidase.

8.2.3.2   Other Carboxylic Acids

Other carboxylic acids are also important intermediates in the synthesis of pharmaceuticals. Researchers have published kilogram-scale resolutions by using proteases (Fig. 169).

subtilisin
renin inhibitors
Mazdiyasni *et al.* (1993)

lactamase
for antiviral nucleoside analog
Taylor *et al.* (1993)

subtilisin
for renin inhibitors
Doswald *et al.* (1994)

subtilisin
for a collagenase inhibitor
Wirz & Soukup (1997)

PGA
for β-lactam antibiotic
Zmijeweski *et al.* (1991)

Fig. 169. Kilogram-scale routes to pharmaceutical precursors involving proteases or amidases.

9 Reactions Catalyzed by Esterases

9.1 Pig Liver Esterase

9.1.1 Biochemical Properties

Pig liver esterase (EC 3.1.1.1) is a serine-type esterase and consists of three subunits. Its physiological role is the hydrolysis of various esters occurring in the pig diet, which might explain its wide substrate tolerance. According to Heymann and Junge (1979) 'most laboratories working on the characterization of highly purified carboxylesterases (especially pig liver esterase) agree that these are highly frustrating enzymes'. Nevertheless, they were able to perform detailed studies of these subunits by isolectric focussing and revealed that they are not identical (Heymann and Junge, 1979). They found up to 16 bands showing esterase activity and identified six major components with isoelectric points ranging from 4.8 to 5.8. Similar observations were reported by Farb and Jencks (1980), who separated seven isozymes. Heymann and Junge reported different substrate specificities of these fractions towards methyl butyrate or butanilicain (*N*-butylglycyl-2-chloro-6-methylanilide hydrochloride). In contrast, no differences in substrate specificity and composition were observed between crude and purified PLE isolated from liver microsomes.

The three major fractions of PLE had apparent molecular weights of 58.2 kDa (α-subunit, *C*-terminus leucine), 59.7 kDa (β-subunit, *C*-terminus glycine) and 61.4 kDa (γ-subunit, *C*-terminus alanine) as determined by SDS-PAGE. Amino acid analysis revealed a lower content of aspartic acid and a higher content of arginine in the α-subunit compared to the γ-subunit (Heymann and Junge, 1979).

Fraction I (γ-subunit) has cholinesterase-like properties (Junge and Heymann, 1979), because it hydrolyzes butyrylcholine and is sensitive to physostigmine and fluoride ions. Fraction V (α-subunit) does not hydrolyze butyrylcholine and acts on short-chain aliphatic esters. Main components of PLE are the trimers γγγ, αγγ, ααγ, and ααα. Fractions I and V differ considerably in their substrate spectra and solvents like acetone, methanol or acetonitrile had different effects on fractions I or V.

A trimeric structure of three identical subunits obtained by electron microscopy was reported for a pig intestinal proline-β-naphthylamidase. At pH 4.5 and 4 °C, this trimer dissociates into active subunits, which show a 2.5-fold lower activity compared to the trimer (Takahashi et al., 1989, 1991).

Later, the same group determined the nucleotide and deduced amino acid sequence and suggested that this proline-β-naphthylamidase is identical with pig liver esterase (Matsushima et al., 1991). This was further undermined by Heymann and Peter (1993), who found that fraction I (γ-subunit) of PLE is capable to hydrolyze proline-β-naphthylamide. However, they claim that the sequence reported by Matsushima et al.

belongs to the main carboxylesterase isozymes but not the γ-subunit. Unfortunately, no one succeeded so far in the expression of the proline-β-naphthylamidase or pig liver esterase genes. This is an important requirement for further verifications of these results.

In contrast to the observations made by Heymann's group and by Öhrner et al. (1990), Jones and co-workers (Lam et al., 1988) claimed that isozymes of commercially available pig liver esterase show essentially the same stereospecificities toward representative monocyclic and acyclic diester substrates. Only differences in activity were observed. However, it cannot be excluded that the stereoselectivity of PLE may vary from batch to batch, because the pig diet probably influences the induction of different PLE isozymes (Seebach and Eberle, 1986).

PLE has the major advantage of easy preparation in form of an acetone powder (Seebach and Eberle, 1986; Adachi et al., 1986). A procedure reported by Adachi et al. (1986) is as follows: "Fresh pig liver (about 4 kg) is homogenized in 18 L cold acetone by using a kitchen juicer. After confirming that the homogenized parts became in a well-powdered state, they were collected by filtration. The residue was further washed with cold acetone (18 L) to remove the fatty material as cleanly as possible. The 'acetone powder' thus obtained was dried at room temperature to afford about 1 kg of the crude pig liver esterase. The fibrous material was removed and about 800 g of fine powder was finally obtained after sieving."

9.1.2   Overview of PLE Substrate Specificity and Models

In contrast to lipases, PLE usually does not accept highly hydrophobic substrates. PLE shows highest activity in aqueous buffered systems (best: pH 7.0), which might be supplemented up to 20 % with polar protic water-miscible solvents like methanol, *t*-butanol, DMSO, acetone or acetonitrile. pH as well as solvent should be carefully choosen, because both can strongly influence the stereoselectivity and rate of reaction. For instance, addition of cosolvents to a PLE-catalyzed hydrolysis of *meso*-cyclohexene diacetate increased the enantioselectivity and decreased the rate (Tab. 20) (Guanti et al., 1986). Others reported similar increases in enantioselectivity (Lam et al., 1986; Björkling et al., 1987; Polla and Frejd, 1991), but in some cases addition of cosolvent decreased the enantioselectivity (for a summary see Faber et al., 1993).

Although PLE converts a huge variety of substrates, only hydrolysis of methylesters of carboxylic acids or acetates of alcohols is recommended. PLE is especially useful in the preparation of optically pure compounds starting from prostereogenic or *meso*-substrates. PLE-catalyzed acylations in organic solvents are extremely slow and therefore have no practical importance. However, activity of PLE in organic solvents could be substantially enhanced after colyophilization of the esterase with methoxypolyethylene glycol (Ruppert and Gais, 1997).

Tab. 20.   Effect of organic solvents on PLE-catalyzed asymmetric hydrolysis (Guanti et al., 1986).

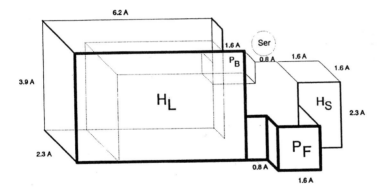

Organic cosolvent	Relative rate (%)	Optical purity (%ee)
None	100	55
20 % DMSO	70	59
40 % DMSO	28	72
20 % DMF	35	84
5 % t-BuOH	70	94
10 % t-BuOH	44	96

Because noone succeeded so far in the crystallization of PLE, no structural data based on x-ray diffraction or NMR measurements is available. However, Jones and co-workers developed a three dimensional cubic model of the active site of PLE (Fig. 170), which allows a prediction of the stereoselectivity and reactivity of PLE towards a given substrate. The model was based on the evaluation of over 100 substrates. The best fit of a substrate is determined by locating the ester group to be hydrolyzed within the locus of the serine residue and then arranging the remaining moieties in the H and P pockets (Toone et al., 1990; Provencher et al., 1993).

Fig. 170.   Active-site model for PLE. H_L is the large and H_S the small hydrophobic binding pocket. P_F is the polar, hydrophilic front pocket and P_B the rear polar pocket. Ser stands for the serine residue of the catalytic triad (Toone et al., 1990; Provencher et al., 1993).

In addition, a substrate model for PLE was proposed (Fig. 171) (Mohr et al., 1983). Optimal selectivity with PLE can be ensured by assigning the α- and β-substituents of methyl carboxylates according to their size (S = small, M = medium, L = large) with the preferably accepted enantiomer oriented as shown in Fig. 171.

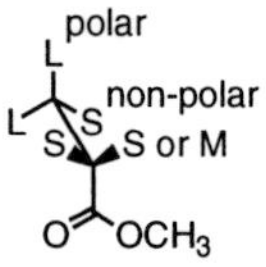

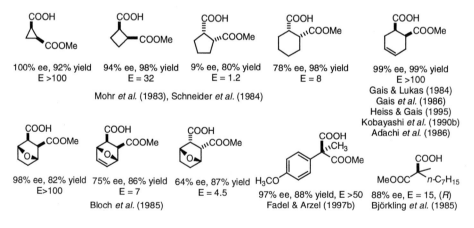

Fig. 171. Substrate model of PLE (Mohr et al., 1983).

In the following survey, various applications of PLE are summarized. Pre-1990 publications dealing with the use of PLE in organic synthesis have been extensively reviewed (Ohno and Otsuka, 1990; Zhu and Tedford, 1990; Tamm, 1992; Jones, 1993; Phytian, 1998) and only a few remarkable examples are included here.

9.1.3   Asymmetrization of Carboxylic Acids with a Stereocenter at the α-Position

The examples summarized in Fig. 172 show that the asymmetrization of carboxylic acids with a stereocenter at the α-position varies considerably. Selectivity was largely affected and even inversed by ring size or presence of a double bond. This changes in enantio-preference could be explained by the active site model of PLE shown in Fig. 170 (Toone and Jones, 1991). In addition, enantiomeric excess differed substantially for hydrolysis of either methyl- (E = 9) or ethyl esters (E > 100) of the acetonide of a bicyclic *meso*-diester (Fig. 172 (continued), first row, first structure, Arita et al., 1983; Adachi et al., 1986).

Fig. 172. PLE-catalyzed asymmetrization of *meso*- or prochiral dicarboxylic acids with a stereocenter at the α-position.

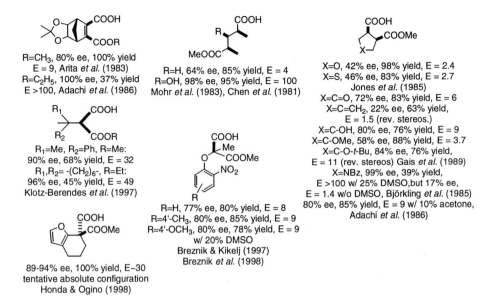

Fig. 172. PLE-catalyzed asymmetrization of *meso-* or prochiral dicarboxylic acids with a stereocenter at the α-position (continued).

9.1.4   Asymmetrization of Carboxylic Acids with Other Stereocenters

Similar tendencies but less examples can be found for the asymmetrization of carboxylic acids with other stereocenters (Fig. 173).

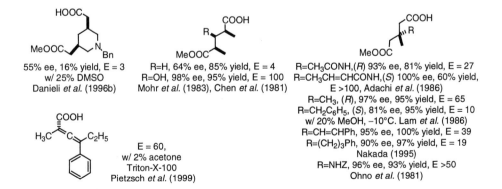

Fig. 173. PLE-catalyzed asymmetrization of *meso-* or prostereogenic dicarboxylic acids with other stereocenters.

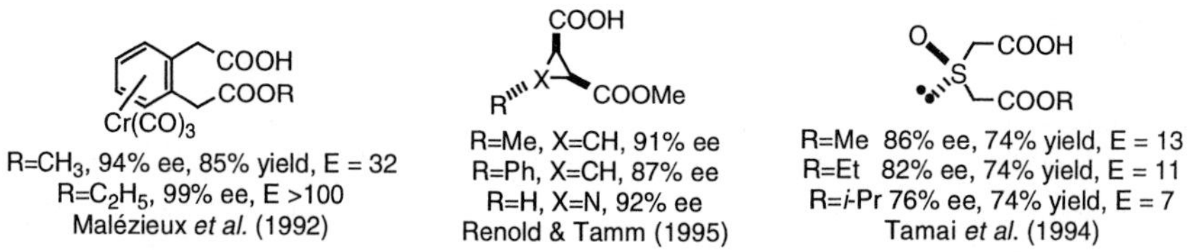

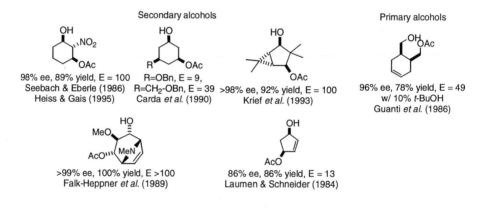

Fig. 173. PLE-catalyzed asymmetrization of *meso-* or prostereogenic dicarboxylic acids with other stereocenters (continued).

9.1.5   Asymmetrization of Primary and Secondary *meso*-Diols

Asymmetrization of primary or secondary *meso*-diols with PLE often proceeded with very high enantioselectivities (Fig. 174).

Fig. 174. PLE-catalyzed asymmetrization of *meso*-diols.

9.1.6   Kinetic Resolution of Alcohols or Lactones

PLE also catalyzed the kinetic resolution of alcohols. Selected examples are summarized in Fig. 175. Note that enantioselectivities are usually high and that in contrast to most lipases (see Sect. 5.1.3.1) even tertiary alcohols reacted. Selectivity towards lactones parallels that of most lipases (see Sect. 5.3).

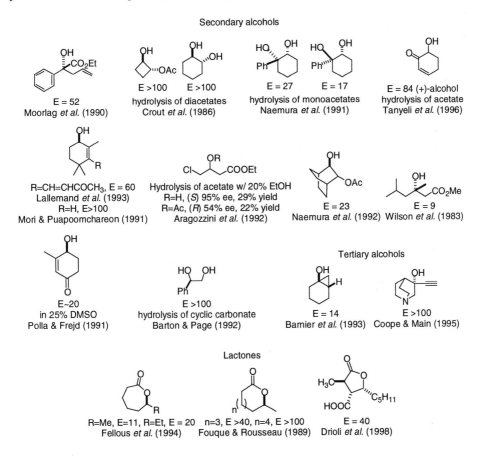

Fig. 175. PLE-catalyzed kinetic resolution of racemic secondary or tertiary alcohols and lactones.

9.1.7   Kinetic Resolution of Carboxylic Acids

Kinetic resolution of racemic carboxylic acids with PLE usually proceeded with high enantioselectivities (Fig. 176) and E-values are in the same range as found for *Candida rugosa* lipase, which represents the most enantioselective lipase for these kind of substrates (see Sect. 5.2.2.2).

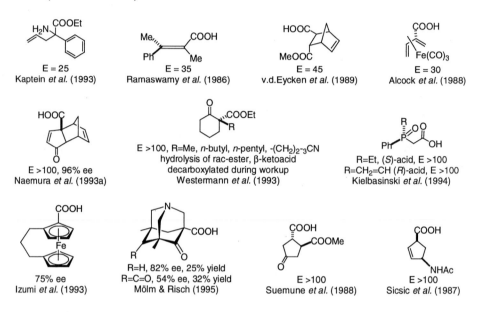

Fig. 176. PLE-catalyzed kinetic resolution of racemic carboxylic acids.

9.1.8   Reactions Involving Miscellaneous Substrates

Selected examples exploiting the regioselectivity of PLE are summarized in Fig. 177.

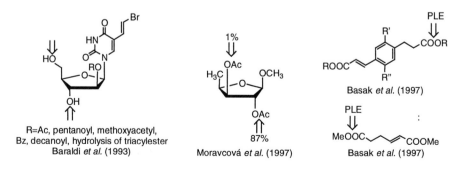

Fig. 177. Other substrates used in PLE-catalyzed biotransformations.

9.2   Acetylcholine Esterase

9.2.1   Biochemical Properties

Acetylcholine esterase (AChE, EC 3.1.1.7) plays a key role in cholinergic neurotransmission. By rapid hydrolysis (25 000 mols per second) of the neurotransmitter acetylcholine, the enzyme effectively terminates the chemical impulse, thereby setting the basis for rapid, repetitive responses and enabling the re-uptake (and recycling) of choline (Maelicke, 1991). Historically, first hints that an esterase is involved in the hydrolysis of acetylcholine, date back to 1914 and the first crude extract of AChE was prepared in 1932 by Stedman's group. Since then, numerous papers on the enzymes' mechanism of action appeared. Cholinesterase from different species differ considerably in their substrate specificity. Currently, the major distinction is based on the substrate specificity towards acetylcholine or butyrylcholine (naming the latter enzyme butyrylcholine esterase, BChE).

AChE from the electric eel *Electrophorus electricus* consists of 12 subunits (75 kDa each), which are assemmbled via disulfide bridges as three tetramers. These subunits are further classified into globulary (G_1, G_2, disulfide-bridged dimers, identical subunits, G_4, tetramer) and asymmetric (A_4, A_8, A_{12}, with 4, 8 or 12 catalytic subunits, resp.) forms.

The structure of AChE from *Torpedo californica* (68 kDa, tetramer) is known (resolution 2.8 Å) (Sussman et al., 1991). It consists of a 12-stranded mixed β-sheet surrounded by 14 α-helices and bears a striking resemblance to several hydrolase structures including dienelactone hydrolase, serine carboxypeptidase-II, three neutral lipases (especially lipases from *Geotrichum candidum* and *Candida rugosa*), and a haloalkane dehalogenase. The active site is unusual because it contains Glu_{327}, not Asp, in the Ser_{200}–His_{440}–acid catalytic triad and because the relation of the triad to the rest of the protein approximates a mirror image of that seen in the serine proteases. Furthermore, the active site lies near the bottom of a 20 Å deep and narrow gorge that reaches halfway into the protein. Modeling of acetylcholine binding to the enzyme suggests that the quaternary ammonium ion is bound not to a negatively charged "anionic" site, but rather to some of the 14 aromatic residues that line the gorge.

AChE is inhibited through formation of (in most cases) irreversible acyl-enzyme complexes. The longest known inhibitor is the natural alkaloid physostigmine, which carbamoylates the active site serine residue. Organophosphorous compounds such as Paraoxon, E 605 and diisopropyl fluorophosphate, which are used as agricultural insecticides or as nerve gases in chemical warfare, react with the active site serine forming a very stable covalent phosphoryl-enzyme complex. This leads to breath paralysis, cardiac arrest and death. Some oximes such as pralidoxim iodine and obidoxime chloride can be used as antidots. A connection between Alzheimer's disease and AChE has been proposed.

BChE cleaves butyryl- and propionylcholine faster than acetylcholine and was found in sera, liver and mammalian pancreas as well as in snake venom. Tetraisopropyldiphosphoramide (*iso*-OMPA) is a specific inhibitor of BChE.

More information about AChE and BChE is collected in databases (e.g., *http://bnlstb. bio.bnl.gov:8000/disk$3/giles/che.html, http://www.ensam.inra.fr/cholinesterase/*).

Activity of AChE is usually determined with acetylthiocholine as substrate and the released thiocholine is determined spectrophotometrically at 410 nm after derivatization with Ellmanns-reagent (5,5-dithio-bis-(2-nitrobenzoic acid), DTNB).

9.2.2   Application of AChE in Organic Syntheses

Only a few biocatalytic applications of AChE can be found, because the enzyme is only commercially available from electric eel, bovine brain, and human or bovine erythrocytes at high prices. Although heterologous expression of recombinant AChE was achieved (Heim et al., 1998), the expression levels are still far to low for the production of AChE on a practical scale. In the examples shown in Fig. 178 only AChE from electric eel was used.

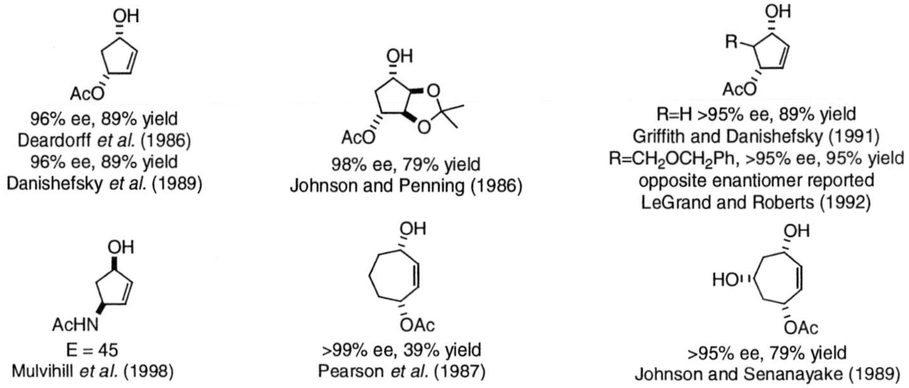

Fig. 178. Products obtained by AChE-catalyzed hydrolysis of *meso*-diacetates.

9.3   Other Mammalian Esterases

Selected examples for the application of other mammalian esterases are summarized in Fig. 179.

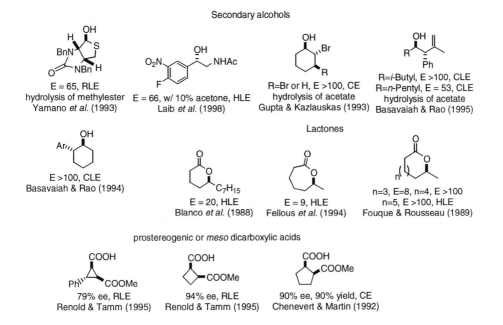

Fig. 179. Application of rabbit liver esterase (RLE), horse liver esterase (HLE), cholesterol esterase (CE) and chicken liver esterase (CLE) in the kinetic resolution of secondary alcohols, lactones or asymmetrization of prostereogenic or *meso*-dicarboxylic acids.

9.4   Microbial Esterases

Although a large number of microbial esterases have been described in literature, only a few of them are synthetically useful enzymes. With the exception of carboxylesterase NP, researchers studied in most cases only one or a few selected substrates. Further reasons are narrow substrate range and low enantioselectivity. In addition, none of the esterases listed in the survey below are commercially available, despite the fact that more than ten microbial esterases have been cloned and overexpressed in common microbial host organisms.

9.4.1   Carboxylesterase NP

Carboxylesterase NP was originally isolated from *Bacillus subtilis* (strain ThaiI-8) and was cloned and expressed in *Bacillus subtilis* (Quax and Broekhuizen, 1994). The esterase shows very high activity and stereoselectivity towards 2-arylpropionic acids, which are e.g., used in the synthesis of (*S*)-naproxen – (+)-(*S*)-2-(6-methoxy-2-naphthyl) propionic acid – a non-steroidal anti-inflammatory drug (Fig. 180). (*S*)-naproxen is ca. 150-times more effective than (*R*)-naproxen, the latter also might promote unwanted gastro-intestinal disorders. Therefore, this esterase is usually abbreviated in most references as carboxylesterase NP. It has a molecular weight of 32 kDa, a pH optimum between pH 8.5–10.5 and a temperature optimum between 35–55 °C. Carboxylesterase NP is produced as intracellular protein; its structure is unknown.

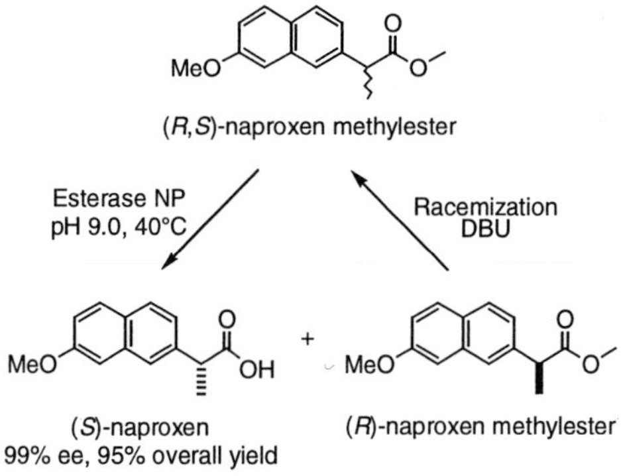

Fig. 180. Synthesis of (*S*)-naproxen by kinetic resolution of the (*R,S*)-methylester with carboxylesterase NP followed by chemical racemization of the (*R*)-naproxen methylester (Quax and Broekhuizen, 1994).

In a pilot-scale process, the (*R,S*)-naproxen methylester is hydrolyzed in the presence of Tween 80 to increase substrate solubility at pH 9.0. The (*S*)-acid is separated from the remaining (*R*)-methylester by filtration and the latter is racemized with DBU. This reaction yields (*S*)-naproxen with excellent optical purity (99 %ee) at an overall yield of 95 %. At concentrations > 20 g/L, an irreversible inactivation of carboxylesterase NP by the acid released was observed. Although this could be circumvented by addition of formaldehyde, the activity dropped by 50 %. A better solution was based on site-directed mutagenesis by replacement of lysine residues with glutamine, thus eliminating the positively charged target prone to the formation of a Schiff base. 11 Lys residues were all successively replaced by Glu and mutant K34Q turned out to be the best choice (Quax and Broekhuizen, 1994). Carboxylesterase NP was also used in the resolution of (*R,S*)-ibuprofen methylester and showed higher selectivity compared to lipase from *Candida rugosa* (Mustranta, 1992).

A detailed investigation of the substrate spectra of carboxylesterase NP towards other chiral acids revealed high enantioselectivity in most cases (Fig. 181) and the (*S*)-enantiomer was usually preferred. All reactions shown have been performed at 20 °C in a phosphate buffer:acetone mixture (9:1) at pH 7.2. Carboxylesterase NP was superior compared to α-chymotrypsin and PCL, which had (*R*)-enantiopreference (Azzolina et al., 1995a, b). Variation of the substituents as well as pH had a strong influence on enantio-selectivity and reaction rate (Azzolina et al., 1995a).

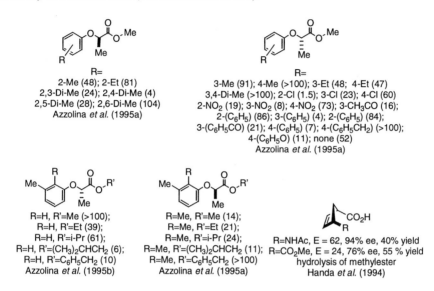

R=
2-Me (48); 2-Et (81)
2,3-Di-Me (24); 2,4-Di-Me (4)
2,5-Di-Me (28); 2,6-Di-Me (104)
Azzolina *et al.* (1995a)

R=
3-Me (91); 4-Me (>100); 3-Et (48; 4-Et (47)
3,4-Di-Me (>100); 2-Cl (1.5); 3-Cl (23); 4-Cl (60)
2-NO$_2$ (19); 3-NO$_2$ (8); 4-NO$_2$ (73); 3-CH$_3$CO (16);
2-(C$_6$H$_5$) (86); 3-(C$_6$H$_5$) (4); 2-(C$_6$H$_5$) (84);
3-(C$_6$H$_5$CO) (21); 4-(C$_6$H$_5$) (7); 4-(C$_6$H$_5$CH$_2$) (>100);
4-(C$_6$H$_5$O) (11); none (52)
Azzolina *et al.* (1995a)

R=H, R'=Me (>100);
R=H, R'=Et (39);
R=H, R'=i-Pr (61);
R=H, R'=(CH$_3$)$_2$CHCH$_2$ (6);
R=H, R'=C$_6$H$_5$CH$_2$ (10)
Azzolina *et al.* (1995b)

R=Me, R'=Me (14);
R=Me, R'=Et (21);
R=Me, R'=i-Pr (24);
R=Me, R'=(CH$_3$)$_2$CHCH$_2$ (11);
R=Me, R'=C$_6$H$_5$CH$_2$ (>100)
Azzolina *et al.* (1995a)

R=NHAc, E = 62, 94% ee, 40% yield
R=CO$_2$Me, E = 24, 76% ee, 55 % yield
hydrolysis of methylester
Handa *et al.* (1994)

Fig. 181. Other examples for carboxylesterase NP-catalyzed resolutions of chiral carboxylic acids. Values in brackets refer to enantioselectivity E.

Carboxylesterase NP was also used in the kinetic resolution of α-chloropropionic acid methylester (Wolff et al., 1994), however the enantioselectivity was rather low (E = 4.7). Performing the reaction in combination with a dehalogenase as a sequential kinetic reso-lution increased E to 15 (Rakels et al., 1994b) (Fig. 182).

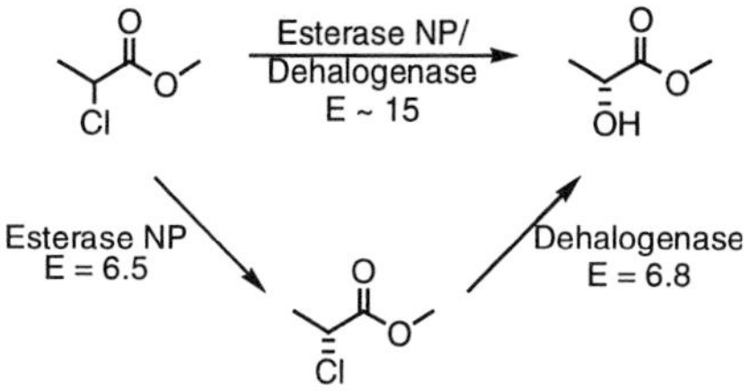

Esterase NP/
Dehalogenase
E ~ 15

Esterase NP
E = 6.5

Dehalogenase
E = 6.8

Fig. 182. Sequential kinetic resolution of α-chloropropionic acid methylester with carboxylester-ase NP and a dehalogenase (Rakels et al., 1994b).

9.4.2   Other Microbial Esterases

Selected examples for the application of other microbial esterases are summarized in Fig. 183. (+)-*trans*-(1*R*,3*R*) chrysanthemic acid (Fig. 183, second row, right structure) is an important precursor of pyrethrin insecticides. An efficient kinetic resolution starting from the (±)-*cis-trans* ethylester was achieved using an esterase from *Arthrobacter globiformis* resulting in the sole formation of the desired enantiomer (> 99 %ee, at 77 % conversion). The enzyme was purified and the gene was cloned in *E. coli* (Nishizawa et al., 1993). In a 160 g scale process, hydrolysis is performed at pH 9.5 at 50 °C. Acid produced is separated through a hollow-fiber membrane module and the esterase was very stable over four cycles of 48 h (Nishizawa et al., 1995).

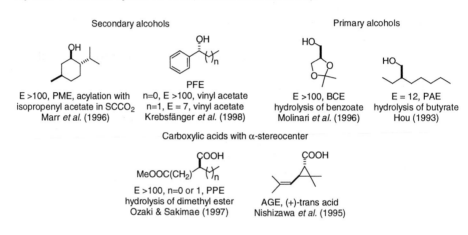

Secondary alcohols · Primary alcohols

PFE

E >100, PME, acylation with isopropenyl acetate in SCCO₂ Marr *et al.* (1996)

n=0, E >100, vinyl acetate n=1, E = 7, vinyl acetate Krebsfänger *et al.* (1998)

E >100, BCE hydrolysis of benzoate Molinari *et al.* (1996)

E = 12, PAE hydrolysis of butyrate Hou (1993)

Carboxylic acids with α-stereocenter

MeOOC(CH₂)ₙ

E >100, n=0 or 1, PPE hydrolysis of dimethyl ester Ozaki & Sakimae (1997)

AGE, (+)-trans acid Nishizawa *et al.* (1995)

Fig. 183. Application of esterases from *Pseudomonas marginata* (PME), *Pseudomonas fluorescens* (PFE), *Pseudomonas aeruginosa* (PAE), *Pseudomonas putida* (PPE), *Bacillus coagulans* (BCE) and *Arthrobacter globiformis* (AGE) in the kinetic resolution of secondary and primary alcohols or carboxylic acids.

Shimizu et al. (1992b) found a novel lactone-hydrolyzing enzyme in the fungus *Fusarium oxysporum*. This lactonase catalyzes the enantioselective ring opening of several aldonate lactones (e.g., D-galactono-γ-lactone, L-mannono-γ-lactone, D-gulono-γ-lactone and D-glucono-δ-lactone) and also D-pantoyl lactone (Eq. 41). A commercial resolution uses fungal mycelia and yields D-pantoyl lactone, a precursor for the synthesis of pantothenic acid (Kataoka et al., 1995).

lactonase E = 35

HOOC

D-pantothenic acid vitamin B₅

(41)

10 Epoxide Hydrolases

10.1 Introduction

Optically pure epoxides are versatile building blocks in organic synthesis. Several chemical ways for the preparation of optically pure epoxides have been developed, such as the methods described by Sharpless or Jacobsen-Katsuki. However, the Sharpless epoxidation (Katsuki and Martin, 1996) is limited to allylic alcohols and the Jacobsen-Katsuki method (Hosoya et al., 1994; Linker, 1997) gives poor enantiomeric excess (≤ 60 %ee) with *gem-* or *trans*-disubstituted olefins.

Direct stereospecific epoxidation of alkenes by isolated monooxygenases (cytochrome P450) and bacterial monooxygenases have been reported (for reviews see de Bont, 1993; Besse and Veschambre, 1994; Onumonu et al., 1994; Pedragosa-Moreau et al., 1995). However, although enantiomeric excess are often high, yields are typically low, most enzymes require cofactors and these biocatalysts accept only a narrow range of substrates making them less attractive for organic synthesis.

Alternatively, epoxides can be resolved by using epoxide hydrolases (EC 3.3.2.3), which catalyze the hydrolysis of an epoxide to furnish the corresponding vicinal diol. The reaction proceeds via an S_N2-specific opening of the epoxide leading to the formation of the corresponding *trans*-configurated 1,2-diols.

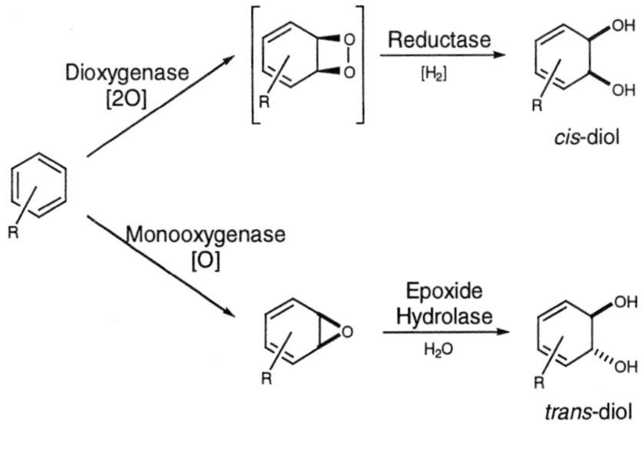

Fig. 184. Degradation of aromatics can proceed via dioxygenase-catalyzed formation of a dioxetane leading after reduction to *cis*-diols or via a monooxygenase-catalyzed epoxidation followed by epoxide hydrolase-catalyzed ring-opening to *trans*-diols.

Epoxide hydrolases (EH's) do not require cofactors, have been found in various sources such as mammals (Bellucci et al., 1991), plants (Blée and Schuber, 1995), insects (Linderman et al., 1995), yeasts (Weijers, 1997), filamentous fungi (Pedragosa-Moreau et al., 1996b; Grogan et al., 1996) and bacteria (Mischitz et al., 1995a; Osprian et al., 1997). They are catalytically active in the presence of organic solvents and often show high regio- and enantioselectivity. In eucaryotes, EH's play a key role in the metabolism of xenobiotics, in particular of aromatic systems. In prokaryotes, EH's hydrolyze epoxides formed by the action of P450-monooxygenases from olefinic or aromatic compounds giving the organisms access to these carbon sources (Fig. 184).

In mammals, microsomal (mEH's) and soluble (sEH's, earlier referred to as cytosolic EH's) epoxide hydrolases have been identified and the most widely studied enzymes are mammalian liver mEH's (for a review see Seidegard and de Pierre, 1983). In addition EH's acting on cholesterol, leukotriene A_4 and hepoxilin have been described (Pinot et al., 1995). Several enzymes have been purified and characterized in detail, the amino acid sequences of some are known and rat mEH was cloned into E. coli (Bell and Kasper, 1993). However, it is unlikely that these enzymes will be widely used as biocatalysts in organic synthesis, since they are not available in reasonable amounts for large-scale biotransformations. The 3D-structures of epoxide hydrolases are still unknown, but due to a high sequence similarity to haloalkane dehalogenase and bromoperoxidase, an α/β-hydrolase fold (see Sect. 2.1.6) is assumed for epoxide hydrolases.

The proposed mechanism of epoxide ring-opening involves the attack of a nucleophilic carboxylate residue at one end of the epoxide which again has been activated by protonation. This leads to an α-hydroxyester intermediate covalently bound to the active site of the enzyme. This intermediate is hydrolyzed by the nucleophilic attack of a water molecule which is activated by a histidine, followed by the release of the diol product and regeneration of the enzyme (Lacourciere and Armstrong, 1993) (Fig. 185). Evidence for the likelihood of this mechanism came from experiments involving ^{18}O-labeled enzyme in unlabeled water and *vice versa*. For soluble rat epoxide hydrolase, Asp_{333}, Asp_{495}, His_{523} (Arand et al., 1996) and for microsomal epoxide hydrolase, Asp_{226}, Asp_{352}, His_{431} were identified to form the catalytic triad (Laughlin et al., 1998). However, in a recent paper it could be demonstrated that indeed not Asp_{352} but Glu_{404} is involved in the catalytic machinery and replacement of Glu_{404} with Asp leads to a strongly increased turnover rate (Arand et al., 1999).

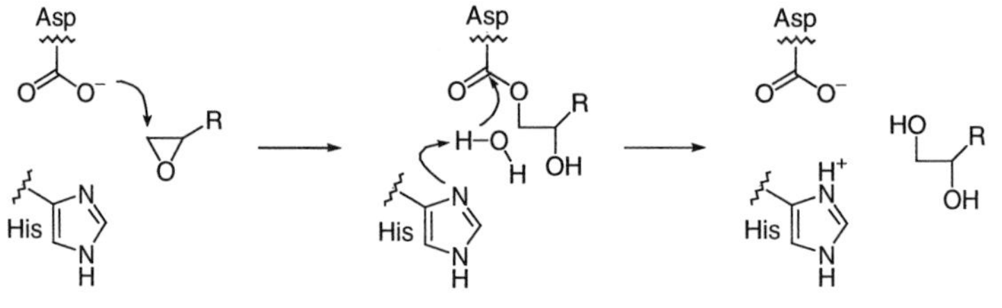

Fig. 185. Mechanism proposed for mammalian epoxide hydrolases.

The enzymatic hydrolysis of terminal epoxides may proceed via attacking either the less shielded oxirane carbon leading to retention of configuration (most common) or at the stereogenic center resulting in inversion of configuration (Faber et al., 1996) (Fig. 186).

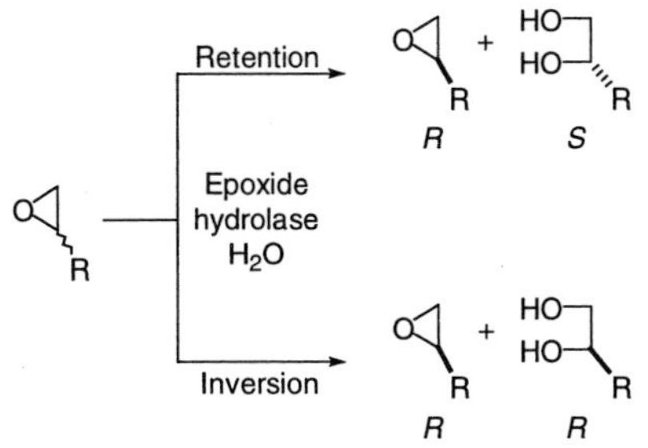

Fig. 186. Hydrolysis of epoxides can proceed with retention or inversion of configuration.

A detailed study of the regioselectivity of EH from *Rhodococcus* sp. NCIMB11216 using ^{18}O-labeled water revealed that also the position of substituents influenced the stereospecifity of the reaction (Fig. 187) (Mischitz et al., 1996). Recently, Mossou et al. (1998a, b) developed equations to determine the regio- and enantioselectivity of epoxide hydrolase-catalyzed ring opening of epoxides without the need for ^{18}O-labelling. By using experimentally obtained data for conversion, %ee diol and %ee epoxide, one can distinguish between (1) total regioselectivity on the same carbon, (2) partial regioselectivity between the two carbon atoms of the oxirane ring, and (3) total but opposite regioselectivity (enantioconvergence) (Moussou et al., 1998a, b).

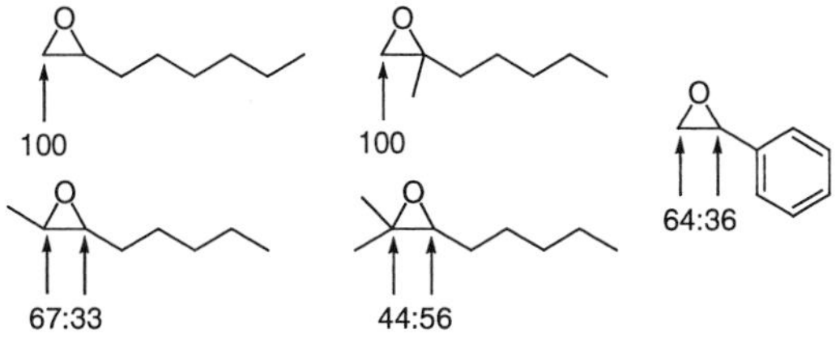

Fig. 187. Regioselectivity of epoxide hydrolase from *Rhodococcus* sp. NCIMB11216 studied using ^{18}O-labeled water revealed that insertion of oxygen usually occurs at the less hindered carbon. Note that in the case of styrene oxide non-enzymatic incorporation took place at the more hindered carbon (Mischitz et al., 1996).

The following sections reviews only selected examples for the application of mammalian and microbial epoxide hydrolases in organic synthesis proceeding with good to excellent enantioselectivities. Further examples can be found in a number of recent reviews (Faber et al., 1996; Archelas and Furstoss, 1998; Orru et al., 1998a; Archelas, 1998; Swaving and de Bont, 1998).

10.2   Mammalian Epoxide Hydrolases

Mammalian epoxide hydrolases have been studied in very much detail with regard to their biological role and mechanism. Less research focussed on the application of these enzymes in organic syntheses. Generally it was found that mEH's show higher enantiose-lectivity than sEH's. For example, Bellucci et al. (1994a) observed remarkable differ-ences in enantio- and regioselectivity between soluble and microsomal EH from rabbit liver in the resolution of phenyl substituted epoxides. In case of alkyl oxiranes, sEH showed only high enantioselectivity for a *t*-butyl derivative. Here attack of the nucleo-phile occurred at the least hindered carbon (Bellucci et al., 1991) (Fig. 188). Stereocon-vergent hydrolyses were reported for 3,4-epoxytetrahydropyran (Bellucci et al., 1981), a series of *cis*-disubstituted aliphatic epoxides (Chiappe et al., 1998) and β-*n*-alkyl-substi-tuted styrene oxides (Bellucci et al., 1996) yielding the corresponding (*R*,*R*)-diols.

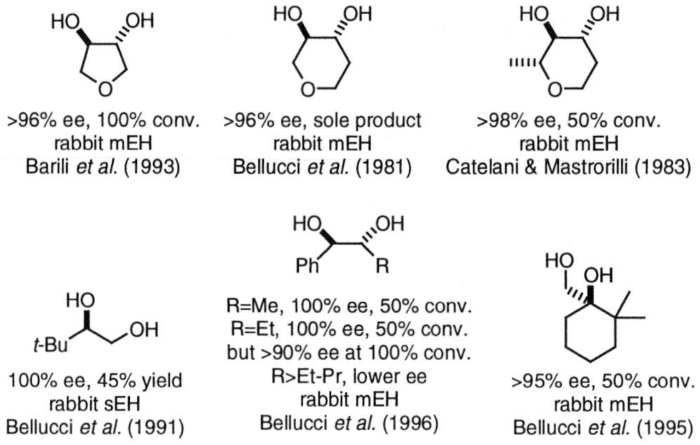

>96% ee, 100% conv.	>96% ee, sole product	>98% ee, 50% conv.
rabbit mEH	rabbit mEH	rabbit mEH
Barili *et al.* (1993)	Bellucci *et al.* (1981)	Catelani & Mastrorilli (1983)

HO	R=Me, 100% ee, 50% conv.	
t-Bu OH	R=Et, 100% ee, 50% conv.	HO OH
	but >90% ee at 100% conv.	
100% ee, 45% yield	R>Et-Pr, lower ee	>95% ee, 50% conv.
rabbit sEH	rabbit mEH	rabbit mEH
Bellucci *et al.* (1991)	Bellucci *et al.* (1996)	Bellucci *et al.* (1995)

Fig. 188. Examples for the resolution of epoxides by mammalian EH's. Note that in the case of 3,4-epoxytetrahydropyran (top row, middle structure) hydrolysis occurred with complete enantio-convergence.

In general, lipophilic aryl- or alkyl chains located close to the oxirane ring enhance the reaction rate in mEH-catalyzed reactions, whereas polar groups have an inhibitory effect, which explains the fact that most (apolar) xenobiotics are converted to more hydrophilic ones prior to their elimination (Jerina and Daly, 1974). In addition, steric hindrance is an important factor, because terminal monosubstituted epoxides are hydrolyzed more rap-idly than their *cis*-1,2-disubstituted analogs. *Trans*-1,2-di-, tri- or tetra-substituted com-pounds are usually not accepted as substrates (Oesch, 1973, 1974).

Some *meso*-epoxides could also be efficiently resolved by mammalian EH's, however optical purity of the resulting diols was affected by the substitution pattern, ring size of the substrates or whether microsomal or soluble EH's were used (Fig. 189).

More examples on the application of mammalian epoxide hydrolases can be found in an excellent recent review (Archer, 1997).

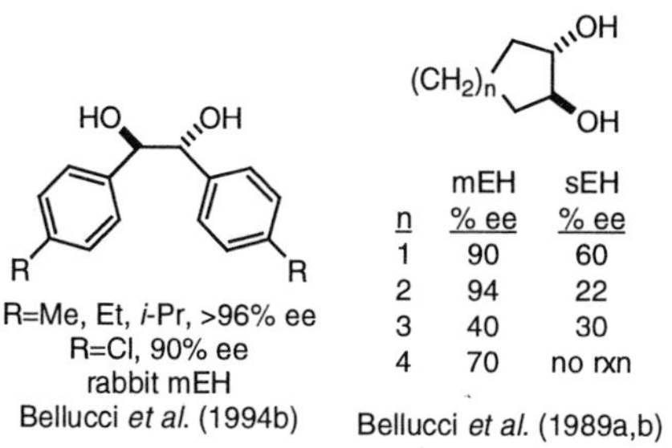

Fig. 189. Asymmetrization of *meso*-epoxides by mammalian EH's. Note that substitution pattern, ring size or origin of EH had a strong influence on enantioselectivity.

10.3   Microbial Epoxide Hydrolases

10.3.1   Bacterial Epoxide Hydrolases

In 1969, Allen reported the first application of an epoxide hydrolase isolated from a *Pseudomonas putida* strain catalyzing the synthesis of L- and *meso*-tartaric acid from an epoxide precursor (Allen and Jacoby, 1969). More than 20 years passed until researchers started to thoroughly investigate the application of epoxide hydrolases of microbial origin. Especially the groups of K. Faber (Austria) and R. Furstoss (France) have shown that a wide variety of microorganisms produce EH's which often exhibit good to excellent enantioselectivity. In contrast to mammalian epoxide hydrolases, conventional fermentation techniques enable the production of EH's from bacteria or fungi in sufficient quantities for large-scale biotransformations.

Initially, Faber's group discovered that the biocatalyst preparation SP409 produced by Novo also exhibits epoxide hydrolase activity. SP409 (from *Rhodococcus* sp.) was already widely used for the hydrolysis of nitriles (see Sect. 11), but was also able to hydrolyze a wide variety of mono- and 1,1-disubstituted epoxides with low to moderate enantioselectivity. In addition, also nucleophiles other than water (e.g., azide) are accepted (Fig. 190) (Mischitz and Faber, 1994).

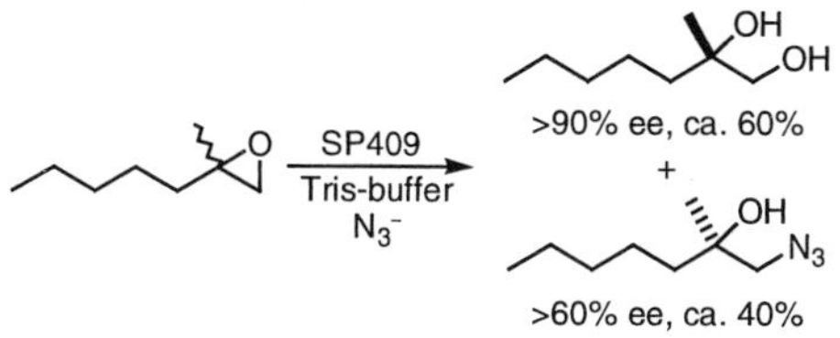

Fig. 190. *Rhodococcus* sp. SP409 also catalyzed epoxide ring-opening and accepts an azide as nucleophile (Mischitz and Faber, 1994).

Stimulated by this discovery, the Faber group screened a wide variety of organisms. Of 43 strains investigated, four bacterial and three fungal strains were found to show epoxide hydrolase activity (Mischitz et al., 1995b). *Rhodococcus* sp. NCIMB11216 and *Corynebacterium* sp. UPT9 preferentially hydrolyzed the (*R*)-epoxide of 1,2-epoxyoctane yielding the (*R*)-1,2-diol, however enantioselectivity was only E = 2.8 and E = 2.6 resp. Similar enantioselectivities were found for an epoxide hydrolase produced by *Corynebacterium* sp. C12 (Carter and Leak, 1995). Only the resolution of 1-methyl-1,2-epoxycyclohexane proceeded with good enantioselectivity (Archer et al., 1996). The purified enzyme is a multimer (probably tetrameric) with a subunit size of 32 kDa. The gene encoding *Corynebacterium* EH was isolated and sequenced. Sequence comparison revealed high similarity to mammalian and plant soluble EH's and to the EH cloned from *Agrobacterium radiobacter* AD1 (see below) (Misawa et al., 1998).

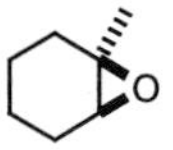

>99% ee, 30% yield
Corynebacterium sp. C12
Archer *et al.* (1996)

Fig. 191. Resolution of 1-methyl-1,2-epoxycyclohexane using *Corynebacterium* sp. C12 proceeded with good enantioselectivity.

Significantly enhanced enantioselectivities were achieved by hydrolyzing 1,1-disubstituted epoxides using the *Rhodococcus* sp. strain, e.g., for 2-methyl-1,2-epoxyheptane enantioselectivity was E > 100 and simultaneously the enantiopreference was inversed. Later work showed that *Rhodococcus* sp. epoxide hydrolases accept a wide range of substrates which are often converted with high enantioselectivity.

Selected examples for the resolution of racemic epoxides are summarized in Fig. 192. Note that in most cases only optical purity and yield of remaining epoxide are given in literature which is probably due to difficult isolation of the more hydrophilic diol. In most cases epoxide hydrolases were not isolated. Instead, substrates were added to the culture medium, to centrifuged and washed whole cells or to lyophilized whole cells.

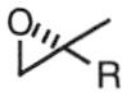

R=*n*-C$_5$H$_{11}$, E >100
Rhodococcus ruber DSM43338
R=CH$_2$-Ph, E >100
Rhodococcus sp. NCIMB11216
Osprian *et al.* (1997)
R=*n*-C$_5$H$_{11}$, E >100; R=*n*-C$_7$H$_{15}$, E >100;
R=*n*-C$_9$H$_{19}$, E>100
Rhodococcus sp. NCIMB11216
(lyophilized cells or purified enzyme)
Mischitz *et al.* (1995a)

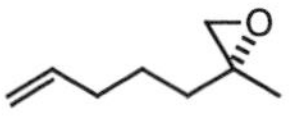

E = 39, 20% ee (*R*)-epoxide,
94% ee (*S*)-diol at 18% conv.
Rhodococcus equi IFO3730
Kroutil *et al.* (1997d)

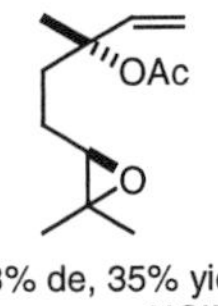

98% de, 35% yield
Rhodococcus sp. NCIMB11216
Mischitz & Faber (1996)

Fig. 192. Examples for the resolution of racemic epoxides using epoxide hydrolases from *Rhodococcus* sp.

In general, EH from *Rhodococcus* sp. NCIMB11216 shows high enantioselectivity for methyl-alkyl substituted oxiranes (Mischitz et al., 1995a; Osprian et al., 1997) but low E for ethyl-alkyl substituted oxiranes (Wandel et al., 1995). The strain was also used in the synthesis of optically pure linalool oxide (Fig. 192, right structure) leading to a product with high diastereomeric excess (98 %de) (Mischitz and Faber, 1996). Epoxide hydrolase activity present in lyophilized cells of *Rhodococcus equi* IFO3730 allowed the synthesis of a precursor of (*S*)-frontalin, a sex pheromone (94 %ee for the produced diol), however conversion was only 18 % (Kroutil et al., 1997d).

Another very enantioselective EH is produced by the strain *Mycobacterium parafinicium* NCIMB10420 (Faber et al., 1996). Botes et al. (1998a) described an EH produced by the bacteria *Chryseomonas luteola* which converted straight-chain aliphatic epoxides with moderate to excellent enantioselectivity. Highest selectivity was found for 1,2-epoxyoctane, where > 98 %ee for the remaining (*S*)-epoxide and 86 %ee for the (*R*)-diol were determined. No enantioselectivity was observed for 2,2-disubstituted epoxides, benzyl glycidyl ether and 2-methyl-1,2-epoxyheptane (Botes et al., 1998a).

Recently, several strains from *Nocardia* sp. (H8, EH1, TB1) have been identified as epoxide hydrolase producers. Especially *Nocardia* sp. EH1 shows high enantioselectivity at 50 % conversion in the resolution of 2-methyl-1,2-epoxyheptane (Fig. 193, middle structure). However, introduction of a phenyl group into the side chain almost destroyed enantioselectivity (E = 5.6). The enzyme was purified to homogeneity via a four-step procedure. It is a monomer with a molecular weight of 34 kDa, a pH optimum of 8–9 and a temperature optimum of 35–40 °C. The pure enzyme is much less stable than a whole cell preparation, but addition of Tween 80 or Triton-X-100 stabilizes it (Kroutil et al., 1998a). Immobilization on DEAE-cellulose doubled specific activity and allowed five repeated batch reactions, however enantioselectivity was slightly lowered (Kroutil et al., 1998b). Using *Nocardia* sp. EH1, the synthesis of naturally occuring (*R*)-(−)-mevalonolactone was achieved by deracemization of 10 g 2-benzyl-2-methyloxirane. The enzymatic reaction gave the corresponding (*S*)-diol, addition of catalytic amounts of sulfuric acid hydrolyzed the remaining (*R*)-epoxide under inversion of configuration, thus allowing the isolation of (*S*)-diol in an overall yield of 94 % at 94 %ee. Subsequent chemical steps afforded (*R*)-(−)-mevalonolactone in a total yield of 55 % (Orru et al., 1997, 1998c).

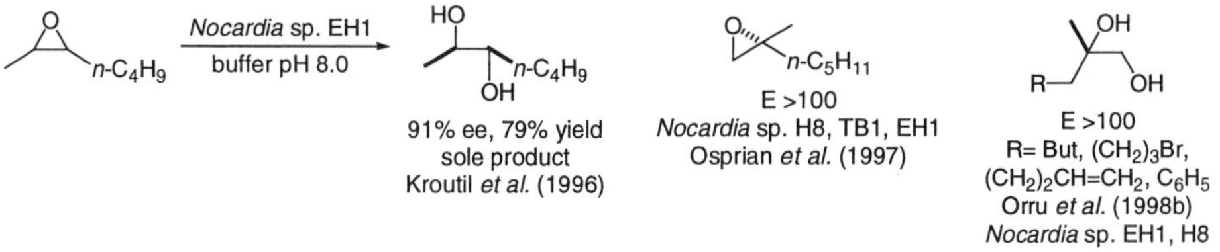

Fig. 193. Examples for the resolution of epoxides using epoxide hydrolases from *Nocardia* sp.

Interestingly, hydrolysis of (±)-*cis*-2,3-epoxyheptane with rehydrated lyophilized cells of *Nocardia* sp. EH1 proceeded in an enantioconvergent fashion and only (2*R*,3*R*)-heptane-2,3-diol was obtained as the sole product (Fig. 193) (Kroutil et al., 1996). Further examples for enantioconvergent reactions can be found in Kroutil et al. (1997c).

The so far only recombinant microbial epoxide hydrolase was cloned from *Agrobacterium radiobacter* AD1. The enzyme was overexpressed in *E. coli* (Rink et al., 1997) and accepts a broad range of styrene oxide derivatives and phenyl glycidyl ether which are converted with excellent enantioselectivity (Fig. 194) (Spelberg et al., 1998). The enzyme has a molecular weight of 34 kDa and the catalytic triad was proposed to consist of Asp_{107}, His_{275} and Asp_{246}, because site-directed mutagenesis of these positions resulted in a dramatic loss of activity toward epichlorohydrin.

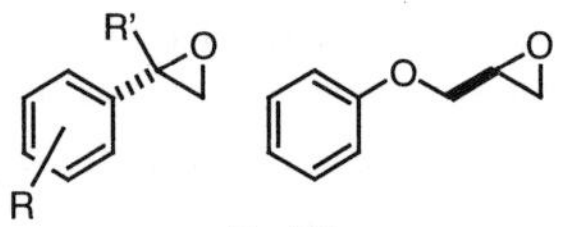

E >100
R'=H, R=H, 4-Me, *o*-Cl, *m*-Cl, *p*-Cl
R'=Me, R=H, Spelberg *et al.* (1998)

Fig. 194. Examples for the resolution of epoxides using recombinant EH from *Agrobacterium radiobacter* AD1.

10.3.2 Fungal and Yeast Epoxide Hydrolases

The first report on the use of a fungal epoxide hydrolase appeared in 1972 by Suzuki and Marumo (1972), who found that *Helminthosporum sativum* catalyzed the enantioselective hydrolysis of 10,11-epoxyfarnesol yielding the corresponding (*S*)-diol in 73 %ee.

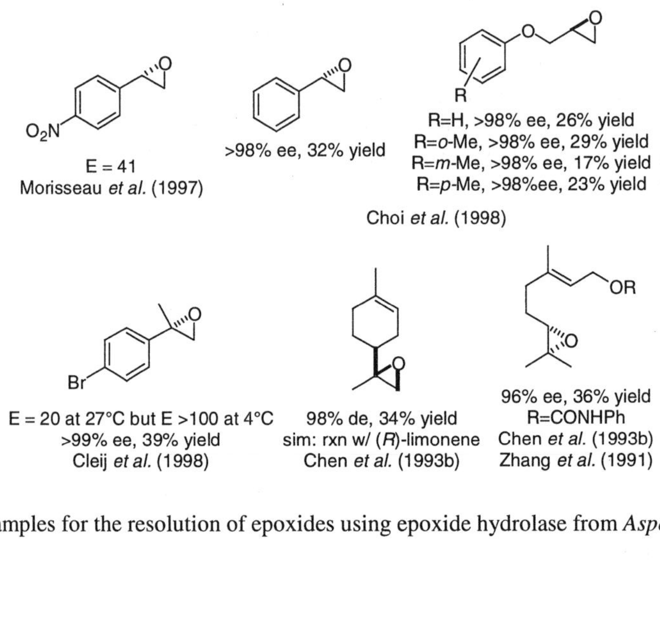

E = 41
Morisseau *et al.* (1997)

>98% ee, 32% yield

R=H, >98% ee, 26% yield
R=*o*-Me, >98% ee, 29% yield
R=*m*-Me, >98% ee, 17% yield
R=*p*-Me, >98%ee, 23% yield

Choi *et al.* (1998)

E = 20 at 27°C but E >100 at 4°C
>99% ee, 39% yield
Cleij *et al.* (1998)

98% de, 34% yield
sim: rxn w/ (*R*)-limonene
Chen *et al.* (1993b)

96% ee, 36% yield
R=CONHPh
Chen *et al.* (1993b)
Zhang *et al.* (1991)

Fig. 195. Examples for the resolution of epoxides using epoxide hydrolase from *Aspergillus niger*.

The first preparative-scale epoxide hydrolysis was reported by the group of Furstoss, who discovered that the fungus *Aspergillus niger* enantioselectively converts geraniol-N-phenylcarbamate yielding the (*S*)-epoxide in high optical purity (96 %ee, Fig. 195, bottom row, right structure), which was further converted to Bower's compound, an analog of insect juvenile hormone (Zhang et al., 1991). Furstoss' group first used the mycelium, but due to several problems, they later used a lyophilized enzyme preparation obtained after concentration and desalting, resulting in a 7-fold higher specific activity. This preparation also showed increased tolerance to higher concentrations of substrate in the resolution of *p*-nitrostyrene oxide. The reaction proceeded with acceptable enantioselectivity (E = 41) and up to 20 % DMSO could be added without significant loss of activity (Nellaiah et al., 1996; Morisseau et al., 1997). Lowering the temperature from 27 °C to 4 °C increased the enantioselectivity 13-fold to E = 260 in the resolution of *p*-bromo-α-methyl styrene oxide (Cleij et al., 1998). A few more substrates hydrolyzed with high enantioselectivity by *Aspergillus niger* are shown in Fig. 195.

Epoxide hydrolase activity was also discovered in *Beauveria sulfurescens* ATCC7159, which converts styrene oxide (Fig. 196) as well as two-membered ring epoxides with high enantioselectivity (Pedragosa-Moreau et al., 1996a).

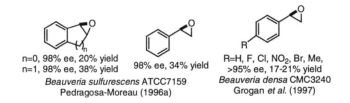

n=0, 98% ee, 20% yield
n=1, 98% ee, 38% yield   98% ee, 34% yield
Beauveria sulfurescens ATCC7159
Pedragosa-Moreau (1996a)

R=H, F, Cl, NO₂, Br, Me,
>95% ee, 17-21% yield
Beauveria densa CMC3240
Grogan *et al.* (1997)

Fig. 196. Examples for the resolution of epoxides using EH from *Beauveria* sp.

Interestingly, *Aspergillus niger* and *Beauveria sulfurescens* produced the (*R*)-diol in the hydrolysis of styrene oxide. Thus, the reaction catalyzed by *A. niger* proceeded with retention of configuration (via attack at C-2), whereas the hydrolysis with *B. sulfurescens* occurred with inversion of configuration (via attack at C-1, benzylic position). Employing a mixture of both organisms permitted the enantioconvergent synthesis of (*R*)-1-phenyl-1,2-dihydroxyethane in 92 % yield and 89 %ee (Fig. 197).

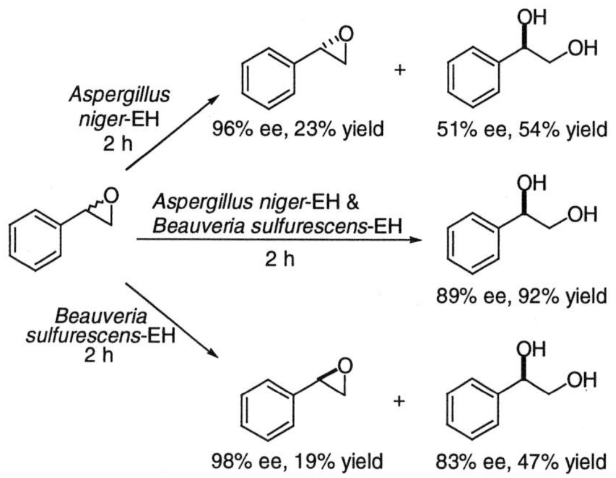

Fig. 197. Resolution of styrene oxide using fungal epoxide hydrolases from *Aspergillus niger* or *Beauveria sulfurescens* or a mixture of both for an enantioconvergent synthesis (Pedragosa-Moreau et al., 1993, 1996c).

In search for an enantioselective EH capable to resolve indene oxide, a precursor to the side chain of HIV protease inhibitor MK639, researchers at Merck found that out of 80 fungal strains investigated, *Diploida gossipina* ATCC16391 and *Lasiodiploida theobromae* MF5215 showed excellent enantioselectivity yielding exclusively the desired (1*S*,2*R*)-enantiomer. Two other strains from *Gilmaniella humicola* MF5363 and from *Altenaria enius* MF4352 showed opposite enantiopreference. Preparative biotransformation using whole cells of *Diploida gossipina* ATCC16391 allowed isolation of optically pure (1*S*,2*R*)-indene oxide in 14 % yield after 4 h reaction time (Fig. 198) (Zhang et al., 1995).

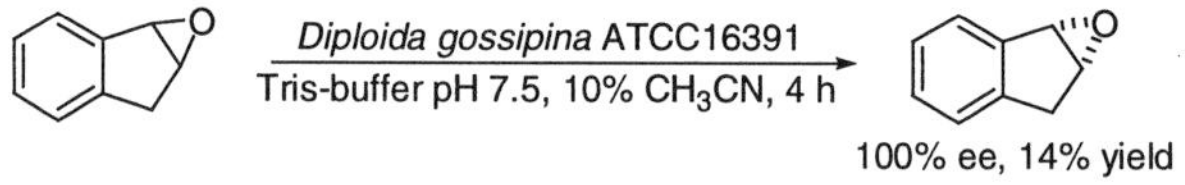

Fig. 198. Resolution of indene oxide catalyzed by EH from *Diploida gossipina* yields a HIV protease inhibitor precursor (Zhang et al., 1995).

Another useful and simple-to-grow strain is the yeast *Rhodotorula glutinis* which converts a wide range of aryl-, alkyl- and alicyclic epoxides with very high enantioselectivity. Best substrates are phenyl substituted epoxides (Weijers, 1997; Weijers et al., 1998) (Fig. 199). The same group recently discovered eight other yeast strains – out of 187 strains investigated – showing good to excellent enantioselectivity (E > 100) in the resolution of 1,2-epoxyoctane yielding the (*R*)-diol. Most stereoselective ones were from *Rhodotorula araucariae* CBS6031 and *Rhodosporidium toruloides* CBS0349 (Botes et al., 1998b).

rac-epoxides

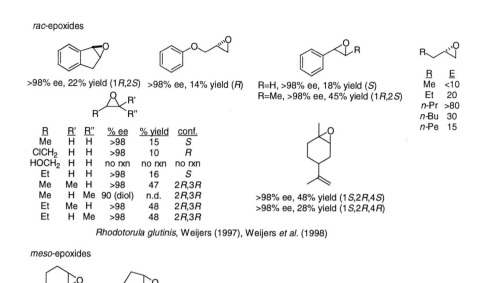

>98% ee, 22% yield (1*R*,2*S*)   >98% ee, 14% yield (*R*)

R=H, >98% ee, 18% yield (*S*)
R=Me, >98% ee, 45% yield (1*R*,2*S*)

R	E
Me	<10
Et	20
n-Pr	>80
n-Bu	30
n-Pe	15

R	R'	R"	% ee	% yield	conf.
Me	H	H	>98	15	*S*
ClCH₂	H	H	>98	10	*R*
HOCH₂	H	H	no rxn	no rxn	no rxn
Et	H	H	>98	16	*S*
Me	Me	H	>98	47	2*R*,3*R*
Me	H	Me	90 (diol)	n.d.	2*R*,3*R*
Et	Me	H	>98	48	2*R*,3*R*
Et	H	Me	>98	48	2*R*,3*R*

>98% ee, 48% yield (1*S*,2*R*,4*S*)
>98% ee, 28% yield (1*S*,2*R*,4*R*)

Rhodotorula glutinis, Weijers (1997), Weijers *et al.* (1998)

meso-epoxides

to *trans*-diol   to *trans*-diol
90% ee (1*R*,2*R*)   >98% ee (1*R*,2*R*)

Rhodotorula glutinis, Weijers (1997)

Fig. 199. Examples for the resolution of racemic and *meso*-epoxides using EH from *Rhodotorula glutinis*.

More examples on the application of microbial (see also Moussou et al., 1998c) as well as plant epoxide hydrolases can be found in a recent review (Archer, 1997).

11 Hydrolysis of Nitriles

11.1 Introduction

Although nitriles can also be hydrolyzed to the corresponding carboxylic acid by strong acid or base at high temperatures, nitrile hydrolyzing enzymes have the advantages that they require mild conditions (Sect. 11.2) and do not produce large amounts of by-products. In addition, during hydrolysis of dinitriles, they are often regioselective so that only one nitrile group is hydrolyzed (Sect. 11.3) and they can be used for the synthesis of optically active substances (Sect. 11.4).

The enzymatic hydrolysis of nitriles follows two different pathways (Fig. 200). Nitrilases (EC 3.5.5.1) directly catalyze the conversion of a nitrile into the corresponding acid plus ammonia. In the other pathway, a nitrile hydratase (NHase, EC 4.2.1.84; a lyase) catalyzes the hydration of a nitrile to the amide, which may be converted to the carboxylic acid and ammonia by an amidase (EC 3.5.1.4).

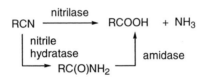

Fig. 200. Hydrolysis of nitriles follows two different pathways.

Both pathways occur in the biosynthesis of the phytohormone indole-3-acetic acid from indole-3-acetonitrile for a nitrilase from *Alcaligenes faecalis* JM3 (Kobayashi et al., 1993) and for the nitrile hydratase/amidase system in *Agrobacterium tumefaciens* and *Rhizobium* sp. (Kobayashi et al., 1995).

Pure nitrilases and nitrile hydratases are usually unstable so most researchers use them in whole cell preparations. Furthermore, the nitrile hydrolyzing activity must be induced first. Common inducers are benzonitrile, isovaleronitrile, crotononitrile, acetonitrile but the inexpensive inducer urea also works. In addition, inducing with ibuprofen or ketoprofen nitriles yields enantioselective enzymes (Layh et al., 1997). After induction, preparative conversion is usually performed by adding the nitriles either during cultivation or by employing resting cells. Most commonly used strains are from *Rhodococcus* sp. and the most important ones are subspecies of *R. rhodochrous*. Probably due to the low stability of isolated enzymes, only one biocatalyst was produced commercially by Novo under the trade names SP361 or SP409. Both contained nitrile hydratase and amidase activity derived from *Rhodococcus* sp. CH5 and were immobilized on ion-exchange resin, however are no more available.

Nitrilases are cysteine hydrolases which act via an enzyme-bound imine intermediate. All nitrilases are inactivated by thiol reagents (e.g., 5,5'-dithiobis(2-nitrobenzoic acid), indicating that they are sulfhydryl enzymes. This was further confirmed by site-directed mutagenesis replacing a conserved cysteine residue by an alanine resulting in a loss of nitrilase activity (Kobayashi and Shimizu, 1994).

Two different groups of nitrile hydratases have been described, which require Fe(III) or Co(III) ions (Sugiura et al., 1987; Nagasawa et al., 1991). NHase from *Pseudomonas chlororaphis* B23, *Brevibacterium* R312 and *Rhodococcus* sp. N-771 contain iron, the enzymes from *R. rhodochrous* J1 (Nagasawa et al., 1991) and from *Pseudomonas putida* NRRL 18668 (Payne et al., 1997) are cobalt-dependent (Shimizu et al., 1997). A NHase from *Agrobacterium tumefaciens* requires both metal ions (Kobayashi et al., 1995). In addition, the *R. rhodochrous* J1 strain produces two kinds of NHases differing in their molecular weight (520 kDa and 130 kDa). The high molecular-weight NHase acts preferentially on aliphatic nitriles, whereas the smaller enzyme also has high affinity toward aromatic nitriles (Wieser et al., 1999). Most nitrile hydratases accept only aliphatic nitriles (e.g., from *Arthrobacter* sp. J-1, *Brevibacterium* R312, *Pseudomonas chlororaphis* B23), however recently strains were described which can act on arylalkylnitriles, arylacetonitriles and heterocyclic nitrilases (Sect. 11.2).

Yamada's group showed by the use of ESR studies, that nitrile hydratases are non-heme ferric iron-containing enzymes. Further spectroscopic studies suggested that the enzyme also binds the cofactor pyrroloquinoline quinone (PQQ) leading to the proposed mechanism shown in Fig. 201 (Sugiura et al., 1987; Nagasawa et al., 1987).

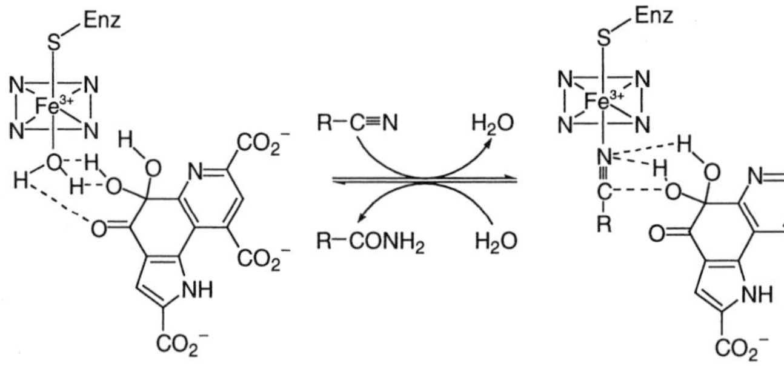

Fig. 201. Proposed mechanism of nitrile hydration catalyzed by a nitrile hydratase involves Fe(III) and PQQ (Sugiura et al., 1987; Nagasawa et al., 1987).

One subset of iron-containing NHases is activated by light in a reaction involving photodetachment of nitric oxide from an endogenous Fe–NO unit (Noguchi et al., 1996). Analysis of the crystal structure suggests that the non-heme iron catalytic center is in the nitrosylated state (Nagashima et al., 1998). In another work, the crystal structure of nitrile hydratase from *Rhodococcus* sp. R312 was resolved at 2.65 Å resolution (Huang et al., 1997). The enzyme is composed of two subunits (α,β) which contain one iron atom per α,β unit. This is in contrast to earlier work reporting a trimeric structure for this NHase, but consistent with the apparently identical enzyme from *Rhodococcus* N-771, which was described as a dimer. The α-subunit is composed of a long *N*-terminal and a *C*-terminal domain that forms a novel fold, which can be described as a four layered α-β-β-α structure. The two subunits form a tight heterodimer that is the functional unit of the enzyme. The active site is located in a cavity at a subunit-subunit interface. The iron center is formed by residues from the α subunit only – three cysteine thiolates and two main chain amide nitrogen atoms are ligands, although the iron center is located between the α and β subunits. Huang et al. (1997) proposed three possible catalytic roles for the metal ion in NHases. In all mechanisms, the metal ion acts as a Lewis acid activating the nitrile for hydration. Further information on structure, regulation and application of metallo nitrile hydratases can be found in a recent review (Kobayashi and Shimizu, 1998).

A single mutation (Gln19Glu) converted papain, a cysteine protease, into a nitrile hydrolase. It converted MeOCO-PheAla-CN to the corresponding amide, which was then hydrolyzed to the carboxylic acid by the amidase activity of papain (Dufour et al., 1995). The mutation places a proton donor in the oxyanion hole to protonate the nitrogen. This change increased the k_{cat}/K_m value for nitrile hydratase activity by at least 4×10^5 at pH 5. Although the wild-type shows some NHase activity, it requires 2-mercaptoethanol as an external nucleophile. The stereoselectivity of nitrile hydratase activity (WT or mutant) is lower than for ester hydrolysis.

Various NHase genes have been cloned and characterized (for reviews see Kobayashi et al., 1992; Shimizu et al., 1997) and it was found that the amidase gene is closely located to the NHase gene supporting the theory that both enzymes are involved in the two-step degradation of nitriles to carboxylic acids.

The properties and applications of nitrile-hydrolyzing enzymes are further summarized in a number of reviews (Kobayashi and Shimizu, 1994; Ohta, 1996; Sugai et al., 1997) or book chapters (Wyatt and Linton, 1988; Ingvorsen et al., 1988; Drauz and Waldmann, 1995; Shimizu et al., 1997; Faber, 1997; Bunch, 1998; Wieser and Nagasawa, 1999).

11.2   Mild Conditions

Two applications based on nitrile hydratase in *Rhodococcus rhodochrous* originally isolated by Yamada's group (Nagasawa and Yamada, 1995) have been commercialized. The large production of the commodity chemical acrylamide is performed by Nitto Chemical (Yokohama, Japan) on a > 30 000 metric tons per year scale. Initially, strains from *Rhodococcus* sp. N-774 or *Pseudomonas chlororaphis* B23 were used, however the current process uses the 10-fold more productive strain *Rhodococcus rhodochrous* J1. The productivity is > 7 000 g acrylamide per g cells at a conversion of acrylonitrile of 99.97 %. Formation of acrylic acid is barely detectable at the reaction temperature of 2–4 °C. In lab-scale experiments with resting cells up to 656 g acrylamide per liter reaction mixtures were achieved (Kobayashi et al., 1992).

One of the most important characteristics of the *R. rhodochrous* strain is its tolerance toward high concentrations of acrylamide (up to 50 %). Induction of NHase activity is performed using urea resulting in more than 50 % of high molecular weight NHase of the total soluble protein. Cobalt ions are essential to get active NHase. Besides acrylamide, also a wide range of other amides can be produced, e.g., acetamide (150 g/L), isobutyramide (100 g/L), methacrylamide (200 g/L), propionamide (560 g/L) and crotonamide (200 g/L) (Kobayashi et al., 1992).

Furthermore, *R. rhodochrous* J1 also accepts aromatic and arylaliphatic nitriles as substrates. For instance, also the conversion of 3-cyanopyridine to nicotinamide (a vitamin in animal feed supplementation) is catalyzed (Nagasawa et al., 1988), which was industrialized by Switzerland's Lonza on a 3 000 metric tons per year scale in their plant in Guangzhou, China. In contrast to the chemical process, no formation of the by-product nicotinic acid is observed using the nitrile hydratase system.

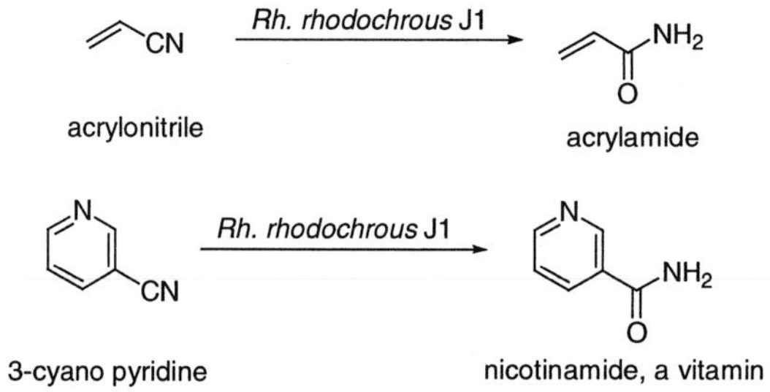

acrylonitrile   acrylamide

3-cyano pyridine   nicotinamide, a vitamin

Fig. 202. Commercial production of acrylamide and nicotinamide using resting cells of *Rhodococcus rhodochrous* J1.

Rhodococcus butanica (~*R. rhodochrous*) ATCC 21197 also catalyzed the mild hydrolysis of a variety of substituted benzonitriles, arylacetonitriles and α- or β-naphthylnitriles. Selected examples are shown in Fig. 203 (Kakeya et al., 1991b).

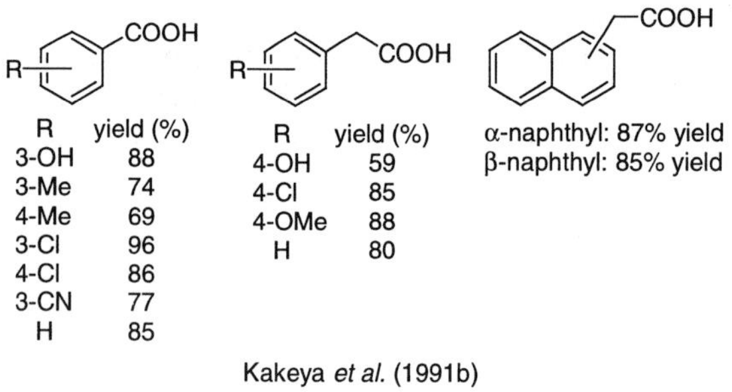

R	yield (%)
3-OH	88
3-Me	74
4-Me	69
3-Cl	96
4-Cl	86
3-CN	77
H	85

R	yield (%)
4-OH	59
4-Cl	85
4-OMe	88
H	80

α-naphthyl: 87% yield
β-naphthyl: 85% yield

Kakeya *et al.* (1991b)

Fig. 203. Acids obtained by mild hydrolysis of substituted aromatic nitriles using cells of *Rhodococcus butanica* ATCC 21197.

Griengl's group used the commercial biocatalyst SP409 (Novo) for the mild hydrolysis of a wide range of aliphatic or alicyclic nitriles either bearing other hydrolyzable groups or functionalities prone to side reactions such as elimination or aldol-type reactions (de Raadt et al., 1992; Klempier et al. 1991). In most cases high yields of acids could be achieved, selected examples are shown in Fig. 204. Note that no enantioselectivity was observed, when racemates were used and in a few cases also the corresponding amides were formed. α-Branched and crystalline substrates did not react.

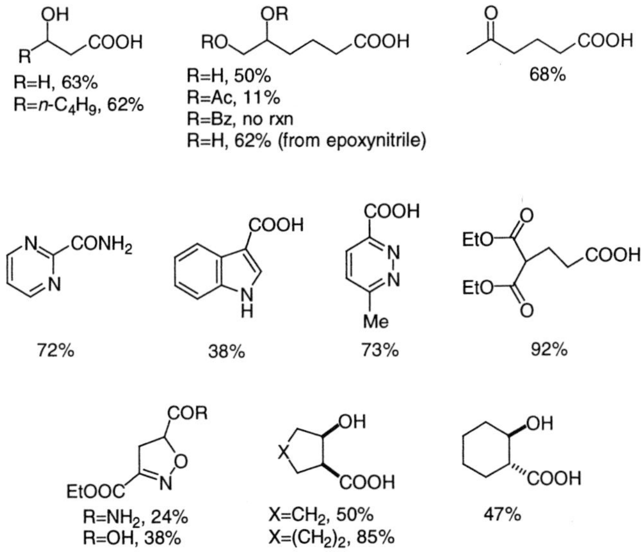

Fig. 204. Selected examples of carboxylic acids obtained by SP409-catalyzed hydrolysis of nitriles. Note that a 1,2-diol-carboxylic acid (R = H) was obtained from the epoxynitrile (top row, middle structure) and in one case the amide could be isolated in substantial amounts (last row, first structure) (Klempier et al., 1991; de Raadt et al., 1992).

Less examples can be found for the use of nitrilases, presumably due to insufficient stability. A nitrilase from *R. rhodochrous* J1 produces nicotinic acid (172 g/L), *p*-aminobenzoic acid (110 g/L), acrylic acid (390 g/L) and metacrylic acid (260 g/L) (Mathew et al., 1988; Nagasawa et al., 1990).

11.3 Regioselective Reactions of Dinitriles

A wide variety of dinitriles can be hydrolyzed with moderate to excellent regioselectivity to the corresponding monocarboxylic acids (Fig. 205). When using NHase from *Rhodococcus* sp. (SP409) monoamides where the major products. Note that poor or no regioselectivity was observed in the hydrolysis of non-fluorinated analogs of the substrates shown in Fig. 205 (last row) (Crosby et al., 1994).

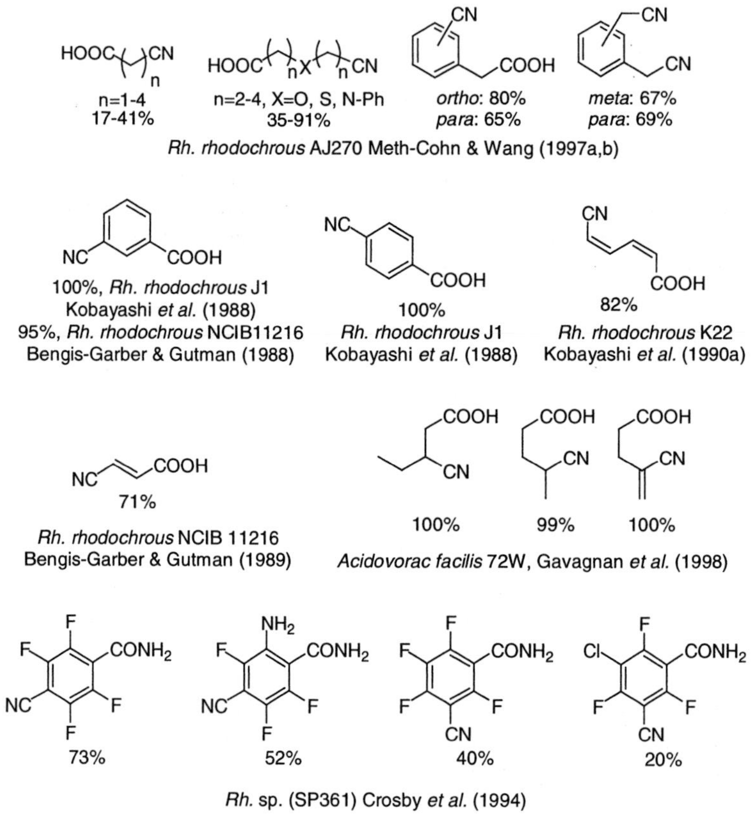
HOOC CN n
n=1-4
17-41%
HOOC n X n CN
n=2-4, X=O, S, N-Ph
35-91%
CN COOH
ortho: 80%
para: 65%
CN CN
meta: 67%
para: 69%
Rh. rhodochrous AJ270 Meth-Cohn & Wang (1997a,b)
NC COOH
100%, Rh. rhodochrous J1
Kobayashi et al. (1988)
95%, Rh. rhodochrous NCIB11216
Bengis-Garber & Gutman (1988)
NC COOH
100%
Rh. rhodochrous J1
Kobayashi et al. (1988)
CN COOH
82%
Rh. rhodochrous K22
Kobayashi et al. (1990a)
NC COOH
71%
Rh. rhodochrous NCIB 11216
Bengis-Garber & Gutman (1989)
COOH CN
100%
COOH CN
99%
COOH CN
100%
Acidovorac facilis 72W, Gavagnan et al. (1998)
F F CONH2 NC F F
73%
NH2 F CONH2 NC F F
52%
F F CONH2 F F CN
40%
F Cl CONH2 F F CN
20%
Rh. sp. (SP361) Crosby et al. (1994)

Fig. 205. Examples for the regioselectivity of NHases and nitrilases.

11.4   Enantioselective Reactions

In contrast to the situation with nitrilases and amidases, only indirect evidence was reported until recently about the enantioselectivity of nitrile hydratases. Recent work using resting cells from *Rhodococcus erythropolis* MP50, *Agrobacterium tumefaciens* d3 or *Pseudomonas putida* NRRL18668 indicates that indeed optically active amides can be formed from racemic nitriles, especially if either the nitrile hydratases are purified or a specific amidase inhibitor such as diethyl phosphoramidate is added (Bauer et al., 1994, 1998; Layh et al., 1994). Note that when the nitrile hydratase/amidase system is used, researchers found that the amidase sometimes shows opposite stereopreference (Ohta, 1996).

In many cases precursors of non-steroidal anti-inflammatory drugs (NSAID) (e.g., ketoprofen, ibuprofen or naproxen) were subjected to nitrile-hydrolase catalyzed kinetic resolutions (Fig. 206). Enantioselectivities differed greatly depending on the strain used and whether nitriles or amides were used as substrates. In addition, the enantioselectivity is influenced by both enzymes in the NHase/amidase system. For instance, in the resolution of naproxen (Fig. 206, top row, right structure) hydrolysis of the nitrile using *R. butanica* proceeded with E ~ 20-40, but the hydrolysis of the amide with only E ~ 10. Reactions in biphasic systems (phosphate buffer/hexane) or immobilization of cells gave no improvement. The enantioselectivity can be enhanced by increasing the reaction temperature (Bauer et al., 1998; Fallon et al., 1997).

The situation becomes more confusing, because some strains produce nitrilases, NHases and amidases, while others produce only nitrilases and in some cases the amidase is not active and the conversion of nitriles stops at the level of carboxylic amides. For example, *Rhodococcus* sp. C3II (Effenberger and Böhme, 1994) and *Rhodococcus equi* A4 (Martínková et al., 1996) contain no nitrilase, but *R. butanica* ATCC 21197 (renamed *R. rhodochrous* (Yokoyama et al., 1993)) exhibits nitrilase, nitrile hydratase and amidase activities. Here, the nitrilase preferentially hydrolyzes the (*S*)-nitrile to give the (*S*)-acid. The stepwise conversion of racemic nitriles proceeds via formation of the (*R*)-enantiomer by the nitrile hydratase, but faster hydrolysis of the (*S*)-amides by the amidase to yield (*R*)-amides in high optical purity (Kakeya et al., 1991b; Effenberger and Böhme, 1994). The strain *Acinetobacter* sp. AK226 produces only a nitrilase, because no amide could be detected as intermediate product (Yamamoto et al., 1990a; Yamamoto and Komatsu, 1991).

The first nitrile hydratase from a gram-negative organism was found in *Pseudomonas putida* NRRL 18668. The NHase yields (*S*)-amides, but the stereoselectivity primarily resides in the amidase (Fallon et al., 1997). A cobalt-containing nitrile hydratase with a very broad substrate spectrum is produced by *Agrobacterium tumefaciens* d3. To ensure that no amidase activity is present, the NHase was purified (Stolz et al., 1998; Bauer et al., 1998).

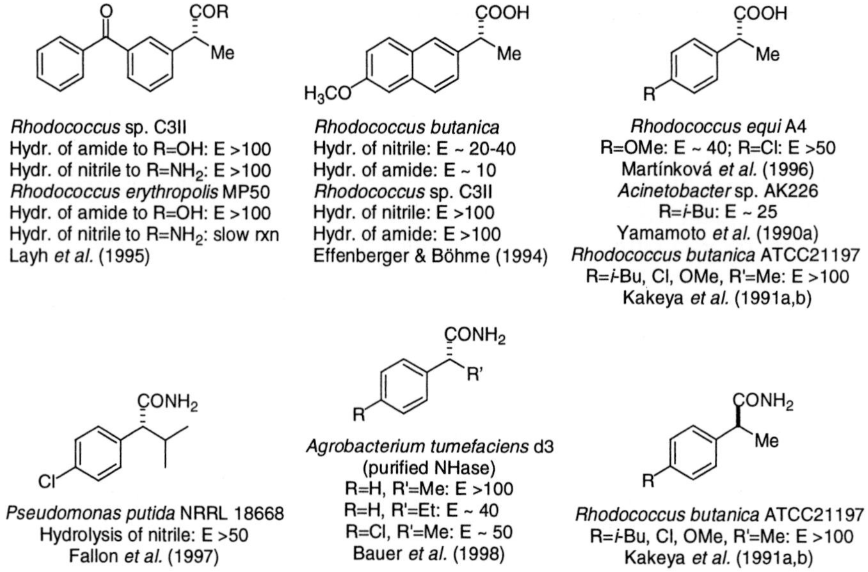
COR
Me
Rhodococcus sp. C3II
Hydr. of amide to R=OH: E >100
Hydr. of nitrile to R=NH2: E >100
Rhodococcus erythropolis MP50
Hydr. of amide to R=OH: E >100
Hydr. of nitrile to R=NH2: slow rxn
Layh et al. (1995)
COOH
Me
H3CO
Rhodococcus butanica
Hydr. of nitrile: E ~ 20-40
Hydr. of amide: E ~ 10
Rhodococcus sp. C3II
Hydr. of nitrile: E >100
Hydr. of amide: E >100
Effenberger & Böhme (1994)
COOH
Me
R
Rhodococcus equi A4
R=OMe: E ~ 40; R=Cl: E >50
Martínková et al. (1996)
Acinetobacter sp. AK226
R=i-Bu: E ~ 25
Yamamoto et al. (1990a)
Rhodococcus butanica ATCC21197
R=i-Bu, Cl, OMe, R'=Me: E >100
Kakeya et al. (1991a,b)
CONH2
Cl
Pseudomonas putida NRRL 18668
Hydrolysis of nitrile: E >50
Fallon et al. (1997)
CONH2
R'
R
Agrobacterium tumefaciens d3
(purified NHase)
R=H, R'=Me: E >100
R=H, R'=Et: E ~ 40
R=Cl, R'=Me: E ~ 50
Bauer et al. (1998)
CONH2
Me
R
Rhodococcus butanica ATCC21197
R=i-Bu, Cl, OMe, R'=Me: E >100
Kakeya et al. (1991a,b)

Fig. 206. Examples for the synthesis of non-steroidal anti-inflammatory drugs by hydrolysis of nitrile or amide precursors. Please note that purified NHase from *Agrobacterium tumefaciens* d3 showed opposite enantiopreference compared to *Rhodococcus butanica* ATCC21197 (last row).

Besides NSAID-precursors, only a few other nitriles were subjected to kinetic resolution (Fig. 207). Using the strain *Rhodococcus rhodochrous* NCIMB11216 (for properties see Hoyle et al., 1998) in the hydrolysis of 2-substituted aliphatic nitriles and aryl-aliphatic nitriles, high enantioselectivity was found only for 2-methylhexanitrile (Gradley et al., 1994; Gradley and Knowles, 1994).

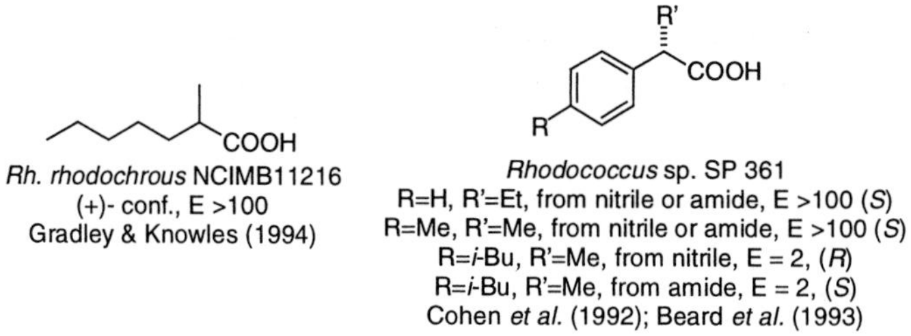
COOH
Rh. rhodochrous NCIMB11216
(+)- conf., E >100
Gradley & Knowles (1994)
R'
COOH
R
Rhodococcus sp. SP 361
R=H, R'=Et, from nitrile or amide, E >100 (S)
R=Me, R'=Me, from nitrile or amide, E >100 (S)
R=i-Bu, R'=Me, from nitrile, E = 2, (R)
R=i-Bu, R'=Me, from amide, E = 2, (S)
Cohen et al. (1992); Beard et al. (1993)

Fig. 207. Other substrates resolved by nitrile-hydrolyzing enzymes.

In the hydrolysis of arylacetonitriles using the Novo catalyst SP361 it was found that the first step to the amide is (*R*)-selective, but the hydrolysis to the acid is (*S*)-selective. Hydrolysis of the corresponding racemic amides revealed that only the amidase is highly enantioselective. Moreover, hydrolysis of 4-*i*-butyl nitrile gave the (*R*)-acid, but the reaction of the amide produced the (S)-acid (but at very low E) (Cohen et al., 1992; Beard et al., 1993).

The strain *Alcaligenes faecalis* ATCC8750 can be used for the efficient production of mandelic acid (Yamamoto et al., 1991). *A. faecalis* contains a nitrilase and an amidase, but no NHase. Using resting cells, (R)-(−)-mandelic acid was formed in excellent optical purity (100 %ee). Moreover, the remaining (S)-mandelonitrile racemized resulting in an overall yield of 91 % mandelonitrile after 6 h reaction time. It was suggested that the rapid racemization is due to an equilibrium between mandelonitrile and benzalde-hyde/HCN, because mandelic acid could also be obtained when only benzaldehyde and HCN are used as substrates (Fig. 208). After partial purification, nitrilase and amidase were investigated separately and it was found that the nitrilase is highly enantioselective and active, but the amidase is not (24 %ee for mandelic acid in the hydration of the amide). Hydrolysis of *O*-acetylmandelonitrile by resting cells of different *Pseudomonas* strains proceeded with lower enantioselectivity (Layh et al., 1992).

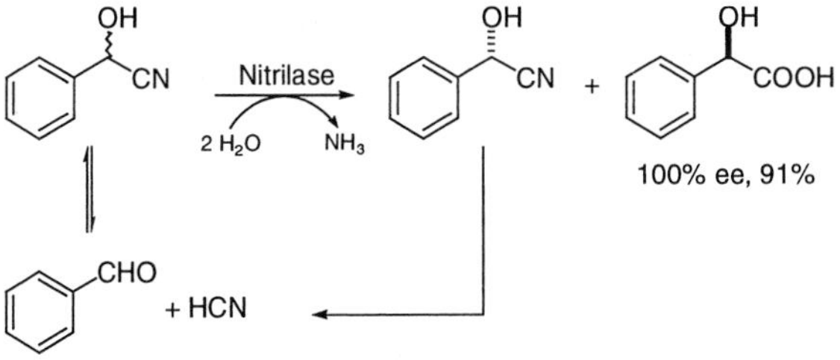

Fig. 208. Synthesis of (R)-(−)-mandelic acid by a nitrilase present in *A. faecalis* resting cells involves *in situ* recycling of (S)-mandelonitrile by disproportion in benzaldehyde and HCN followed by formation of (R,S)-mandelonitrile (Yamamoto et al., 1991, 1992).

A few groups investigated the hydrolysis of prochiral dinitriles (Fig. 209). Hydrolysis of 3-hydroxyglutaronitriles revealed that only with Bn- or Bz-proctecting groups acceptable enantiomeric excess for the monocarboxylic acid could be obtained. Using SP361, the dinitrile is hydrolyzed initially by a *pro*-(S)-selective NHase followed by a fast, non-selective amidase-catalyzed reaction (Crosby et al., 1992).

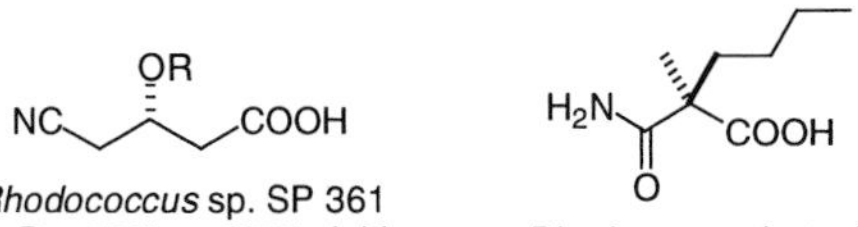

Rhodococcus sp. SP 361
R=Bn, 83% ee, 73% yield
R=Bz, 84% ee, 25% yield
Crosby *et al.* (1992)
Rhodococcus butanica ATCC21197
R=Bn, 90% ee, 68% yield
R=Bz, 99% ee, 71% yield
Kakeya *et al.* (1991a)

Rhodococcus butanica
ATCC21197, 96% ee, 92% yield
Yokoyama et al. (1993)

Fig. 209. Resolution of prochiral dinitriles proceeds with good enantioselectivity.

In contrast, hydrolysis of a disubstituted malononitrile with *R. butanica* resulted in the formation of 2-carbamoyl-2-methylhexanoic acid in high purity and yield. The reaction is presumably catalyzed by a fast non-selective nitrilhydratase forming the prochiral diamide followed by slow *pro*-(*R*)-enantiotopic selective hydrolysis catalyzed by an amidase (Yokoyama et al., 1993).

12 Glycosidases

Glycosidases (EC 3.2) are cofactor-independent enzymes catalyzing the hydrolysis of glycosidic bonds. Glycosidases also catalyze the reverse reaction, formation of a glycosidic link, and this ability is their most important synthetic application. Major advantages of glycosidase-catalyzed glycosyl transfers are that there is no – or minimal need – for protection and that the stereochemistry at the newly formed anomeric center can be controlled by the use of either α- or β-glycosidases.

Most glycosidases are specific for both the glycosyl moiety and the nature of the glycosidic linkage, e.g., α-D-galactosidase only accepts derivatives of α-D-galactose (Fig. 212). However, most glycosidases show low specificity toward the leaving group. This is usally a sugar, but can also be a primary or secondary alcohol. The regioselectivity towards the nucleophile is also low. Thus, a synthesis involving a sugar nucleophile may give a mixture where different hydroxyl groups reacted.

Studies using β-glucosidase from *Agrobacterium* revealed that catalysis proceeds via a double displacement. A glucosyl-enzyme intermediate forms by displacing the leaving group, then a nucleophile displaces the enzyme. The overall reaction is a net retention of the configuration at the anomeric center. Kinetic analysis identified Glu170 as active site residue, because replacement with Gly resulted in a 10^6-fold drop in activity. Surprisingly, activity of the Glu170Gly-mutant could be restored by using carboxyphenyl-β-D-glycosides bearing the missing acid catalytic group (Wang and Withers, 1995).

Glycosides can be synthesized by either thermodynamic or kinetic control – similar to the formation of peptide bonds (Sect. 4.2.5) – of the reaction (Fig. 210). For thermodynamic control, researchers shift the reaction equilibrium toward synthesis by increasing the concentrations of donor and nucleophile. Since the free energy of hydrolysis of a glycosyl link (–3.8 kcal/mol) is large, the required concentrations are quite high usually yielding a thick syrup. The concentration of water cannot be reduced below 10 % because glycosidases requires > 10 % water to remain active. Yields of glycoside from the thermodynamic approach are usually below 15 %. By comparison, the free energy of hydrolysis of a peptide is smaller (–2.2 kcal/mol) so the thermodynamic approach is more successful.

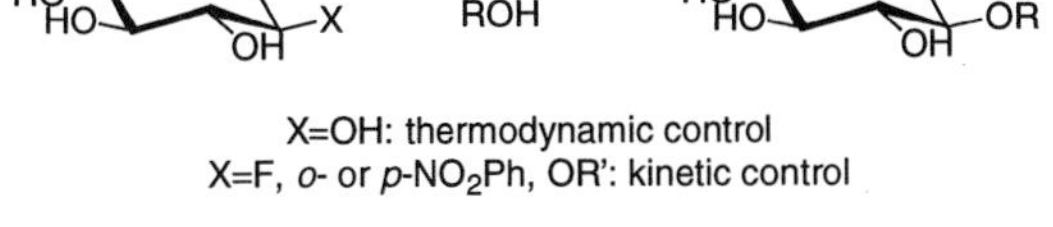

X=OH: thermodynamic control
X=F, *o*- or *p*-NO$_2$Ph, OR': kinetic control

Fig. 210. Principle of glycoside synthesis using glycosidases. For thermodynamic control, researchers shift the equilibrium toward synthesis by increasing the concentrations of donor or nucleophile. For kinetic control researchers use an activated donor and stop the reaction before hydrolysis of product takes place.

Moderate yields (21 %) could be obtained by a combination of ß-glucosidase-catalyzed synthesis of *n*-butyl glucose under thermodynamic control followed by lipase-catalyzed acylation using 4-phenylbutyric acid (Fig. 211). However, with *n*-octanol yields of *n*-octyl glucose were < 10 % (Otto et al., 1998b). Yields of *n*-alkyl-glycosides are also affected by water activity (Chahid et al., 1992).

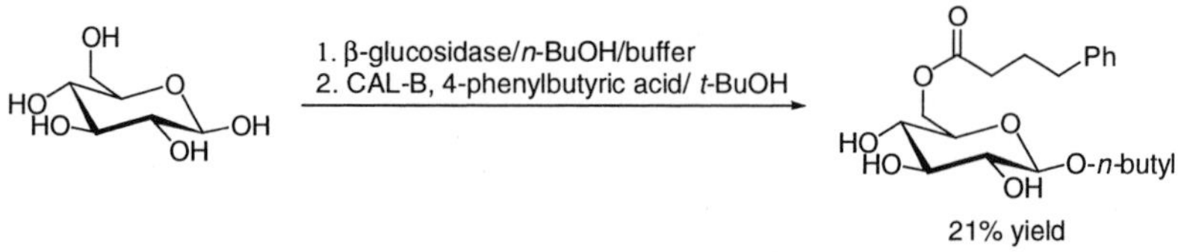

Fig. 211. Synthesis of 1-*n*-butyl-6-(4'-phenylbutyryl)-β-D-glucose in a coupled β-glucosidase and lipase reaction.

The alternative kinetically-controlled synthesis starts from an activated glycoside (e.g., fluoride, *o*- or *p*-nitrophenyl derivative, azides) which reacts with the nucleophile (Fig. 212). This transglycosidation gives acceptable yields but requires activated donors. Two competing reactions are hydrolysis of the glycosyl donor and hydrolysis of the product glycoside. For instance, Gopalan et al. (1992) found that the rate of alcoholysis is only 24-fold greater than the rate of competing hydrolysis in the formation of octyl-β-D-glycoside from octanol and D-glucose using guinea pig liver cytosolic β-glucosidase.

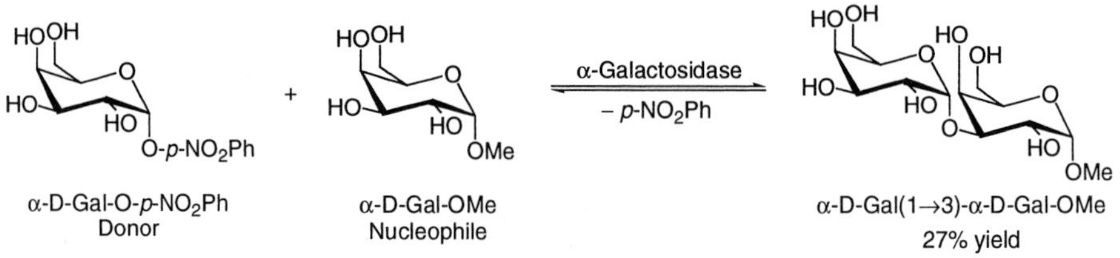

Fig. 212. Kinetically-controlled synthesis of α-D-Gal(1→3)-α-D-Gal-OMe using α-D-galactosidase from coffee beans (Nilsson, 1987).

Despite these difficults, glycosidases are often preferred over glycosyl transferases, which require not readily available phosphorylated glycosyl donors (e.g., uridine diphosphate, UDP) and are less available than glycosidases. The application of glycosyl transferases is beyond the scope of this book and readers are referred to a number of books (Wong and Whitesides, 1994; Drauz and Waldmann, 1995; Faber, 1997).

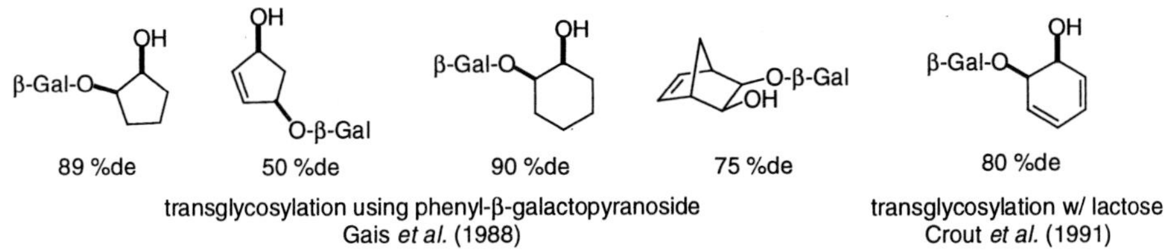
β-Gal-O
OH
OH
O-β-Gal
β-Gal-O
OH
O-β-Gal
OH
β-Gal-O
OH
89 %de
50 %de
90 %de
75 %de
80 %de
transglycosylation using phenyl-β-galactopyranoside
Gais et al. (1988)
transglycosylation w/ lactose
Crout et al. (1991)

Fig. 213. Examples for the asymmetrization of *meso*-diols using β-galactosidase from *E. coli*.

Beside the synthesis of natural glycosides, glycosidases have been also used to obtain sugar derivatives based on non-natural alcohols (Huber et al., 1984; Ooi et al., 1985). In some cases, also the desymmetrization of *meso*-diols (Fig. 213) or kinetic resolution of racemic alcohols (Fig. 214) was possible. However, in general enantioselectivity of glycosidases is to low to obtain optically pure compounds. Hydrolysis of β-galactopyranoside derivatives bearing racemic alcohol moieties proceeded with e.g., E = 7 for the isopropylidene glycerol derivative (Werschkun et al., 1995) and similar trends were observed in the reverse reaction using related substrates (Crout et al., 1990).

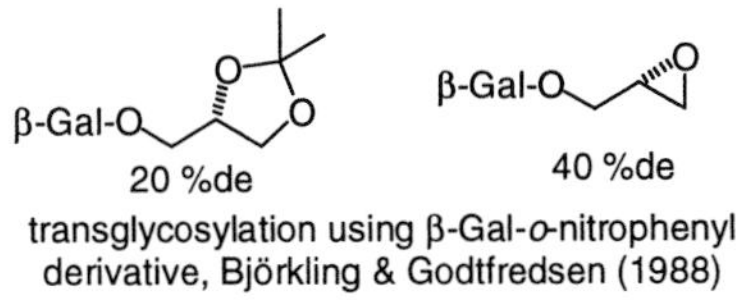
β-Gal-O
O
O
β-Gal-O
O
20 %de
40 %de
transglycosylation using β-Gal-o-nitrophenyl
derivative, Björkling & Godtfredsen (1988)

Fig. 214. Examples for the resolution of racemic alcohols using β-galactosidase from *E. coli*.

Further examples on the use of glycosidases – especially for the synthesis of oligosaccharides – can be found in recent publications (Hashimoto et al., 1995; Vic et al., 1996) or books (Wong and Whitesides, 1994; Drauz and Waldmann, 1995; Faber, 1997).

Abbreviations

7-ACA	7-Aminocephalosporanic acid	DNA	desoxyribonucleic acid
AChE	acetylcholine esterase	DTNB	5,5-dithio-bis-(2-nitrobenzoic acid
AGE	*Arthrobacter globiformis* esterase		
Alloc	allyloxycarbonyl	E	enantioselectivity / enantiomeric ratio
ANL	*Aspergillus niger* lipase		
6-APA	6-aminopenicillanic acid	*E.*	*Escherichia*
AOT	bis(2-ethylhexyl)sodium sulfo-succinate	ee	enantiomeric excess
		EH	epoxide hydrolase
Ar	aryl	EPA	eicosapentaenoic acid
ATTC	American Type Culture Collection (Manassas, Virginia)	Eq.	equation
		ESR	electron spin resonance
a_w	water activity	Et	ethyl
BCE	*Bacillus coagulans* esterase	Fmoc	9-fluorenylmethoxycarbonyl
BChE	butyrylcholine esterase	GC	gas chromatography
Bn	benzyl	GCL	*Geotrichum candidum* lipase
Boc	*t*-butyloxycarbonyl	HEPES	*N*-(2-hydroxyethyl)-piperazine-*N'*-ethanesulfonic acid
bp	base pair(s), boiling point		
BTL2	*Bacillus thermocatenulatus* lipase	HLE	horse liver esterase
Bu	butyl	HLL	*Humicola lanuginosa* lipase
But	butyryl	HPLC	high performance liquid chromatography
Bz	benzoyl		
c	conversion	Hx	hexyl
CAL	*Candida antarctica* lipase	IPG	isopropylidene glycerol, solketal
Cbz	*N*-carbobenzyloxy	kDa	kilo Dalton
CE	cholesterol esterase	Me	methyl
cEH	cytosolic epoxide hydrolase	MEE	(methoxyethoxy)ethyl
cHX	cyclohexyl	mEH	microsomal epoxide hydrolase
CLE	chicken liver esterase	MG	monoglyceride
CLEC	Cross-linked enzyme crystals	MLM	M = medium, L = long chain fatty acid in triglycerides
CLL	*Candida lipolytica* lipase		
CRL	*Candida rugosa* lipase	MOM	methoxymethyl
CT	Chymotrypsin	MPGM	*trans*-3-(4-methoxy- phenyl)-glycidic acid methylester
CTAB	cetyltrimethyl ammonium bromide		
		MTBE	methyl-*t*-butylether
CVL	*Chromobacterium viscosum* lipase	NCIMB	National Collections of Industrial and Marine Bacteria (Aberdeen, Scotland)
DG	diglyceride		
DHA	docosahexaenoic acid	NHase	nitrile hydratase
DMF	*N,N'*-dimethylformamide		
DMSO	dimethylsulfoxid		

NRRL	Northern Regional Research Laboratory, now called National Center For Agricultural Utilization Research (Peoria, Illinois)	PPE	*Pseudomonas putida* esterase
		PPL	porcine pancreatic lipase
		PQQ	pyrroloquinoline quinone
		Pr	Propyl
NSAID	non-steroidal anti-inflammatory drugs	ProqL	*Penicillium roquefortii* lipase
		pTol	p-toluyl
PAE	*Pseudomonas aeruginosa* esterase	PUFA	polyunsaturated fatty acid
		RDL	*Rhizopus delemar* lipase
PAGE	polyacrylamide gelelectrophoresis	RJL	*Rhizopus javanicus* lipase
		RLE	rabbit liver esterase
PBA	phenylboronic acid	RML	*Rhizomucor miehei* lipase
PC	phosphatidyl choline	RNL	*Rhizopus niveus* lipase
PcamL	*Penicillium camembertii* lipase	ROL	*Rhizopus oryzae* lipase
PCL	*Pseudomonas cepacia* lipase	SCCO$_2$	supercritical carbon dioxide
PCR	polymerase chain reaction	SDS	sodium dodecylsulfate
pdb	brookhaven data base	sEH	soluble epoxide hydrolase
PEG	polyethylene glycol	*sn*	stereochemical numbering
PFE	*Pseudomonas fluorescens* esterase	ST	structured triglyceride
		StEP	staggered extension process
PFL	*Pseudomonas fluorescens* lipase	*t.*	tertiary
PfragiL	*Pseudomonas flragi* lipase	TFEA	2,2,2-trifluoroethyl acetate
PGA	Penicillin G amidase	TG	triglyceride
PL	phospholipase	THF	tetrahydrofuran
PLE	pig liver esterase	TLC	thin layer chromatography
PME	*Pseudomonas marginata* esterase	Tr	trityl
PMSF	phenylmethylsulfonyl fluoride	U	Units
pNPA	p-nitrophenyl acetate	Z	t-butyloxycarbonyl
pNPP	p-nitrophenyl palmitate		

References

Aaltonen, O., Rantakylae, M. (1991), Biocatalysis in supercritical carbon dioxide, *Chemtech.* **21**, 240-248.

Adachi, K., Kobayashi, S., Ohno, M. (1986), Chiral synthons by enantioselective hydrolysis of *meso*-diesters with pig liver esterase: substrate-stereoselectivity relationships, *Chimia* **40**, 311-314.

Adachi, T., Ishii, M., Ohta, Y., Ota, T., Ogawa, T., Hanada, K. (1993), Chemoenzymatic synthesis of optically active 1,4-dihydropyridine derivatives via enantioselective hydrolysis and trans-esterification, *Tetrahedron: Asymmetry* **4**, 2061-2068.

Adam, W., Diaz, M. T., Fell, R. T., Sahamoller, C. R. (1996), Kinetic resolution of racemic α-hydroxy ketones by lipase-catalyzed irreversible transesterification, *Tetrahedron: Asymmetry* **7**, 2207-2210.

Adam, W., Groer, P., Saha-Moeller, C. R. (1997a), Enzymic preparation of optically active α-methylene β-lactones by lipase-catalyzed kinetic resolution through asymmetric transesterification, *Tetrahedron: Asymmetry* **8**, 833-836.

Adam, W., Mock-Knoblauch, C., Saha-Möller, C. R. (1997b), Kinetic resolution of hydroxy vinylsilanes by lipase-catalyzed enantioselective acetylation, *Tetrahedron: Asymmetry* **8**, 1441-1444.

Adamczyk, M., Gebler, J. C., Mattingly, P. G. (1994), Lipase mediated hydrolysis of Rapamycin 42-hemisuccinate benzyl and methyl esters, *Tetrahedron Lett.* **35**, 1019-1022.

Adamczyk, M., Gebler, J. C., Grote, J. (1995), Chemo-enzymic transformations of sensitive systems. Preparation of digoxigenin haptens via regioselective lipase mediated hydrolysis, *Tetrahedron Lett.* **36**, 6987-6990.

Adamczyk, M., Grote, J. (1996), *Pseudomonas cepacia* lipase mediated amidation of benzyl esters, *Tetrahedron Lett.* **37**, 7913-7916.

Adelhorst, K., Björkling, F., Godtfredsen, S. E., Kirk, O. (1990), Enzyme catalysed preparation of 6-*O*-acylglucopyranosides, *Synthesis*, 112-115.

Ader, U., Schneider, M. P. (1992), Enzyme assisted preparation of enantiomerically pure β-andrenergic blockers III. Optically active chlorohydrin derivatives and their conversion, *Tetrahedron: Asymmetry* **3**, 521-524.

Ader, U., Andersch, P., Berger, M., Goergens, U., Haase, B., Hermann, J., Laumen, K., Seemayer, R., Waldinger, C., Schneider, M. P. (1997), Screening techniques for lipase catalyst selection, *Methods Enzymol.* **286**, 351-386.

Adjé, N., Breuilles, P., Uguen, D. (1993), Desymmetrisation of *meso* propargylic diols, *Tetrahedron Lett.* **34**, 4631-4634.

Adlercreutz, P. (1991), On the importance of the support material for enzymatic synthesis in organic media. Support effects at controlled water activity, *Eur. J. Biochem.* **199**, 609-614.

Adlercreutz, P. (1994), Enzyme-catalyzed lipid modification, *Biotechnol. Genet. Eng. Rev.* **12**, 231-254.

Adshiri, T., Akiya, H., Chin, L. C., Arai, K., Fujimoto, K. (1992), Lipase-catalyzed interesterification of triglyceride with supercritical carbon dioxide extraction, *J. Chem. Eng. Jpn.* **25**, 104-105.

Affleck, R., Xu, Z. F., Suzawa, V., Focht, K., Clark, D. S., Dordick, J. S. (1992), Enzymatic catalysis and dynamics in low-water environments, *Proc. Natl. Acad. Sci. USA* **89**, 1100-1104.

Ahmed, S. N., Kazlauskas, R. J., Morinville, A. H., Grochulski, P., Schrag, J. D., Cygler, M. (1994), Enantioselectivity of *Candida rugosa* lipase toward carboxylic acids: a predictive rule from substrate mapping and X-ray crystallography, *Biocatalysis* **9**, 209-225.

Ajima, A., Yoshimoto, T., K. Takahashi, Tamaura, Y., Saito, Y., Inada, Y. (1985), Polymerization of 10-hydroxydecanoic acid in benzene with polyethylene glycol-modified lipase, *Biotechnol. Lett.* **7**, 303-306.

Akai, S., Naka, T., Takebe, Y., Kita, Y. (1997), Enzyme-catalyzed asymmetrization of 2,2-disubstituted 1,3-propanediols using 1-ethoxyvinyl esters, *Tetrahedron Lett.* **38**, 4243-4246.

Akita, H. (1996), Recent advances in the use of immobilized lipases directed toward the asymmetric syntheses of complex molecules, *Biocatal. Biotransform.* **13**, 141-156.

Akita, H., Umezawa, I., Matsukura, H., Oishi, T. (1991), A lipid- lipase aggregate as a new type of immobilized enzyme, *Chem. Pharm. Bull.* **39**, 1632-1633.

Akita, H., Umezawa, I., Tisnadjaja, D., Matsukura, H., Oishi, T. (1993), Enantioselective acetylation of an α-hydroxy ester by using ether-linked lipid- lipase aggregates in organic solvents, *Chem. Pharm. Bull.* **41**, 16-20.

Akita, H., Nozawa, M., Futagami, Y., Miyamoto, M., Saotome, C. (1997a), Simple approach to optically active drimane sesquiterpenes based on enzymatic resolution, *Chem. Pharm. Bull.* **45**, 824-831.

Akita, H., Umezawa, I., Matsukura, H. (1997b), Enzymatic hydrolysis in organic solvents for kinetic resolution of water-insoluble α-acyloxy esters with immobilized lipases, *Chem. Pharm. Bull.* **45**, 272-278.

Akoh, C. C. (1993), Lipase-catalyzed synthesis of partial glycerides, *Biotechnol. Lett.* **15**, 949-954.

Akoh, C. C. (1995), Structured lipids - enzymatic approach, *Inform* **6**, 1055-1061.

Akoh, C. C., Cooper, C., Nwosu, C. V. (1992), Lipase G-catalyzed synthesis of monoglycerides in organic solvent and analysis by HPLC, *J. Am. Oil Chem. Soc.* **69**, 257-260.

Akoh, C. C., Jennings, B. H., Lillard, D. A. (1995), Enzymic modification of trilinolein: incorporation of n-3 polyunsaturated fatty acids, *J. Am. Oil Chem. Soc.* **72**, 1317-1321.

Alberghina, L., Schmid, R. D., Verger, R. (Eds.) (1991), *Lipases: Structure, Mechanism and Genetic Engineering, GBF Monographs* **Vol. 16**, Weinheim: Wiley-VCH.

Alcock, N. W., Crout, D. H. G., Henderson, C. M., Thomas, S. E. (1988), Enzymatic resolution of a chiral organometallic ester - enantioselective hydrolysis of 2-ethoxycarbonylbuta-1,3-dienetricarbonyliron by pig liver esterase, *J. Chem. Soc., Chem. Commun.*, 746-747.

Alfonso, I., Astorga, C., Rebolledo, F., Gotor, V. (1996), Sequential biocatalytic resolution of (±)-*trans*-cyclohexane-1,2- diamine. Chemoenzymic synthesis of an optically active polyamine, *Chem. Commun.*, 2471-2472.

Allen, R. H., Jacoby, W. B. (1969), Tartaric acid metabolism. IX. Synthesis with tartrate epoxidase, *J. Biol. Chem.* **244**, 2078-2084.

Allen, J. V., Williams, J. M. J. (1996), Dynamic kinetic resolution with enzyme and palladium combinations, *Tetrahedron Lett.* **37**, 1859-1862.

Allenmark, S., Ohlsson, A. (1992a), Enantioselectivity of lipase-catalyzed hydrolysis of some 2-chloroethyl 2-arylpropanoates studied by chiral reversed-phase liquid chromatography, *Chirality* **4**, 98-102.

Allenmark, S., Ohlsson, A. (1992b), Studies of the heterogeneity of a *Candida cylindracea (rugosa)* lipase: monitoring of esterolytic activity and enantioselectivity by chiral liquid chromatography, *Biocatalysis* **6**, 211-221.

Allenmark, S. G., Andersson, A. C. (1993), Lipase-catalyzed kinetic resolution of a series of esters having a sulfoxide group as the stereogenic center, *Tetrahedron: Asymmetry* **4**, 2371-2376.

Allevi, P., Anastasia, M., Cajone, F., Ciuffreda, P., Sanvito, A. M. (1993), Enzymatic resolution of (*R*)- and (*S*)-(*E*)-4-hydroxyalk-2-enals related to lipid peroxidation., *J. Org. Chem.* **58**, 5000-5002.

Allevi, P., Ciuffreda, P., Tarocco, G., Anastasia, M. (1996), Enzymatic resolution of (*R*)- and (*S*)-2-(1-hydroxyalkyl)thiazoles, synthetic equivalents of (*R*)- and (*S*)-2-hydroxy aldehydes, *J. Org. Chem.* **61**, 4144-4147.

Ampon, K., Basri, M., Salleh, A. B., Yunus, W. M. Z. W., Razak, C. N. A. (1994), Immobilization by adsorption of hydrophobic lipase derivatives to porous polymer beads for use in ester synthesis, *Biocatalysis* **10**, 341-351.

Anantharamaiah, G. M., Roeske, R. W. (1982), Resolution of α-methyl amino esters by chymotrypsin, *Tetrahedron Lett.* **23**, 3335-3336.

Andersch, P., Schneider, M. P. (1993), Enzyme assisted synthesis of enantiomerically pure myo-inositol derivatives. Chiral building blocks for inositol polyphosphates, *Tetrahedron: Asymmetry* **4**, 2135-2138.

Anderson, E. M., Larsson, K. M., Kirk, O. (1998), One biocatalyst - many applications: The use of *Candida antarctica* B-lipase in organic synthesis, *Biocatal. Biotransform.* **16**, 181-204.

Anthonsen, H. W., Hoff, B. H., Anthonsen, T. (1995), A simple method for calculating enantiomer ratio and equilibrium constants in biocatalytic resolutions, *Tetrahedron: Asymmetry* **6**, 3015-3022.

Aouf, N. E., Djerourou, A. H., Blanco, L. (1994), Preparation et essais de dedoublement enzymatique d'acetoxymethyl- et hydroxymethylsilanes chiraux, *Phosphorus, Sulfur, and Silicon* **88**, 207-215.

Aragozzini, F., Valenti, M., Santaniello, E., Ferraboschi, P., Grisenti, P. (1992), Biocatalytic, enantioselective preparations of (R)- and (S)-ethyl 4-chloro-3-hydroxybutanoate, a useful chiral synthon, *Biocatalysis* **5**, 325-332.

Arand, M., Wagner, H., Oesch, F. (1996), Asp[333], Asp[495], and His[523] from the catalytic triad of rat soluble epoxide hydrolase, *J. Biol. Chem.* **271**, 4223-4229.

Arand, M., Müller, F., Mecky, A., Hinz, W., Urban, P., Pompon, D., Kellner, R., Oesch, F. (1999), Catalytic triad of microsomal epoxide hydrolase: replacement of Glu[404] with Asp leads to a strongly increased turnover rate, *Biochem. J.* **337**, 37-43.

Archelas, A. (1998), Epoxide hydrolases: new tools for the synthesis of enantiopure epoxides and diols, *J. Mol. Catal. B: Enzymatic* **5**, 79-85.

Archelas, A., Furstoss, R. (1998), Epoxide hydrolases: new tools for the synthesis of fine chemicals, *Trends Biotechnol.* **16**, 108-116.

Archer, I. V. J. (1997), Epoxide hydrolases as asymmetric catalysts, *Tetrahedron* **53**, 15617-15662.

Archer, I. V. J., Leak, D. J., Widdowson, D. A. (1996), Chemoenzymic resolution and deracemization of (±)-1-methyl-1,2-epoxycyclohexane: the synthesis of (1-*S*,2-*S*)-1-methylcyclohexane-1,2-diol, *Tetrahedron Lett.* **37**, 8819-8822.

Arita, M., Adachi, K., Ito, Y., Sawai, H., Ohno, M. (1983), Enantioselective synthesis of the carbocyclic nucleosides (-)-aristeromycin and (-)-neplanocin A by a chemicoenzymatic approach, *J. Am. Chem. Soc.* **105**, 4049-4055.

Arnold, F. H. (1998), Design by directed evolution, *Acc. Chem. Res.* **31**, 125-131.

Arnold, F. H., Moore, J. C. (1997), Optimizing industrial enzymes by directed evolution, *Adv. Biochem. Eng./Biotechnol.* **58**, 1-14.

Arroyo, M., Sinisterra, J. V. (1994), High enantioselective esterification of 2-arylpropionic acids catalyzed by immobilized lipase from *Candida antarctica*: A mechanistic approach, *J. Org. Chem.* **59**, 4410-4417.

Arroyo, M., Sinisterra, J. V. (1995), Influence of chiral carvones on selectivity of pure lipase-B from *Candida antarctica*, *Biotechnol. Lett.* **17**, 525-530.

Astorga, C., Rebolledo, F., Gotor, V. (1991), Synthesis of hydrazides through an enzymatic hydrazinolysis reaction, *Synthesis*, 350-352.

Astorga, C., Rebolledo, F., Gotor, V. (1993), Enzymic hydrazinolysis of diesters and synthesis of N-aminosuccinimide derivatives, *Synthesis*, 287-289.

Athawale, V. D., Gaonkar, S. R. (1994), Enzymatic synthesis of polyesters by lipase catalysed polytransesterification, *Biotechnol. Lett.* **16**, 149-154.

Atomi, H., Bornscheuer, U., Soumanou, M. M., Beer, H. D., Wohlfahrt, G., Schmid, R. D. (1996), Microbial lipases - from screening to design, in.: *Oils-Fats and Lipids*, **Vol. 1**, pp. 49-50. Bridgwater: PJ Barnes & Associates.

Azzolina, O., Collina, S., Vercesi, D. (1995a), Stereoselective hydrolysis by esterase: a strategy for resolving 2-(R,R'-phenoxy)propionyl ester racemates, *Farmaco* **50**, 725-733.

Azzolina, O., Vercesi, D., Collina, S., Ghislandi, V. (1995b), Chiral resolution of methyl 2-aryloxypropionates by biocatalytic stereospecific hydrolysis, *Farmaco* **50**, 221-226.

Baba, N., Mimura, M., Hiratake, J., Uchida, K., Oda, J. (1988), Enzymic resolution of racemic hydroperoxides in organic solvent, *Agric. Biol. Chem.* **52**, 2685-2687.

Baba, N., Mimura, M., Oda, J., Iwasa, J. (1990a), Lipase-catalyzed stereoselective hydrolysis of thiol acetate, *Bull. Inst. Chem. Res. Kyoto Univ.* **68**, 208-212.

Baba, N., Tateno, K., Iwasa, J., Oda, J. (1990b), Lipase-catalyzed kinetic resolution of racemic methyl 13-hydroperoxy-9Z,11E-octadecadienoate in an organic solvent, *Agric. Biol. Chem.* **54**, 3349-3350.

Baba, N., Yoneda, K., Tahara, S., Iwasa, J., Kaneko, T., Matsuo, M. (1990c), A regioselective, stereoselective synthesis of a diacylglycerophosphocholine hydroperoxide by use of lipoxygenase and lipase, *J. Chem. Soc., Chem. Commun.*, 1281-1282.

Babayan, V. K. (1987), Medium chain triglycerides and structured lipids, *Lipids* **22**, 417-420.

Babayan, V. K., Rosenau, J. R. (1991), Medium-chain triglyceride cheese, *Food Technol.* **45**, 111-114.

Babiak, K. A., Ng, J. S., Dygos, J. H., Weyker, C. L., Wang, Y.-F., Wong, C. H. (1990), Lipase-catalyzed irreversible transesterification using enol esters: Resolution of prostaglandin synthons 4-hydroxy-2-alkyl-2-cyclopentenones and inversion of the 4S enantiomer to the 4R enantiomer, *J. Org. Chem.* **55**, 3377-3381.

Backlund, S., Eriksson, F., Kanerva, L. T., Rantala, M. (1995), Selective enzymic reactions using microemulsion-based gels, *Colloids Surf., B* **4**, 121-127.

Baczko, K., Liu, W. Q., Roques, B. P., Garbay-Jaureguiberry, C. (1996), New synthesis of D,L-Fmoc protected 4-phosphonomethylphenylalanine derivatives and their enzymatic resolution, *Tetrahedron* **52**, 2021-2030.

Baillargeon, M. W., McCarthy, S. G. (1991), *Geotrichum candidum* NRRL Y-553 lipase: purification, characterization and fatty acid specificity, *Lipids* **26**, 831-836.

Bakke, M., Takizawa, M., Sugai, T., Ohta, H. (1998), Lipase-catalyzed enantiomeric resolution of ceramides, *J. Org. Chem.* **63**, 6929-6938.

Balcao, V. M., Paiva, A. L., Malcata, F. X. (1996), Bioreactors with immobilized lipases: state of the art, *Enzyme Microb. Technol.* **18**, 392-416.

Baldaro, E., Fuganti, C., Servi, S., Tagliani, A., Terreni, M. (1992), The use of immobilized penicillin G acylase in organic synthesis, in.: *Microbial Reagents in Organic Synthesis* (Servi, S.; Ed.), pp. 175-188. Dordrecht: Kluwer Academic.

Baldaro, E., D'Arrigo, P., Pedrocchi-Fantoni, G., Rosell, C. M., Servi, S., Tagliani, A., Terreni, M. (1993), Pen G acylase catalyzed resolution of phenyl acetate esters of secondary alcohols, *Tetrahedron: Asymmetry* **4**, 1031-1034.

Baldessari, A., Iglesias, L. E., Gros, E. G. (1994), An improved procedure for chemospecific acylation of 2-mercaptoethanol by lipase-catalyzed transesterification, *Biotechnol. Lett.* **16**, 479-484.

Balkenhohl, F., Hauer, B., Ladner, W., Pressler, U., Rettenmaier, H., Adam, G. (1993a), A method to increase the activity of hydrolytic enzymes in organic solvents using surfactants, *German Patent* DE 4 344 211 (Chem. Abstr. 123:137 435).

Balkenhohl, F., Hauer, B., Ladner, W., Schnell, U., Pressler, U., Staudenmaier, H. R. (1993b), Lipase-catalyzed acylation of alcohols with diketenes, *German Patent* DE 4 329 293 (Chem. Abstr. 122:212 267).

Balkenhohl, F., Ditrich, K., Hauer, B., Ladner, W. (1997), Optically active amines via lipase-catalyzed methoxyacetylation, *J. Prakt. Chem.* **339**, 381-384.

Ballesteros, A., Bernabé, M., Cruzado, C., Martin-Lomas, M., Otero, C. (1989), Regioselective deacylation of 1,6-anhydro-β-D-galactopyranose derivatives catalyzed by soluble and immobilized lipases, *Tetrahedron* **45**, 7077-7082.

Ballesteros, A., Bornscheuer, U., Capewell, A., Combes, D., Condorét, J. S., König, K., Kolisis, F. N., Marty, A., Menge, U., Scheper, T., Stamatis, H., Xenakis, A. (1995), Enzymes in non-conventional phases, *Biocatal. Biotransform.* **13**, 1-42.

Banfi, L., Guanti, G., Riva, R. (1995), On the optimization of pig pancreatic lipase catalyzed monoacetylation of prochiral diols, *Tetrahedron: Asymmetry* **6**, 1345-1356.

Bänziger, M., Griffiths, G. J., McGarrity, J. F. (1993a), A facile synthesis of (2R,3E)-4-iodobut-3-en-2-ol and (2S,3E)-4-iodobut-3-en-2-yl chloroacetate, *Tetrahedron: Asymmetry* **4**, 723-726.

Bänziger, M., McGarrity, J. F., Meul, T. (1993b), A facile synthesis of N-protected statine and analogues via a lipase-catalyzes kinetic resolution, *J. Org. Chem.* **58**, 4010-4012.

Baraldi, P. G., Bazzanini, R., Manfredini, S., Simoni, D., Robins, M. J. (1993), Facile access to 2'-O-acyl prodrugs of 1-(β-D-arabinofuranosyl)-5(E)-(2-bromovinyl)uracil (BVARAU) via regioselective esterase-catalyzed hydrolysis of 2',3',5'-triester, *Tetrahedron Lett.* **48**, 7731-7734.

Barco, A., Benetti, S., Risi, C. D., Pollini, G. P., Romagnoli, R., Zanirato, V. (1994), A chemoenzymic approach to chiral phenylisoserinates using 4-isopropyl-2-oxazolin-5-one as masked umpoled synthon for hydroxycarbonyl anion, *Tetrahedron Lett.* **35**, 9289-9292.

Barili, P. L., Berti, G., Mastrorilli, E. (1993), Regio- and stereochemistry of the acid catalyzed and of a highly enantioselective enzymatic hydrolysis of some epoxytetrahydrofurans, *Tetrahedron* **49**, 6263-6276.

Barnier, J. P., Blanco, L., Guibe-Jampel, E., Rousseau, G. (1989), Preparation of (R)-Veratryl- and (R)-(3-Methoxybenzyl)succinates, *Tetrahedron* **45**, 5051-5058.

Barnier, J. P., Blanco, L., Rousseau, G., Guibé-Jampel, E., Fresse, I. (1993), Enzymic resolution of cyclopropanols. An easy access to optically active cyclohexanones possessing an α-quaternary chiral carbon, *J. Org. Chem.* **58**, 1570-1574.

Barton, P., Page, M. I. (1992), The resolution of racemic 1,2-diols by the esterase-catalyzed hydrolysis of the corresponding cyclic carbonate, *Tetrahedron* **48**, 7731-7734.

Barz, M., Herdtweck, E., Thiel, W. R. (1996), Kinetic resolution of *trans*-2-(1-pyrazolyl)cyclohexan-1-ol catalyzed by lipase B from *Candida antarctica*, *Tetrahedron: Asymmetry* **7**, 1717-1722.

Basak, A., Bhattacharya, G., Palit, S. K. (1997), Novel regioselective ester hydrolysis by pig-liver esterase, *Bull. Chem. Soc. Jpn.* **70**, 2509-2513.

Basavaiah, D., Krishna, P. R. (1994), Pig liver acetone powder (PLAP) as biocatalyst: enantioselective synthesis of *trans*-2-alkoxycyclohexan-1-ols, *Tetrahedron* **50**, 10521-10530.

Basavaiah, D., Rao, P. D. (1994), Enzymatic resolution of *trans*-2-arylcyclohexan-1-ols using crude chicken liver esterase (CCLE) as biocatalyst, *Tetrahedron: Asymmetry* **5**, 223-234.

Basavaiah, D., Rao, P. D. (1995), Synthesis of enantiomerically enriched *anti*-homoallyl alcohols mediated by crude chicken liver esterase (CCLE), *Tetrahedron: Asymmetry* **6**, 789-800.

Basheer, S., Mogi, K.-i., Nakajima, M. (1995a), Interesterification kinetics of triglycerides and fatty acids with modified lipase in n-hexane, *J. Am. Oil Chem. Soc.* **72**, 511-518.

Basheer, S., Mogi, K.-i., Nakajima, M. (1995b), Surfactant-modified lipase for the catalysis of the interesterification of triglycerides and fatty acids, *Biotechnol. Bioeng.* **45**, 187-195.

Bashir, N. B., Phythian, S. J., Roberts, S. M., (1995), Enzymatic regioselective acylation and deacylation of carbohydrates, in.: *Preparative Biotransformations* (Roberts, S. M., Ed.), pp. 0:0.43-40:40.74. New York: Wiley.

Basri, M., Ampon, K., Yunus, W. M. Z. W., Razak, C. N. A., Salleh, A. B. (1995), Synthesis of fatty esters by polyethylene glycol-modified lipase, *J. Chem.Technol. Biotechnol.* **64**, 10-16.

Battistel, E., Bianchi, D., Cesti, P., Pina, C. (1991), Enzymatic resolution of (*S*)-(+)-naproxen in a continuous reactor, *Biotechnol. Bioeng.* **38**, 659-664.

Bauer, R., Hirrlinger, B., Layh, N., Stolz, A., Knackmuss, H.-J. (1994), *Appl. Microbiol. Biotechnol.*

Bauer, R., Knackmuss, H.-J., Stolz, A. (1998), Enantioselective hydration of 2-arylpropionitriles by a nitrile hydratase from *Agrobacterium tumefaciens* strain d3, *Appl. Microbiol. Biotechnol.* **49**, 89-95.

Baumann, H., Bühler, M., Fochem, H., Hirsinger, F., Zoebelein, H., Falbe, J. (1988), Natural fats and oils - Renewable raw materials for the chemical industry, *Angew. Chem. Int. Ed. Engl.* **27**, 41-62.

Beard, T., Cohen, M. A., Paratt, J. S., Turner, N. J., Crosby, J., Moilliet, J. (1993), Stereoselective hydrolysis of nitriles and amides under mild conditions using a whole cell catalyst, *Tetrahedron: Asymmetry* **4**, 1085-1104.

Beer, H. D., Wohlfahrt, G., McCarthy, J. E. G., Schomburg, D., Schmid, R. D. (1996), Analysis of the catalytic mechanism of a fungal lipase using computer-aided design and structural mutants, *Protein Eng.* **9**, 507-517.

Beer, H. D., Bornscheuer, U. T., McCarthy, J. E. G., Schmid, R. D. (1998), Cloning, expression, characterization and role of the leader sequence of a lipase from *Rhizopus oryzae*, *Biochim. Biophys. Acta* **1399**, 173-180.

Bell, P. A., Kasper, C. B. (1993), Expression of rat microsomal epoxide hydrolase in *Escherichia coli*. Identification of a histidyl residue essential for catalysis, *J. Biol. Chem.* **268**, 14011-14017.

Bell, G., Halling, P. J., Moore, B. D., Partidge, J., Rees, D. G. (1995), Biocatalyst behaviour in low-water systems, *Trends Biotechnol.* **13**, 468-473.

Bello, M., Thomas, D., Legoy, M. D. (1987), Interesterification and synthesis by *Candida cylindracea* lipase in microemulsions, *Biochem. Biophys. Res. Commun.*, **146**, 361-367.

Bellucci, G., Berti, G., Catelani, C., Mastrorilli, E. (1981), Unusual steric course of the epoxide hydrolase catalyzed hydrolysis of (±)-3,4-epoxytetrahydropyran. A case of complete stereoconvergence, *J. Org. Chem.* **46**, 5148-5150.

Bellucci, G., Capitani, I., Chiappe, C., Marioni, F. (1989a), Product enantioselectivity of the microsomal and cytosolic epoxide hydrolase catalysed hydrolysis of *meso* epoxides, *J. Chem. Soc., Chem. Commun.*, 1170-1171.

Bellucci, G., Chiappe, C., Marioni, F. (1989b), Enantioselectivity of the enzymatic hydrolysis of cyclohexene oxide and (±)-1-methylcyclohexene oxide: A comparison between microsomal and cytosolic epoxide hydrolases, *J. Chem. Soc., Perkin Trans. 1*, 2369-2373.

Bellucci, G., Chiappe, C., Marioni, F., Benetti, M. (1991), Regio- and enantio-selectivity of the cytosolic epoxide hydrolase-catalyzed hydrolysis of racemic monosubstituted alkyloxiranes, *J. Chem. Soc., Perkin Trans. 1*, 361-363.

Bellucci, G., Chiappe, C., Cordoni, A., Marioni, F. (1994a), Different enantioselectivity and regioselectivity of the cytosolic and microsomal epoxide hydrolase catalyzed hydrolysis of simple phenyl substituted epoxides, *Tetrahedron Lett.* **35**, 4219-4222.

Bellucci, G., Chiappe, C., Ingrosso, G. (1994b), Kinetics and stereochemistry of the microsomal epoxide hydrolase-catalyzed hydrolysis of *cis*-stilbene oxides, *Chirality* **6**, 577-582.

Bellucci, G., Chiappe, C., Ingrosso, G., Rosini, C. (1995), Kinetic resolution by epoxide hydrolase catalyzed hydrolysis of racemic methyl substituted methylenecyclohexene oxides, *Tetrahedron: Asymmetry* **6**, 1911-1918.

Bellucci, G., Chiappe, C., Cordoni, A. (1996), Enantioconvergent transformation of racemic *cis*-β-alkyl substituted styrene oxides to (*R,R*) threo diols by microsomal epoxide hydrolase catalysed hydrolysis, *Tetrahedron: Asymmetry* **7**, 197-202.

Bengis-Garber, C., Gutman, A. L. (1988), Bacteria in organic synthesis: Selective conversion of 1,3-dicyanobenzene into 3-cyanobenzoic acid, *Tetrahedron Lett.* **29**, 2589-2590.

Bengis-Garber, C., Gutman, A. L. (1989), Selective hydrolysis of dinitriles into cyano-carboxylic acids by *Rhodococcus rhodochrous* N.C.I.B. 11216, *Appl. Microbiol. Biotechnol.* **32**, 11-16.

Berger, B., Faber, K. (1991), 'Immunization' of lipase against acetaldehyde emerging in acyl transfer reactions from vinyl acetate, *J. Chem. Soc., Chem. Commun.*, 1198-1200.

Berger, M., Schneider, M. P. (1991a), Lipases in organic solvents: The fatty acid chain length profile, *Biotechnol. Lett.* **13**, 641-645.

Berger, M., Schneider, M. P. (1991b), Regioselectivity of lipases in organic solvents, *Biotechnol. Lett.* **13**, 333-338.

Berger, M., Schneider, M. P. (1992), Enzymatic esterification of glycerol II. Lipase-catalyzed synthesis of regioisomerically pure 1(3)-*rac*-monoacylglycerols, *J. Am. Oil. Chem. Soc.* **69**, 961-965.

Berger, M., Schneider, M. (1993), Regioisomerically pure mono- and diglycerols as synthetic building blocks, *Fat Sci. Technol.* **95**, 169-175.

Berger, A., Smolarsky, M., Kurn, N., Bosshard, H. R. (1973), A new method for the synthesis of optically active α-amino acids and their Na derivatives via acylamino malonates, *J. Org. Chem* **38**, 457-460.

Berger, B., Rabiller, C. G., Königsberger, K., Faber, K., Griengl, H. (1990), Enzymatic acylation using acid anhydrides: crucial removal of acid, *Tetrahedron: Asymmetry* **1**, 541-546.

Berger, M., Laumen, K., Schneider, M. P. (1992), Enzymatic esterification of glycerol I. Lipase-catalyzed synthesis of regioisomerically pure 1,3-*sn*-diacylglycerols, *J. Am. Oil Chem. Soc.* **69**, 955-960.

Berglund, P., Voerde, C., Hoegberg, H.-E. (1994), Esterification of 2-methylalkanoic acids catalyzed by lipase from *Candida rugosa*: enantioselectivity as a function of water activity and alcohol chain length, *Biocatalysis* **9**, 123-130.

Berglund, P., Desantis, G., Stabile, M. R., Shang, X., Gold, M., Bott, R. R., Graycar, T. P., Lau, T. H., Mitchinson, C., Jones, J. B. (1997), Chemical modification of cysteine mutants of subtilisin *Bacillus lentus* can create better catalysts than the wild–type enzyme, *J. Am. Chem. Soc.* **119**, 5265–5266.

Berkowitz, D. B., Maeng, J. H. (1996), Enantioselective entry into benzoxabicyclo[2.2.1]heptyl systems via enzymatic desymmetrization - toward chiral building blocks for lignan synthesis, *Tetrahedron: Asymmetry* **7**, 1577-1580.

Berkowitz, D. B., Danishefsky, S. J., Schulte, G. K. (1992), A route to artificial glycoconjugates and oligosaccharides via enzymatically resolved glycals: Dramatic effects of the handedness of the sugar domain upon the properties of an anthracycline drug, *J. Am. Chem. Soc.* **114**, 4518-4529.

Berkowitz, D. B., Pumphrey, J. A., Shen, Q. (1994), Enantiomerically enriched α-vinyl amino acids via lipase-mediated reverse transesterification, *Tetrahedron Lett.* **35**, 8743-8746.

Bernard, P., Barth, D. (1995), Internal mass transfer limitation during enzymic esterification in supercritical carbon dioxide and hexane, *Biocatal. Biotransform.* **12**, 299-308.

Bernard, P., Barth, D., Perrut, M. (1992), The integration of biocatalysis and downstream processing in supercritical carbon dioxide, in.: *High Press. Biotechnol.* (Balny, C., Hayashi, R., Heremans, K. , Masson, P.; Eds.), **Vol. 224**, pp. 451-455.

Berry, D. R., Paterson, A. (1990), Enzymes in the food industry, in.: *Enzyme Chemistry* (Suckling, C. J.; Ed.), **2nd Ed.** pp. 306-351. London: Chapman and Hall.

Besse, P., Veschambre, H. (1994), Chemical and biological synthesis of chiral epoxides, *Tetrahedron* **50**, 8885-8927.

Betzel, C., Klupsch, S., Papendorf, G., Hastrup, S., Branner, S., Wilson, K. S. (1992), Crystal structure of the alkaline proteinase Savinase from *Bacillus lentus* at 1.4 A resolution, *J. Mol. Biol.* **223**, 427-445.

Bevinakatti, H. S., Banerji, A. A. (1991), Practical chemoenzymic synthesis of both enantiomers of propranolol, *J. Org. Chem.* **56**, 5372-5375.

Bevinakatti, H. S., Banerji, A. A., Newadkar, R. V., Mokashi, A. A. (1992), Enzymic synthesis of optically active amino acids. Effect of solvent on the enantioselectivity of lipase-catalyzed ring-opening of oxazolin-5-ones, *Tetrahedron: Asymmetry* **3**, 1505-1508.

Bevinakatti, H. S., Newadkar, R. V. (1993), Lipase catalysis in organic solvents. In search of practical derivatizing agents for the kinetic resolution of alcohols, *Tetrahedron: Asymmetry* **4**, 773-776.

Bevinakatti, H. S., Newadkar, R. V., Banerji, A. A. (1990), Lipase-catalyzed enantioselective ring-opening of oxazol-5(4H)-ones coupled with partial in situ racemization of the less reactive isomer, *J. Chem. Soc., Chem. Commun.*, 1091-1092.

Bhalerao, U. T., Dasaradhi, L., Neelankantan, P., Fadnavis, N. W. (1991), Lipase-catalyzed regio- and enantioselective hydrolysis: molecular recognition phenomenon and synthesis of *R*-dimorphecolic acid, *J. Chem. Soc., Chem. Commun.*, 1197-1198.

Bhaskar-Rao, A., Rehman, H., Krishnakumari, B., Yadav, J. S. (1994), Lipase catalyzed kinetic resolution of racemic (±)-2,2-dimethyl-3-(2-methyl-1-propenyl)cyclopropane carboxyl esters, *Tetrahedron Lett.* **35**, 2611-2614.

Biadatti, T., Esker, J. L., Johnson, C. R. (1996), Chemoenzymatic synthesis of a versatile cyclopentenone: (+)-(3aS,6aS)-2,2-dimethyl-3aβ,6aβ-dihydro-4H-cyclopenta-1,3-dioxol-4-one, *Tetrahedron: Asymmetry* **7**, 2313-2320.

Bianchi, D., Cesti, P. (1990), Lipase-catalyzed stereoselective thiotransesterification of mercapto esters, *J. Org. Chem.* **55**, 5657-5659.

Bianchi, D., Cabri, W., Cesti, P., Francalanci, F., Rama, F. (1988a), Enzymatic resolution of 2,3-epoxyalcohols, intermediates in the synthesis of the gypsy moth sex pheromone, *Tetrahedron Lett.* **29**, 2455-2458.

Bianchi, D., Cabri, W., Cesti, P., Francalanci, F., Ricci, M. (1988b), Enzymic hydrolysis of alkyl 3,4-epoxybutyrates. A new route to (R)-(−)-carnitine chloride, *J. Org. Chem.* **53**, 104-107.

Bianchi, D., Cesti, P., Battistel, E. (1988c), Anhydrides as acylating agents in lipase-catalyzed stereoselective esterification of racemic alcohols, *J. Org. Chem.* **53**, 5531-5534.

Bianchi, D., Bosetti, A., Golini, P., Cesti, P., Pina, C. (1997), Resolution of isopropylidene-glycerol benzoate by sequential enzymatic hydrolysis and preferential crystallization, *Tetrahedron: Asymmetry* **8**, 817-819.

Binns, F., Roberts, S. M., Taylor, A., Williams, C. F. (1993), Enzymic polymerization of an unactivated diol/diacid system, *J. Chem. Soc., Perkin Trans. 1*, 899-904.

Bis, S. J., Whitaker, D. T., Johnson, C. R. (1993), Lipase asymmetrization of *cis*-3,7-dihydroxycycloheptene derivatives in organic and aqueous media, *Tetrahedron: Asymmetry* **4**, 875-878.

Bisht, K. S., Parmar, V. S. (1993), Diastereo- and enantioselective esterification of butane-2,3-diol catalyzed by the lipase from *Pseudomonas fluorescens*, *Tetrahedron: Asymmetry* **4**, 957-958.

Bisht, K. S., Kumar, A., Kumar, N., Parmar, V. S. (1996), Preparative and mechanistic aspects of interesterification reactions on diols and peracetylated polyphenolic compounds catalysed by lipases, *Pure Appl. Chem.* **68**, 749-752.

Björkling, F., Godtfredsen, S. E. (1988), New enzyme catalyzed synthesis of monoacyl galactoglycerides, *Tetrahedron* **44**, 2957-2962.

Björkling, F., Boutelje, J., Gatenbeck, S., Hult, K., Norin, T. (1985), Enzyme catalysed hydrolysis of dialkylated propanedioic acid diesters, synthesis of optically pure (S)-α-methylphenylalanine, (S)-α-methyltyrosine, and (S)-α-methyl-3,4-dihydroxyphenylalanine, *Tetrahedron Lett.* **26**, 4957-4958.

Björkling, F., Boutelje, J., Hjalmarsson, M., Hult, K., Norin, T. (1987), Highly enantioselective route to (R)-proline derivates via enzyme catalysed hydrolysis of *cis*-N-benzyl-2,5-bismethoxycarbonylpyrrolidine in an aqueous dimethyl sulphoxide medium, *J. Chem. Soc., Chem. Commun.*, 1041-1042.

Björkling, F., Godtfredsen, S. E., Kirk, O. (1989), A highly selective enzyme-catalyzed esterification of simple glucosides, *J. Chem. Soc., Chem. Commun.*, 934-935.

Björkling, F., Godtfredsen, S. E., Kirk, O. (1991), The future impact of industrial lipases, *Trends. Biotechnol.* **9**, 360-363.

Björkling, F., Frykman, H., Godtfredsen, S. E., Kirk, O. (1992), Lipase-catalyzed synthesis of peroxycarboxylic acids and lipase-mediated oxidations, *Tetrahedron* **48**, 4587-4592.

Blackman, R. L., Spence, J. M., Field, L. M., Devonshire, A. L. (1995), Chromosomal location of the amplified esterase genes conferring resistance to insecticides in *Mycus persicae* (Homoptera: Aphidae), *Heredity* **75**, 297-302.

Blackwood, A. D., Curran, L. J., Moore, B. D., Halling, P. J. (1994), Organic phase buffers control biocatalyst activity independent of initial aqueous pH, *Biochim. Biophys. Acta* **1206**, 161-165.

Blanco, L., Guibé-Jampel, E., Rousseau, G. (1988), Enzymatic resolution of racemic lactones, *Tetrahedron Lett.* **29**, 1915-1918.

Blanco, L., Rousseau, G., Barnier, J. P., Guibé-Jampel, E. (1993), Enzymic resolution of 3-substituted-4-oxoesters, *Tetrahedron: Asymmetry* **4**, 783-792.

Blée, E., Schuber, F. (1995), Stereocontrolled hydrolysis of the linoleic acid monoepoxide regioisomers catalyzed by soy-bean epoxide hydrolase, *Eur. J. Biochem.* **230**, 229-234.

Bloch, R., Guibé-Jampel, E., Girard, G. (1985), Stereoselective pig-liver esterase-catalyzed hydrolysis of rigid bicyclic *meso*-diester - preparation of optically pure 4,7-epoxytetra-hydrophthalides and hexa-hydrophthalides, *Tetrahedron Lett.* **26**, 4087-4090.

Bloomer, S., Adlercreutz, P., Mattiasson, B. (1991), Triglyceride interesterification by lipases. 2. Reaction parameters for the reduction of trisaturated impurities and diglycerides in batch reactions, *Biocatalysis* **5**, 145-162.

Bloomer, S., Adlercreutz, P., Mattiasson, B. (1992), Facile synthesis of fatty acid esters in high yields, *Enzyme Microb. Technol.* **14**, 546-552.

Blow, D. M. (1976), Structure and mechanism of chymotrypsin, *Acc. Chem. Res.* **9**, 145-152.

Boaz, N. (1989), Enzymatic esterification of 1-ferrocenylethanol: an alternate approach to chiral ferrocenyl bis-phosphines, *Tetrahedron Lett.* **30**, 2061-2064.

Boaz, N. W. (1992), Enzymic hydrolysis of ethyl 3-hydroxy-3-phenylpropanoate: Observations on an enzyme active site model, *J. Org. Chem.* **57**, 4289-4292.

Bogenstätter, M., Steglich, W. (1997), Synthesis of optically pure α-hydroxyglycine peptides, *Tetrahedron* **53**, 7267-7274.

Boland, W., Frößl, C., Lorenz, M. (1991), Esterolytic and lipolytic enzymes in organic synthesis, *Synthesis*, 1049-1072.

Bommarius, A. S., Drauz, K., Groeger, U., Wandrey, C. (1992), Membrane bioreactors for the production of enantiomerically pure α-amino acids, in.: *Chirality in Industry* (Collins, A. N., Sheldrake, G. N. , Crosby, J.; Eds.), pp. 371-397. New York: Wiley.

Bommarius, A. S., Hatton, T. A., Wang, D. I. C. (1995), Xanthine oxidase reactivity in reversed micellular systems: A contribution to the prediction of enzymatic activity in organized media, *J. Am. Chem. Soc.* **117**, 4515-4523.

Bommarius, A. S., Drauz, K., Günther, K., G., K., Schwarm, M. (1997), L-Methionine related L-amino acids by acylase cleavage of their corresponding N-acetyl-DL-derivatives, *Tetrahedron: Asymmetry* **8**, 3197-3200.

Bonini, C., Racioppi, R., Righi, G., Viggiani, L. (1993), Polyhydroxylated chiral building block by enzymic desymmetrization of *meso* 1,3 syn diols, *J. Org. Chem.* **58**, 802-803.

Bonini, C., Racioppi, R., Viggiani, L. (1997), Unprecedented biocatalytic desymmetrization of hexitols, *Tetrahedron: Asymmetry* **8**, 353-354.

Bonneau, P. R., Eyer, M., Graycar, T. P., Estelle, D. A., Jones, J. B. (1993), The effects of organic solvents on wild-type and mutant subtilisin-catalyzed hydrolyses, *Bioorg. Chem.* **21**, 431-438.

Borgstrom, B., Brockman, H. L. (Eds.) (1984), *Lipases*, New York: Elsevier.

Bornemann, S., Crout, D. H. G., Dalton, H., Hutchinson, D. W. (1992), Activities in crude porcine pancreatic lipase: enantioselectivity in hydrolysis of the diacetate of 2-phenylpropane-1,3-diol, *Biocatalysis* **5**, 297-303.

Bornscheuer, U. T. (1995), Lipase-catalyzed syntheses of monoacylglycerols, *Enzyme Microb. Technol.* **17**, 578-586.

Bornscheuer, U. T. (1998), Directed evolution of enzymes, *Angew. Chem. Int. Ed. Engl.* **37**, 3105-3108.

Bornscheuer, U. T., Yamane, T. (1994), Activity and stability of lipase in the solid-phase glycerolysis of triolein, *Enzyme Microb. Technol.* **16**, 864-869.

Bornscheuer, U. T., Yamane, T. (1995), Fatty acid vinyl esters as acylating agents: a new method for the enzymic synthesis of monoacylglycerols, *J. Am. Oil Chem. Soc.* **72**, 193-197.

Bornscheuer, U., Herar, A., Kreye, L., Wendel, V., Capewell, A., Meyer, H. H., Scheper, T., Kolisis, F. N. (1993), Factors affecting the lipase catalyzed transesterification reactions of 3-hydroxy esters in organic solvents, *Tetrahedron: Asymmetry* **4**, 1007-1016.

Bornscheuer, U., Reif, O.-W., Lausch, R., Freitag, R., Scheper, T., Kolisis, F. N., Menge, U. (1994a), Lipase of *Pseudomonas cepacia* for biotechnological purposes: purification, crystallization and characterization, *Biochim. Biophys. Acta* **1201**, 55-60.

Bornscheuer, U., Stamatis, H., Xenakis, A., Yamane, T., Kolisis, F. N. (1994b), A comparison of different strategies for lipase-catalyzed synthesis of partial glycerides, *Biotechnol. Lett.* **16**, 697-702.

Bornscheuer, U., Capewell, A., Wendel, V., Scheper, T. (1996), On-line determination of the conversion in a lipase-catalyzed kinetic resolution in supercritical carbon dioxide, *J. Biotechnol.* **46**, 139-143.

Bornscheuer, U. T., Altenbuchner, J., Meyer, H. H. (1998a), Directed evolution of an esterase for the stereoselective resolution of a key intermediate in the synthesis of Epothilones, *Biotechnol. Bioeng.* **58**, 554-559.

Bornscheuer, U. T., Soumanou, M. M., Schmid, R. D., Schmid, U. (1998b), Preparation of symmetrical triglycerides ABA, *European Patent* EP 88 27 92 (Unilever N.V.) (Chem. Abstr. 130:51410).

Bornscheuer, U. T., Altenbuchner, J., Meyer, H. H. (1999), Directed evolution of an esterase: Screening of enzyme libraries based on pH-indicators and a growth assay, *Bioorg. Med. Chem.*, in press.

Borrelly, S., Paquette, L. A. (1993), An intramolecular cyclization approach to optically-active cyclopentenyl bromides, *J. Org. Chem.* **58**, 2714-2717.

Borzeix, F., Monot, F., Vandecasteele, J.-P. (1992), Strategies for enzymatic esterification in organic solvents: Comparison of microaqueous, biphasic, and micellar systems, *Enzyme Microb. Technol.* **14**, 791-797.

Bosetti, A., Bianchi, D., Cesti, P., Golini, P. (1994), Lipase-catalyzed resolution of isopropylidene glycerol: effect of co-solvents on enantioselectivity, *Biocatalysis* **9**, 71-77.

Bosley, J. (1997), Turning lipases into industrial biocatalysts, *Biochem. Soc. Trans.* **25**, 174-178.

Botes, A. L., Steenkamp, J. A., Letloenyane, M. Z., Van Dyk, M. S. (1998a), Epoxide hydrolase activity of *Chryseomonas luteola* for the asymmetric hydrolysis of aliphatic mono-substituted epoxides, *Biotechnol. Lett.* **20**, 427-430.

Botes, A. L., Weijers, C. A. G. M., Van Dyk, M. S. (1998b), Biocatalytic resolution of 1,2-epoxyoctane using resting cells of different yeast strains with novel epoxide hydrolase activities, *Biotechnol. Lett.* **20**, 421-426.

Bott, R., Dauberman, J., Wilson, L., Ganshaw, G., Sagar, H., Graycar, T., Estell, D. (1996), Structural changes leading to increased enzymic activity in an engineered variant of *Bacillus lentus* subtilisin., *Adv. Exp. Med. Biol.* **379**, 277-283.

Botta, M., Cernia, E., Corelli, F., Manetti, F., Soro, S. (1997), Probing the substrate specificity for lipases. 2. Kinetic and modeling studies on the molecular recognition of 2-arylpropionic esters by *Candida rugosa* and *Rhizomucor miehei* lipases, *Biochim. Biophys. Acta* **1337**, 302-310.

Bourne, Y., Martinez, C., Kerfelec, B., Lombardo, D., Chapus, C., Cambillau, C. (1994), Horse pancreatic lipase - the crystal structure refined at 2.3 Å resolution, *J. Mol. Biol.* **238**, 709-732.

Bovara, R., Carrea, G., Ferrara, L., Riva, S. (1991), Resolution of *trans*-Sobrerol by lipase PS-catalyzed transesterification and effects of organic solvents on enantioselectivity, *Tetrahedron: Asymmetry* **2**, 931-938.

Brackenridge, I., McCague, R., Roberts, S. M., Turner, N. J. (1993), Enzymatic resolution of oxalate esters of a tertiary alcohol using porcine pancreatic lipase, *J. Chem. Soc., Perkin Trans. 1*, 1093-1094.

Brady, L., Brzozowski, A. M., Derewenda, Z. S., Dodson, E., Dodson, G., Tolley, S., Turkenburg, J. P., Christiansen, L., Huge-Jensen, B., Norskov, L., Thim, L., Menge, U. (1990), A serine protease triad forms the catalytic centre of a triacylglycerol lipase, *Nature* **343**, 767-770.

Brand, S., Jones, M. F., Rayner, C. M. (1995), The first examples of dynamic kinetic resolution by enantioselective acetylation of hemithioacetals: an efficient synthesis of homochiral α-acetoxysulfides, *Tetrahedron Lett.* **36**, 8493-8496.

Branden, C., Tooze, J. (1991), An example of enzyme catalysis: serine proteinases, in.: *Introduction to Protein Structure*, pp. 231-246. New York: Garland.

Braun, P., Waldmann, H., Vogt, W., Kunz, H. (1990), Selective enzymatic removal of protecting functions: heptyl esters as carboxy protecting groups in peptide synthesis, *Synthesis*, 105-107.

Braun, P., Waldmann, H., Vogt, W., Kunz, H. (1991), Selective enzymatic removal of protecting groups: *n*-heptyl esters as carboxy protecting functions in peptide synthesis, *Liebigs Ann. Chem.*, 165-170.

Braun, P., Waldmann, H., Kunz, H. (1992), Selective enzymatic removal of protecting functions: heptyl esters as carboxy protecting groups in glycopeptide synthesis, *Synlett*, 39-40.

Braun, P., Waldmann, H., Kunz, H. (1993), Chemoenzymic synthesis of *O*-glycopeptides carrying the tumor associated TN-antigen structure, *Bioorg. Med. Chem.* **1**, 197-207.

Brazwell, E. M., Filos, D. Y., Morrow, C. J. (1995), Biocatalytic synthesis of polymers. III. Formation of a high molecular weight polyester through limitation of hydrolysis by enzyme-bound water and through equilibrium control, *J. Polym. Sci., Part A: Polym. Chem.*, **33**, 89-95.

Breitgoff, D., Laumen, K., Schneider, M. P. (1986), Enzymatic differentiation of the enantiotopic hydroxymethyl groups of glycerol; synthesis of chiral building blocks, *J. Chem. Soc., Chem. Commun.*, 1523-1524.

Breznik, M., Kikelj, D. (1997), Pig liver esterase catalyzed hydrolysis of dimethyl and diethyl 2-methyl-2-(*o*-nitrophenoxy)malonates, *Tetrahedron:Asymmetry* **8**, 425-434.

Breznik, M., Mrcina, A., Kikelj, D. (1998), Enantioselective synthesis of (R)- and (S)-2-methyl-3-oxo-3,4-dihydro-2H-1,4-benzoxazine-2-carboxamides, *Tetrahedron:Asymmetry* **9**, 1115-1116.

Briand, D., Dubreucq, E., Galzy, P. (1995), Functioning and regioselectivity of the lipase of *Candida parapsilosis* (Ashford) Langeron and Talice in aqueous medium. New interpretation of regioselectivity taking acyl migration into account, *Eur. J. Biochem.* **228**, 169-175.

Brieva, R., Rebolledo, F., Gotor, V. (1990), Enzymatic synthesis of amides with two chiral centres, *J. Chem. Soc., Chem. Commun.*, 1386-1387.

Brocca, S., Schmidt-Dannert, C., Lotti, M., Alberghina, L., Schmid, R. D. (1998), Design, total synthesis, and functional overexpression of the *Candida rugosa lip1* gene encoding for a major industrial lipase, *Protein Sci.* **7**, 1415-1422.

Broos, J., Engbersen, J. F. J., Verboom, W., Reinhoudt, D. N. (1995a), Inversion of enantioselectivity of serine proteases, *Recl. Trav. Chim. Pays-Bas* **114**, 255-257.

Broos, J., Visser, A. J. W. G., Engbersen, J. F. J., Verboom, W., van Hoek, A., Reinhoudt, D. N. (1995b), Flexibility of enzymes suspended in organic solvents probed by time-resolved fluorescence anisotropy. Evidence that enzyme activity and enantioselectivity are directly related to enzyme flexibility, *J. Am. Chem. Soc.* **117**, 12657-12663.

Brown, S. M., Davies, S. G., de Sousa, J. A. A. (1993), Kinetic resolution strategies II: Enhanced enantiomeric excesses and yields for the faster reacting enantiomer in lipase mediated kinetic resolutions, *Tetrahedron: Asymmetry* **4**, 813-822.

Bruce, M. A., St. Laurent, D. R., Poindexter, G. S., Monkovic, I., Huang, S., Balasubramanian, N. (1995), Kinetic resolution of piperazine-2-carboxamide by leucine aminopeptidase. An application in the synthesis of the nucleoside transport blocker (–) draflazine, *Synth. Commun.*, **25**, 2673-2684.

Bruzik, K. S., Guan, Z., Riddle, S., Tsai, M.-D. (1996), Synthesis of inositol phosphodiesters by phospholipase C-catalyzed transesterification, *J. Am. Chem. Soc.* **118**, 7679-7688.

Brzozowski, A. M., Derewenda, U., Derewenda, G. G., Dodson, D. M., Lawson, J., Turkenburg, P., Bjorkling, F., Huge-Jensen, B., Patkar, S. A., Thim, L. (1991), A model for interfacial activation in lipases from the structure of a fungal lipase-inhibitor complex, *Nature* **351**, 491-494.

Bucciarelli, M., Formi, A., Moretti, I., Prati, F. (1988), Enzymatic resolution of chiral *N*-alkyloxaziridine-3,3-dicarboxylic esters, *J. Chem. Soc., Chem. Commun.*, 1614-1615.

Bühler, M., Wandrey, C. (1987a), Continuous use of lipases in fat hydrolysis, *Fat Sci. Technol.* **89**, 598-605.

Bühler, M., Wandrey, C. (1987b), Enzymatic fat splitting, *Fat Sci. Technol.* **89**, 156-164.

Bunch, A. W. (1998), Nitriles, in.: *Biotechnology-Series* (Rehm, H. J., Reed, G., Pühler, A., Stadler, P. J. W. , Kelly, D. R.; Eds.), **Vol. 8a**, pp. 277-324. Weinheim: Wiley-VCH.

Burgess, K., Henderson, I. (1989), A facile route to homochiral sulfoxides, *Tetrahedron Lett.* **30**, 3633-3636.

Burgess, K., Henderson, I. (1990), Lipase-mediated resolution of SPAC reaction products, *Tetrahedron: Asymmetry* **1**, 57-60.

Burgess, K., Henderson, I. (1991), Biocatalytic desymmetrizations of pentitol derivatives, *Tetrahedron Lett.* **32**, 5701-5704.

Burgess, K., Ho, K.-K. (1992), Asymmetric syntheses of all four stereoisomers of 2,3-methanomethionine, *J. Org. Chem.* **57**, 5931-5936.

Burgess, K., Jennings, L. D. (1991), Enantioselective esterifications of unsaturated alcohols mediated by a lipase prepared from *Pseudomonas* sp., *J. Am. Chem. Soc.* **113**, 6129-6139.

Burgess, K., Henderson, I., Ho, K.-K. (1992), Biocatalytic resolutions of sulfinylalkanoates: a facile route to optically active sulfoxides, *J. Org. Chem.* **57**, 1290-1295.

Cadwell, R. C., Joyce, G. F. (1992), Randomization of genes by PCR mutagenesis, *PCR Meth. Appl.* **2**, 28-33.

Cainelli, G., Giacomini, D., Galletti, P., DaCol, M. (1997), Penicillin G acylase mediated synthesis of the enantiopure (*S*)-3-amino- azetidin-2-one, *Tetrahedron: Asymmetry* **8**, 3231-3235.

Cambillau, C., Tilbeurgh, H. van (1993), Structure of hydrolases - lipases and cellulases, *Curr. Opin. Struct. Biol.* **3**, 885-895.

Cambou, B., Klibanov, A. M. (1984), Preparative production of optically-active esters and alcohols using esterase-catalyzed stereospecific transesterification in organic media, *J. Am. Chem. Soc.* **106**, 2687-2692.

Cao, L., Bornscheuer, U. T., Schmid, R. D. (1996), Lipase-catalyzed solid-phase synthesis of sugar esters, *Fett/Lipid* **98**, 332-335.

Cao, L., Fischer, A., Bornscheuer, U. T., Schmid, R. D. (1997), Lipase-catalyzed solid phase preparation of sugar fatty acid esters, *Biocatal. Biotransform.* **14**, 269-283.

Capewell, A., Wendel, V., Bornscheuer, U., Meyer, H. H., Scheper, T. (1996), Lipase-catalyzed kinetic resolution of 3-hydroxy esters in organic solvents and supercritical carbon dioxide, *Enzyme Microb. Technol.* **19**, 181-186.

Carda, M., van der Eycken, J., Vandewalle, M. (1990), Enantiotoposelective PLE-catalyzed hydrolysis of *cis*-5-substituted-1,3-diacyloxycyclohexanes. Preparation of some useful chiral building blocks, *Tetrahedron: Asymmetry* **1**, 17-20.

Cardellicchio, C., Naso, F., Scilimati, A. (1994), An efficient biocatalyzed kinetic resolution of methyl (Z)-3-(arylsulfinyl)propenoates, *Tetrahedron Lett.* **35**, 4635-4638.

Caron, G., Kazlauskas, R. J. (1991), An optimized sequential kinetic resolution of *trans*-1,2- cyclohexanediol, *J. Org. Chem.* **56**, 7251-7256.

Caron, G., Kazlauskas, R. J. (1993), Sequential kinetic resolution of (±)-2,3-butanediol using lipase from *Pseudomonas cepacia*, *Tetrahedron: Asymmetry* **4**, 1995-2000.

Caron, G., Kazlauskas, R. J. (1994), Isolation of racemic 2,4-pentanediol and 2,5-hexanediol from commercial mixtures of *meso* and racemic isomers by way of cyclic sulfites, *Tetrahedron: Asymmetry* **5**, 657-664.

Carrea, G., Riva, S., Secundo, F., Danieli, B. (1989), Enzymatic synthesis of various 1-*O*-sucrose and 1-*O*-fructose esters, *J. Chem. Soc., Perkin Trans. 1*, 1057-1061.

Carrea, G., Danieli, B., Palmisano, G., Riva, S., Santagostino, M. (1992), Lipase mediated resolution of 2-cyclohexen-1-ols as chiral building blocks en route to eburane alkaloids, *Tetrahedron: Asymmetry* **3**, 775-784.

Carrea, G., Ottolina, G., Riva, S. (1995), Role of solvents in the control of enzyme selectivity in organic media, *Trends Biotechnol.* **13**, 63-70.

Carrillo-Munoz, J. R., Bouvet, D., Guibé-Jampel, E., Loupy, A., Petit, A. (1996), Microwave-promoted lipase-catalyzed reactions. Resolution of (±)-1-phenylethanol, *J. Org. Chem.* **61**, 7746-7749.

Carter, S. F., Leak, D. J. (1995), The isolation and characterization of a carbocyclic epoxide-degrading *Corynebacterium* sp., *Biocatal. Biotransform.* **13**, 111-129.

Casey, J., Macrae, A. R. (1992), Biotechnology and the oleochemical industry, *Inform* **3**, 203-207.

Castillo, E., Marty, A., Combes, D., Condoret, J. S. (1994), Polar substrates for enzymatic reactions in supercritical CO_2: How to overcome the solubility limitation, *Biotechnol. Lett.* **16**, 169-174.

Castillo, E., Dossat, V., Marty, A., Condoret, J. S., Combes, D. (1997), The role of silica gel in lipase-catalyzed esterification reactions of high-polar substrates, *J. Am. Oil Chem. Soc.* **74**, 77-85.

Catelani, C., Mastrorilli, E. (1983), Acid-catalyzed and enzymatic hydrolysis of *trans*- and *cis*-2-methyl-3,4-epoxytetrahydropyran, *J. Chem. Soc., Perkin Trans. 1*, 2717-2721.

Cernia, E., Palocci, C., Gasparrini, F., Misiti, D. (1994a), *Pseudomonas* lipase catalytic activity in supercritical carbon dioxide, *Chem. Biochem. Eng. Q.* **8**, 1-4.

Cernia, E., Palocci, C., Gasparrini, F., Misiti, D., Fagnano, N. (1994b), Enantioselectivity and reactivity of immobilized lipase in supercritical carbon dioxide, *J. Mol. Catal.* **89**, L11-L18.

Chadha, A., Manohar, M. (1995), Enzymic resolution of 2-hydroxy-4-phenylbutanoic acid and 2-hydroxy-4-phenylbutenoic acid, *Tetrahedron: Asymmetry* **6**, 651-652.

Chahid, Z., Montet, D., Pina, M., Graille, J. (1992), Effect of water activity on enzymatic synthesis of alkylglycosides, *Biotechnol. Lett.* **14**, 281-284.

Chamorro, C., Gonzalez-Muniz, R., Conde, S. (1995), Regio- and enantioselectivity of the *Candida antarctica* lipase catalyzed amidations of Cbz-L- and Cbz-D-glutamic acid diesters, *Tetrahedron: Asymmetry* **6**, 2343-2352.

Chan, C., Cox, P. B., Roberts, S. M. (1990), Chemo-enzymatic synthesis of 13-(*S*)-hydroxyoctadecadienoic acid (13-*S*-HODE), *Biocatalysis* **3**, 111-118.

Chang, M. K., Abraham, G., John, V. T. (1990), Production of cocoa butter-like fat from interesterification of vegetable oils, *J. Am. Oil Chem. Soc.* **67**, 832-834.

Chang, P. S., Rhee, J. S., Kim, J. J. (1991), Continuous glycerolysis of olive oil by *Chromobacterium viscosum* lipase immobilized on liposome in reversed micelles, *Biotechnol. Bioeng.* **38**, 1159-1165.

Chang, R. C., Chou, S. J., Shaw, J. F. (1994), Multiple forms and functions of *Candida rugosa* lipase, *Biotechnol. Appl. Biochem.* **19**, 93-97.

Chapman, D. T., Crout, D. H. G., Mahmoudian, M., Scopes, D. I. C., Smith, P. W. (1996), Enantiomerically pure amines by a new method: biotransformation of oxalamic esters using the lipase from *Candida antarctica*, *Chem. Commun.*, 2415-2416.

Charton, E., Macrae, A. R. (1993), Specificities of immobilized *Geotrichum candidum* CMICC 335426 lipases A and B in hydrolysis and ester synthesis reactions in organic solvents, *Enzyme Microb. Technol.* **15**, 489-493.

Chartrain, M., Katz, L., Marcin, C., Thien, M., Smith, S., Fisher, E., Goklen, K., Salmon, P., al., T. B. e. (1993), Purification and characterization of a novel bioconverting lipase from *Pseudomonas aeruginosa* MB 5001, *Enzyme Microb. Technol.* **15**, 575-580.

Chattapadhyay, S., Mamdapur, V. R. (1993), Enzymic esterification of 3-hydroxybutryic acid, *Biotechnol. Lett.* **15**, 245-250.

Chaudhary, A., Beckman, E. J., Russell, A. J. (1995), Rational control of polymer molecular weight and dispersity during enzyme-catalyzed polyester synthesis in supercritical fluids, *J. Am. Chem. Soc.* **117**, 3728-3733.

Cheetham, P. S. J. (1993), The use of biotransformations for the production of flavours and fragrances, *Trends Biotechnol.* **11**, 478-488.

Cheetham, P. S. J. (1997), Combining the technical push and the business pull for natural flavors, *Adv. Biochem. Eng./Biotechnol.* **55**, 1-49.

Chen, S. T., Fang, J. M. (1997), Preparation of optically active tertiary alcohols by enzymatic methods. Application to the synthesis of drugs and natural products, *J. Org. Chem.* **62**, 4349-4357.

Chen, C. S., Liu, Y. C. (1991), Amplification of enantioselectivity in biocatalyzed kinetic resolution of racemic alcohols, *J. Org. Chem.* **56**, 1966-1968.

Chen, C. S., Sih, C. J. (1989), Enantioselective biocatalysis in organic solvents. Lipase catalyzed reactions, *Angew. Chem. Int. Ed. Engl.* **28**, 695-707.

Chen, C.-A., Sih, C. J. (1998), Chemoenzymatic synthesis of conjugated linoleic acid, *J. Org. Chem.* **63**, 9620-9621.

Chen, C. S., Fujimoto, Y., Sih, C. J. (1981), Bifunctional chiral synthons via microbiological methods. 1. Optically active 2,4-dimethylglutaric acid monomethyl esters, *J. Am. Chem. Soc.* **103**, 3580-3582.

Chen, C. S., Fujimoto, Y., Girdaukas, G., Sih, C. J. (1982), Quantitative analyses of biochemical kinetic resolutions of enantiomers, *J. Am. Chem. Soc.* **104**, 7294-7299.

Chen, S.-T., Wang, K.-T., Wong, C.-H. (1986), Chirally selective hydrolysis of D,L-amino acid esters by alkaline protease, *J. Chem. Soc., Chem. Commun.*, 1514-1516.

Chen, C. S., Wu, S. H., Girdaukas, G., Sih, C. J. (1987), Quantitative analyses of biochemical kinetic resolution of enantiomers. 2. Enzyme-catalyzed esterifications in water-organic solvent biphasic systems, *J. Am. Chem. Soc.* **109**, 2812-2817.

Chen, X., Siddiqi, S. M., Schneller, S. W. (1992), Chemoenzymatic synthesis of (–)-carbocyclic 7-deazaoxetanocin G, *Tetrahedron Lett.* **33**, 2249-2252.

Chen, P.-Y., Wu, S.-H., Wang, K.-T. (1993a), Double enantioselective esterification of racemic acids and alcohols by lipase from *Candida cylindracea, Biotechnol. Lett.* **15**, 181-184.

Chen, X.-J., Archelas, A., Furstoss, R. (1993b), Microbiological transformations. 27. The first examples for preparative-scale enantioselective or diastereoselective epoxide hydrolyses using microorganisms. An unequivocal access to all four Bisabolol stereoisomers, *J. Org. Chem.* **58**, 5528-5532.

Chenault, H. K., Dahmer, J., Whitesides, G. M. (1989), Kinetic resolution of unnatural and rarely occurring amino acids: enantioselective hydrolysis of N-acyl amino acids catalyzed by acylase I, *J. Am. Chem. Soc.* **111**, 6354-6364.

Chênevert, R., Courchesne, G. (1995), Enzymatic desymmetrization of *meso(anti-anti)*-2,4-dimethyl-1,3,5-pentanetriol, *Tetrahedron: Asymmetry* **6**, 2093-2096.

Chênevert, R., D'Astous, L. (1988), Enzyme-catalyzed hydrolysis of chlorophenoxypropionates, *Can. J. Chem.* **66**, 1219-1222.

Chênevert, R., Desjardins, M. (1996), Enzymatic desymmetrization of *meso*-2,6-dimethyl-1,7-heptanediol. Enantioselective formal synthesis of the vitamin E side chain and the insect pheromone tribolure, *J. Org. Chem.* **61**, 1219-1222.

Chênevert, R., Dickman, M. (1992), Enzyme-catalyzed hydrolysis of *N*-benzyloxycarbonyl-*cis*-2,6-(acetoxymethyl)piperidine. A facile route to optically active piperidines, *Tetrahedron: Asymmetry* **3**, 1021-1024.

Chênevert, R., Dickman, M. (1996), Enzymatic route to chiral, nonracemic *cis*-2,6- and *cis,cis*-2,4,6-substituted piperidines. Synthesis of (+)-dihydropinidine and dendrobate alkaloid (+)-241D, *J. Org. Chem.* **61**, 3332-3341.

Chênevert, R., Gagnon, R. (1993), Lipase-catalyzed enantioselective esterification or hydrolysis of 1-*O*-alkyl-3-*O*-tosylglycerol derivatives. Practical synthesis of (*S*)-(+)-1-*O*-hexadecyl-2,3-di-*O*-hexadecanoylglycerol, a marine natural product, *J. Org. Chem.* **58**, 1054-1057.

Chênevert, R., Martin, R. (1992), Enantioselective synthesis of (+) and (-)-*cis*-3-aminocyclopentane carboxylic acids by enzymatic asymmetrization, *Tetrahedron: Asymmetry* **3**, 199-200.

Chênevert, R., Thiboutot, S. (1989), Synthesis of chloramphenicol via an enzymatic enantioselective hydrolysis, *Synthesis*, 444-446.

Chênevert, R., Desjardins, M., Gagnon, R. (1990), Enzyme catalyzed asymmetric hydrolysis of chloral acetyl methyl acetal, *Chem. Lett.* 33-34.

Chênevert, R., Rhlid, R. B., Letourneau, M., Gagnon, R., D'Astous, L. (1993), Selectivity in carbonic anhydrase catalyzed hydrolysis of standard N-acetyl-DL-amino acid methyl esters, *Tetrahedron: Asymmetry* **4**, 1137-1140.

Chênevert, R., Lavoie, M., Courchesne, G., Martin, R. (1994a), Chemoenzymatic enantioselective synthesis of amidinomycin, *Chem. Lett.* 93-96.

Chênevert, R., Pouliot, R., Bureau, P. (1994b), Enantioselective hydrolysis of (±)-chloramphenicol palmitate by hydrolases, *Bioorg. Med. Chem. Lett.* **4**, 2941-2944.

Cheong, C. S., Im, D. S., Kim, J., Kim, I. O. (1996), Lipase-catalyzed resolution of primary alcohol containing quaternary chiral carbon, *Biotechnol. Lett.* **18**, 1419-1422.

Chi, Y. M., Nakamura, K., Yano, T. (1988), Enzymic interesterification in supercritical carbon dioxide, *Agric. Biol. Chem.* **52**, 1541-1550.

Chiappe, C., Cordoni, A., Moro, G. L., Palese, C. D. (1998), Deracemization of (±)-*cis*-dialkyl substituted oxides via enantioconvergent hydrolysis catalyzed by microsomal epoxide hydrolase, *Tetrahedron: Asymmetry* **9**, 341-350.

Chibata, I., Tosa, T., Shibatani, T. (1992), The industrial production of optically active compounds by immobilized biocatalysts, in.: *Chirality in Industry* (Collins, A. N., Sheldrake, G. N. , Crosby, J.; Eds.), pp. 352-370. New York: Wiley.

Chinn, M. J., Iacazio, G., Spackman, D. G., Turner, N. J., Roberts, S. M. (1992), Regioselective enzymic acylation of methyl 4,6-*O*-benzylidene-α- and β-D-glucopyranoside, *J. Chem. Soc., Perkin Trans. 1*, 661-662.

Chiou, A. J., Wu, S. H., Wang, K. T. (1992), Enantioselective hydrolysis of hydrophobic amino acid derivatives by lipases, *Biotechnol. Lett.* **14**, 461-464.

Cho, S. W., Rhee, J. S. (1993), Immobilization of lipase for effective interesterification of fats and oils in organic solvent, *Biotechnol. Bioeng.* **41**, 204-210.

Cho, F., de Man, J. M., Allen, O. B. (1994), Modification of hydrogenated canola oil/ palm stearin/ canola oil blens by continuous enzymatic interesterification, *Elaeis* **6**, 25-38.

Choi, W. J., Huh, E. C., Park, H. J., Lee, E. Y., Choi, C. Y. (1998), Kinetic resolution for optically active epoxides by microbial enantioselective hydrolysis, *Biotechnol. Techn.* **12**, 225-228.

Chong, J. M., Mar, E. K. (1991), Preparation of enantiomerically enriched α-hydroxystannanes via enzymatic resolution, *Tetrahedron Lett.* **32**, 5683-5686.

Christophe, A. B. (Ed.) (1998), Structural Modified Food Fats: Synthesis, Biochemistry, and Use, Champaign: AOCS Press.

Chulalaksananukul, W., Condorét, J. S., Combes, D. (1993), Geranyl acetate synthesis by lipase-catalyzed transesterification in supercritical carbon dioxide, *Enzyme Microb. Technol.* **15**, 691-698.

Cipiciani, A., Fringuelli, F., Scappini, A. M. (1996), Enzymatic hydrolyses of acetoxy- and phenethylbenzoates by *Candida cylindracea* lipase, *Tetrahedron* **52**, 9869-9876.

Ciuffreda, P., Colombo, D., Ronchetti, F., Toma, L. (1990), Regioselective acylation of 6-deoxy-L- and -D-hexosides through lipase-catalyzed transesterification. Enhanced reactivity of the 4-hydroxy function in the L series, *J. Org. Chem.* **55**, 4187-4190.

Cleij, M., Archelas, A., Furstoss, R. (1998), Microbiological transformations. part 42: A two-liquid-phase preparative scale process for an epoxide hydrolase catalyzed resolution of *para*-bromo-α-methyl styrene oxide. Occurrence of a surprising enantioselectivity enhancement, *Tetrahedron: Asymmetry* **9**, 1839-1842.

Cohen, S. G., Crossley, J. (1964), Kinetics of hydrolysis of dicarboxylic esters and their a-acetamido dderivatives by α-chymotrypsin, *J. Am. Chem. Soc.* **86**, 4999-5003.

Cohen, S. G., Lo, L. W. (1970), On the active site of α-chymotrypsin. Cyclized and noncyclized substrates with tetrasubstituted α-carbon atoms, *J. Biol. Chem.* **245**, 5718-5727.

Cohen, S. G., Crossley, J., Khedouri, E. (1963), Action of α-chymotrypsin on diethyl N-acetylaspartate and on diethyl N-methyl-N-acetylaspartate, *Biochemistry* **2**, 820-823.

Cohen, S. G., Weinstein, S. Y. (1964), Hydrolysis of D(−)-ethyl β-phenyl-β-hydroxypropionate and D(−)-ethyl β-phenyl-β-acetamidopropionate by α-chymotrypsin, *J. Am. Chem. Soc.* **86**, 725-728.

Cohen, S. G., Schultz, R. M., Weinstein, S. Y. (1966), Stereospecificity in hydrolysis by α-chymotrypsin of esters of α,α-disubstituted acetic and β,β-disubstituted propionic acids, *J. Am. Chem. Soc.* **88**, 5315-5319.

Cohen, S. G., Milanovic, A., Schultz, R. M., Weinstein, S. Y. (1969), On the active site of α-chymotrypsin. Absolute configurations and kinetics of hydrolysis of cyclized and noncyclized substrates, *J. Biol. Chem.* **244**, 2664-2674.

Cohen, M. A., Paratt, J. S., Turner, N. J. (1992), Enantioselective hydrolysis of nitriles and amides using an immobilized whole cell system, *Tetrahedron: Asymmetry* **3**, 1543-1546.

Coleman, M. H., Macrae, A. R. (1977), Rearrangement of fatty acid esters in fat reaction reactants, *German Patent* DE 2 705 608 (Unilever N.V.) (Chem. Abstr. 87: 166 366).

Collins, A. N., Sheldrake, G. N., Crosby, J. (1992), Chirality in industry - the commercial manufacture and application of optically-active compounds, New York: Wiley

Colton, I. J., Ahmed, S. N., Kazlauskas, R. J. (1995), Isopropanol treatment increases enantioselectivity of *Candida rugosa* lipase toward carboxylic acid esters, *J. Org. Chem.* **60**, 212-217.

Comins, D. L., Salvador, J. M. (1993), Efficient synthesis and resolution of *trans*-2-(1-aryl-1-methylethyl)cyclohexanols: practical alternatives to 8-phenylmenthol, *J. Org. Chem.* **58**, 4656-4661.

Coope, J. F., Main, B. G. (1995), Biocatalytic resolution of a tertiary quinuclidinol ester using pig liver esterase, *Tetrahedron: Asymmetry* **6**, 1393-1398.

Cotterill, I. C., Finch, H., Reynolds, D. P., Roberts, S. M., Rzepa, H. S., Short, K. M., Slawin, A. M. Z., Wallis, C. J., Williams, D. J. (1988a), Enzymatic resolution of bicyclo[4.2.0]oct-2-en-7-ol and the preparation of some polysubstituted bicyclo[3.3.0]octan-2-ones via highly strained tricyclo[4.2.0.01,5]1octan-8-ones, *J. Chem. Soc., Chem. Commun.*, 470-472.

Cotterill, I. C., Macfarlane, E. L. A., Roberts, S. M. (1988b), Resolution of bicyclo[3.2.0]hept-2-en-6-ols and bicyclo[4.2.0]oct-2-en-endo-7-ol using lipases, *J. Chem. Soc., Perkin Trans. 1*, 3387-3389.

Cotterill, I. C., Dorman, G., Faber, K., Jaouhari, R., Roberts, S. M., Scheinmann, F., Spreitz, J., Sutherland, A. G., Winders, J. A., Wakefield, B. J. (1990), Chemoenzymic, enantiocomplementary, total asymmetric synthesis of leukotrienes-B3 and -B4, *J. Chem. Soc., Chem. Commun.*, 1661-1663.

Cotterill, I. C., Sutherland, A. G., Roberts, S. M., Grobbauer, R., Spreitz, J., Faber, K. (1991), Enzymatic resolution of sterically demanding bicyclo[3.2.0]heptanes: evidence for a novel hydrolase in crude porcine pancreatic lipase and the advantages of using organic media for some of the biotransformations, *J. Chem. Soc., Perkin Trans. 1*, 1365-1368.

Cousins, R. P. C., Mahmoudian, M., Youds, P. M. (1995), Enzymic resolution of oxathiolane intermediates - an alternative approach to the anti-viral agent lamivudine (3TC), *Tetrahedron: Asymmetry* **6**, 393-396.

Crameri, A., Whitehorn, E. A., Tate, E., Stemmer, W. P. C. (1996), Improved green fluorescent protein by molecular evolution using DNA shuffling, *Nature Biotechnol.* **14**, 315-319.

Crich, J. Z., Brieva, R., Marquart, P., Gu, R. L., Flemming, S., Sih, C. J. (1993), Enzymic asymmetric synthesis of α-amino acids. Enantioselective cleavage of 4-substituted oxazolin-5-ones and thiazolin-5-ones, *J. Org. Chem.* **58**, 3252-3258.

Crosby, J. A., Parratt, J. S., Turner, N. J. (1992), Enzymic hydrolysis of prochiral dinitriles, *Tetrahedron: Asymmetry* **3**, 1547-1550.

Crosby, J., Moilliet, J., Paratt, J. S., Turner, N. J. (1994), Regioselective hydrolysis of aromatic dinitriles using a whole cell catalyst, *J. Chem. Soc., Perkin Trans. 1*, 1679-1687.

Crotti, P., Dibussolo, V., Favero, L., Minutolo, F., Pineschi, M. (1996), Efficient application of lipase-catalyzed transesterification to the resolution of γ-hydroxy ketones, *Tetrahedron: Asymmetry* **7**, 1347-1356.

Crout, D. H. G., Gaudet, V. S. B., Laumen, K., Schneider, M. P. (1986), Enzymatic hydrolysis of (±)-trans-1,2-diacetoxycycloalkanes - a facile route to optically-active cycloalkane-1,2-diols, *J. Chem. Soc., Chem. Commun.*, 808-810.

Crout, D. H. G., MacManus, D. A., Critchley, P. (1990), Enzymatic synthesis of glycosides using the β-galactosidase of *Escherichia coli*: regio- and stereo-chemical studies, *J. Chem. Soc., Perkin Trans. 1*, 1865-1868.

Crout, D. H. G., MacManus, D. A., Critchley, P. (1991), Stereoselective galactosyl transfer to *cis*-cyclohexa-3,5-diene-1,2-diol, *J. Chem. Soc., Chem. Commun.*, 376-378.

Crout, D. H. G., Gaudet, V. S. B., Hallinan, K. O. (1993), Application of biotransformations in organic synthesis: preparation of enantiomerically pure esters of *cis*- and *trans*-epoxysuccinic acid, *J. Chem. Soc., Perkin Trans. 1*, 805-812.

Csomós, P., Kanerva, L. T., Bernáth, G., Fülöp, F. (1996), Biocatalysis for the preparation of optically active β-lactam precursors of amino acids, *Tetrahedron: Asymmetry* **7**, 1789-1796.

Csuk, R., Dorr, P. (1994), Enantioselective enzymatic hydrolyses of building blocks for the synthesis of carbocyclic nucleosides, *Tetrahedron: Asymmetry* **5**, 269-276.

Cuperus, F. P., Bouwer, S. T., Kramer, G. F. H., Derksen, J. T. P. (1994), Lipases used for the production of peroxycarboxylic acids, *Biocatalysis* **9**, 89-96.

Cvetovich, R. J., Chartrain, M., Hartner, F. W., Roberge, C., Amato, J. S., Grabowski, E. J. (1996), An asymmetric synthesis of L-694,458, a human leukocyte elastase inhibitor, via novel enzyme resolution of β-lactam esters, *J. Org. Chem.* **61**, 6575-6580.

Cygler, M., Schrag, J. D., Ergan, F. (1992), Advances in structural understanding of lipases, *Biotechnol. Genet. Eng. Rev.* **10**, 143-184.

Cygler, M., Schrag, J. D., Sussman, J. L., Harel, M., Silman, I., Gentry, M. K., Doctor, B. P. (1993), Relationship between sequence conservation and three-dimensional structure in a large family of esterases, lipases, and related proteins, *Protein Sci.* **2**, 366-382.

Cygler, M., Grochulski, P., Kazlauskas, R. J., Schrag, J. D., Bouthillier, F., Rubin, B., Serreqi, A. N., Gupta, A. K. (1994), Molecular basis for the chiral preferences of lipases, *J. Am. Chem. Soc.* **116**, 3180-3186.

Dabulis, K., Klibanov, A. M. (1993), Dramatic enhancement of enzymatic activity in organic solvents by lyoprotectants, *Biotechnol. Bioeng.* **41**, 566-571.

Daffe, V., Fastrez, J. (1980), Enantiomeric enrichment in the hydrolysis of oxazolones catalyzed by cyclodextrins or proteolytic enzymes, *J. Am. Chem. Soc.* **102**, 3601-3605.

Dalbøge, H., Lange, L. (1998), Using molecular techniques to identify new microbial biocatalysts, *Trends Biotechnol.* **16**, 265-272.

Dalrymple, B. P., Swadling, Y., Cybinski, D. H., Xue, G. P. (1996), Cloning of a gene encoding cinnamoyl ester hydrolase from the ruminal bacterium *Butyrivibrio fibrisolvens* E14 by a novel method, *FEMS Lett.* **143**, 115-120.

Danieli, B., Luisetti, M., Riva, S., Bertinotti, A., Ragg, E., Scaglioni, L., Bombardelli, E. (1995), Regioselective enzyme-mediated acylation of polyhydroxy natural compounds. A remarkable, highly efficient preparation of 6'-*O*-acetyl and 6'-*O*-carboxyacetyl Ginsenoside Rg1, *J. Org. Chem.* **60**, 3637-3642.

Danieli, B., Barra, C., Carrea, G., Riva, S. (1996a), Oxynitrilase-catalyzed transformation of substituted aldehydes. The case of (±)-2-phenylpropionaldehyde and of (±)-3-phenylbutyraldehyde, *Tetrahedron: Asymmetry* **7**, 1675-1682.

Danieli, B., Lesma, G., Passarella, D., Silvani, A. (1996b), Stereoselective enzymatic hydrolysis of dimethyl *meso*-piperidine-3,5-dicarboxylates, *Tetrahedron: Asymmetry* **7**, 345-348.

Danishefsky, S. J., Cabal, M. P., Chow, K. (1989), Novel stereospecific silyl group transfer reaction - practical routes to the prostaglandins, *J. Am. Chem. Soc.* **111**, 3456-3457.

D'Arrigo, P., de Ferra, L., Piergianni, V., Selva, A., Servi, S., Strini, A. (1996), Preparative transformation of natural phospholipids catalysed by phospholipase D from *Streptomyces*, *J. Chem. Soc., Perkin Trans. 1*, 2651-2656.

D'Arrigo, P., de Ferra, L., Piergianni, V., Ricci, A., Scarcelli, D., Servi, S. (1994), Phospholipase D from *Streptomyces* catalyses the transfer of secondary alcohols, *J. Chem. Soc., Chem. Commun.*,, 1709-1710.

De Amici, M., de Micheli, C., Carrea, G., Spezia, S. (1989), Chemoenzymatic synthesis of chiral isoxazole derivatives, *J. Org. Chem.* **54**, 2646-2650.

De Amici, M., de Micheli, C., Carrea, G., Riva, S. (1996), Nitrile oxides in medicinal chemistry. 6. Enzymatic resolution of a set of bicyclic D2-isoxazolines, *Tetrahedron: Asymmetry* **7**, 787-796.

De Bont, J. A. M. (1993), Bioformation of optically pure epoxides, *Tetrahedron: Asymmetry* **4**, 1331-1340.

De Jesus, P. C., Rezende, M. C., Nascimento, M. G. (1995), Enzymic resolution of alcohols via lipases immobilized in microemulsion-based gels, *Tetrahedron: Asymmetry* **6**, 63-66.

De Raadt, A., Klempier, N., Faber, K., Griengl, H. (1992), Chemoselective enzymatic hydrolysis of aliphatic and alicyclic nitriles, *J. Chem. Soc., Perkin Trans. 1*, 137-140.

Deardorff, D. A., Matthews, A. J., McMeekin, D. S., Craney, C. L. (1986), A highly enantioselective hydrolysis of *cis*-3,5-diacetoxycyclopent-1-ene - an enzymatic preparation of 3(*R*)-acetoxy-5(*S*)-hydroxycyclopent-1-ene, *Tetrahedron Lett.* **27**, 1255-1256.

Degueil-Castaing, M., de Jeso, B., Drouillard, S., Maillard, B. (1987), Enzymatic reactions in organic synthesis. 2. Ester interchange of vinyl esters, *Tetrahedron Lett.* **28**, 953-954.

Derewenda, Z. S. (1994), Structure and function of lipases, *Adv. Prot. Chem.* **45**, 1-52.

Derewenda, U., Brzozowski, A. M., Lawson, D. M., Derewenda, Z. S. (1992), Catalysis at the interface: the anatomy of a conformational change in a triglyceride lipase, *Biochemistry* **31**, 1532-1541.

Derewenda, U., Swenson, L., Green, R., Wei, Y., Dodson, G. G., Yamaguchi, S., Haas, M. J., Derewenda, Z. S. (1994a), An unsusal buried polar cluster in a family of fungal lipases, *Nature Struct. Biol.* **1**, 36-47.

Derewenda, U., Swenson, L., Green, R., Wei, Y., Yamaguchi, S., Joerger, R., Haas, M. J., Derewenda, Z. S. (1994b), Current progress in crystallographic studies of new lipases from filamentous fungi, *Protein Engineer.* **7**, 551-557.

Derewenda, U., Swenson, L., Wei, Y., Green, R., Kobos, P. M., Joerger, R., Haas, M. J., Derewenda, Z. S. (1994c), Conformational lability of lipases observed in the absence of an oil-water interface: crystallografic studies of enzymes from the fungi *Humicola lanuginosa* and *Rhizopus delemar*, *J. Lipid Res.* **35**, 524-535.

Desai, S. B., Argade, N. P., Ganesh, K. N. (1996), Remarkable chemo-, regio-, and enantioselectivity in lipase-catalyzed hydrolysis - efficient resolution of (±)-*threo*-ethyl 3-(4-methoxyphenyl)-2,3-diacetoxypropionate leading to chiral intermediates of (+)-diltiazem, *J. Org. Chem.* **61**, 6730-6732.

DeTar, D. F. (1981), Computation of peptide-protein interactions. Catalysis by chymotrypsin: prediction of relative substrate reactivities, *J. Am. Chem. Soc.* **103**, 107-110.

Dinh, P. M., Howarth, J. A., Hudnott, A. R., Williams, J. M. J., Harris, W. (1996), Catalytic racemization of alcohols: applications to enzymic resolution reactions, *Tetrahedron Lett.* **37**, 7623-7626.

Dirlam, N. L., Moore, B. S., Urban, F. J. (1987), Novel synthesis of the aldose reductase inhibitor sorbinil via amidoalkylation, intramolecular oxazolidin-5-one alkylation, and chymotrypsin resolution, *J. Org. Chem.* **52**, 3587-3591.

Djerourou, A.-H., Blanco, L. (1991), Synthesis of optically active 2-sila-1,3-propanediols derivatives by enzymatic transesterification, *Tetrahedron Lett.* **32**, 6325-6326.

Doddema, H. J., Janssens, R. J. J., de Jong, J. P. J., van der Lugt, J. P., Oostrom, H. H. M. (1990), Enzymatic reactions in supercritical carbon dioxide and integrated product-recovery, in.: *Proc. 5th Eur. Congr. Biotechnol.* (Christiansen, L., Munck,. L., Villadsen, J., Eds.), pp. 239-242. Copenhagen: Munksgaard.

Dodson, G., Wlodawer, A. (1998), Catalytic triads and their relatives, *Trends Biochem. Sci.* **23**, 347-352.

Dodson, G. G., Lawson, D. M., Winkler, F. K. (1992), Structural and evolutionary relationships in lipase mechanism and activation, *Faraday Discuss.* 95-105.

Dorset, D. L. (1987), Is the initial solid state acyl shift of 1,2-diglycerides intermolecular?, *Chem. Phys. Lipids* **43**, 179-191.

Doswald, S., Estermann, H., Kupfer, E., Stadler, H., Walther, W., Weisbrod, T., Wirz, B., Wostl, W. (1994), Large scale preparation of chiral building blocks for the P3 site of renin inhibitors, *Bioorg. Med. Chem.* **2**, 403-410.

Drauz, K., Waldmann, H. (Eds.) (1995), *Enzyme catalysis in organic synthesis*, **Vol. 1 & 2**, Weinheim: Wiley-VCH.

Drenth, J., Hol, W. G., Jansonius, J. N., Koekoek, R. (1972), Subtilisin Novo. The three-dimensional structure and its comparison with subtilisin BPN', *Eur. J. Biochem.* **26**, 177-181.

Drioli, S., Felluga, F., Forzato, C., Nitti, P., Pitacco, G. (1996), A facile route to (+)- and (−)-*trans*-tetrahydro-5-oxo-2-pentylfuran-3-carboxylic acid, precursors of (+)- and (−)-methylenolactocin, *Chem. Commun.*, 1289-1290.

Drioli, S., Felluga, F., Forzato, C., Nitti, P., Pitacco, G., Valentin, E. (1998), Synthesis of (+)- and (−)-Phaseolinic acid by combination of enzymatic hydrolysis and chemical transformations with revision of the absolute configuration of the natural product, *J. Org. Chem.* **63**, 2385-2388.

Ducret, A., Giroux, A., Trani, M., Lortie, R. (1995), Enzymatic preparation of biosurfactants from sugars or sugar alcohols and fatty acids in organic media under reduced pressure, *Biotechnol. Bioeng.* **48**, 214-221.

Dudal, Y., Lortie, R. (1995), Influence of water activity on the synthesis of triolein catalyzed by immobilized *Mucor miehei* lipase, *Biotechnol. Bioeng.* **45**, 129-134.

Dufour, E., Storer, A. C., Ménard, R. (1995), Engineering nitrile hydratase activity into a cysteine protease be a single mutation, *Biochemistry* **34**, 16382-16388.

Duggleby, H. J., Tolley, S. P., Hill, C. P., Dodson, E. J., Dodson, G., Moody, P. C. (1995), Penicillin acylase has a single-amino-acid catalytic centre, *Nature* **373**, 264-268.

Duhamel, P., Renouf, P., Cahard, D., Yebga, A., Poirier, J.-M. (1993), Asymmetric synthesis by enzymatic hydrolysis of prochiral dienol diacetate, *Tetrahedron: Asymmetry* **4**, 2447-2450.

Dumont, T., Barth, D., Corbier, C., Branlant, G., Perrut, M. (1992), Enzymatic reaction kinetic: Comparison in an organic solvent and in supercritical carbon dioxide, *Biotechnol. Bioeng.* **39**, 329-333.

Dumortier, L., Liu, P., Dobbelaere, S., van der Eycken, J., Vandewalle, M. (1992), Lipase-catalyzed enantiotoposelective hydrolysis of *meso*-compounds derived from 2-cyclohexene-1,4-diol: synthesis of (–)-conduritol C, (–)-conduritol E, (–)-conduritol F and a homochiral derivative of conduritol A, *Synlett*, 243-245.

Eberling, J., Braun, P., Kowalczyk, D., Schultz, M., Kunz, H. (1996), Chemoselective removal of protecting groups from *O*-glycosyl amino acid and peptide (methoxyethoxy)ethyl esters using lipases and papain, *J. Org. Chem.* **61**, 2638-2646.

Ebiike, H., Terao, Y., Achiwa, K. (1991), Acyloxymethyl as an activating group in lipase-catalyzed enantioselective hydrolysis. A versatile approach to chiral 4-aryl-1,4-dihydro-2,6-dimethyl-3,5-pyridinedicarboxylates, *Tetrahedron Lett.* **32**, 5805-5808.

Edwards, M. L., Matt, J. E., Wenstrup, D. L., Kemper, C. A., Persichetti, P. A., Margolin, A. L. (1996), Synthesis and enzymatic resolution of a carbocyclic analogue of ribofuranosylamine, *Org. Prep. Proc. Intl.* **28**, 193-201.

Effenberger, F., Böhme, J. (1994), Enzyme-catalyzed enantioselective hydrolysis of racemic naproxen nitrile, *Bioorg. Med. Chem.* **2**, 715-721.

Effenberger, F., Gutterer, B., Ziegler, T., Eckhardt, E., Aichholz, R. (1991), Enantioselective esterification of racemic cyanohydrins and enantioselective hydrolysis or transesterification of cyanohydrin esters by lipases, *Liebigs Ann. Chem.* 47-54.

Egloff, M. P., Marguet, F., Bueno, G., Verger, R., Cambillau, C., Tilbeurgh, H. van (1995a), The 2.46 Å resolution structure of the pancreatic lipase-colipase complex inhibited by a C11 alkyl phosphonate, *Biochemistry* **34**, 2751-2762.

Egloff, M. P., Sarda, L., Verger, R., Cambillau, C., Tilbeurgh, H. van (1995b), Crystallographic study of the structure of colipase and of the interaction with pancreatic lipase, *Protein Sci.* **4**, 44-57.

Egri, G., Baitzgacs, E., Poppe, L. (1996), Kinetic resolution of 2-acylated-1,2-diols by lipase-catalyzed enantiomer selective acylation, *Tetrahedron: Asymmetry* **7**, 1437-1448.

Ehrler, J., Seebach, D. (1990), Enantioselective saponification of substituted, achiral 3-acyloxypropionates with lipases: synthesis of chiral derivatives of "Tris(hydroxymethyl)methane", *Liebigs Ann. Chem.*, 379-388.

Eichhorn, E., Roduit, J.-P., Shaw, N., Heinzmann, K., Kiener, A. (1997), Preparation of (*S*)-piperazine-2-carboxylic acid, (*R*)-piperazine-2-carboxylic acid, and (*S*)-piperidine-2-carboxylic acid by kinetic resolution of the corresponding racemic carboxamides with stereoselective amidases in whole bacterial cells, *Tetrahedron: Asymmetry* **8**, 2533-2536.

Eigen, M., Rigler, R. (1994), Sorting single molecules: Application to diagnostics and evolutionary biotechnology, *Proc. Natl. Acad. Sci. USA* **91**, 5740-5747.

Ema, T., Maeno, S., Takaya, Y., Sakai, T., Utaka, M. (1996), Kinetic resolution of racemic 2-substituted 3-cyclopenten-1-ols by lipase-catalyzed transesterifications - a rational strategy to improve enantioselectivity, *J. Org. Chem.* **61**, 8610-8616.

Enzelberger, M. M., Bornscheuer, U. T., Gatfield, I., Schmid, R. D. (1997), Lipase-catalysed resolution of γ- and δ-lactones, *J. Biotechnol.* **56**, 129-133.

Ergan, F., Trani, M., Andre, G. (1990), Production of glycerides from glycerol and fatty acid by immobilized lipases in non-aqueous media, *Biotechnol. Bioeng.* **35**, 195-200.

Ergan, F., Trani, M., Andre, G. (1991), Use of lipases to discriminate against fatty acids of industrial interest, *Inform* **2**, 357.

Erickson, J. C., Schyns, P., Cooney, C. L. (1990), Effect of pressure on an enzymic reaction in a supercritical fluid, *AIChE J.* **36**, 299-301.

Esteban, A. I., Juanes, O., Conde, S., Goya, P., de Clercq, E., Martinez, A. (1995), New 1,2,6-thiadiazine dioxide acylonucleosides: synthesis and antiviral evalutation, *Bioorg. Med. Chem.* **3**, 1527-1535.

Esteban, A. I., Juanes, O., Martinez, A., Conde, S. (1994), Regioselective lipase-mediated acylation-deacylation in thiadiazine diacyclonucleosides, *Tetrahedron* **50**, 13865-13870.

Estell, D. A., Graycar, T. P., Miller, J. V., Powers, D. B., Burnier, J. P., Ng, P. G., Wells, J. A. (1986), Probing steric and hydrophobic effects on enzyme-substrate interactions by protein engineering, *Science* **233**, 659-663.

Estermann, H., Prasad, K., Shapiro, M. J., Bolsterli, J. J., Walkinshaw, M. D. (1990), Enzyme-catalyzed asymmetrization of bis(hydroxymethyl)-tetrahydrofurans, *Tetrahedron Lett.* **31**, 445-448.

Evans, C. T., Roberts, S. M., Shoberu, K. A., Sutherland, A. G. (1992), Potential use of carbocyclic nucleosides for the treatment of AIDS: chemo-enzymic syntheses of the enantiomers of carbovir, *J. Chem. Soc., Perkin Trans. 1*, 589-592.

Exl, C., Hoenig, H., Renner, G., Rogi-Kohlenprath, R., Seebauer, V., Seufer-Wasserthal, P. (1992), How large are the active sites of the lipases from *Candida rugosa* and from *Pseudomonas cepacia*?, *Tetrahedron: Asymmetry* **3**, 1391-1394.

Faber, K. (1997), *Biotransformations in Organic Chemistry*, 3rd Ed., Heidelberg: Springer.

Faber, K., Riva, S. (1992), Enzyme-catalyzed irreversible acyl transfer, *Synthesis*, 895-910.

Faber, K., Ottolina, G., Riva, S. (1993), Selectivity-enhancement of hydrolase reactions, *Biocatalysis* **8**, 91-132.

Faber, K., Mischitz, M., Kroutil, W. (1996), Microbial epoxide hydrolases, *Acta Chem. Scand.* **50**, 249-258.

Fabre, J., Betbeder, D., Paul, F., Monsan, P., Périé, J. (1993a), Regiospecific enzymic acylation of butyl-α-D-glucopyranoside, *Carbohydr. Res.* **243**, 407-411.

Fabre, J., Betbeder, D., Paul, F., Monsan, P., Périé, J. (1993b), Versatile enzymatic diacid ester synthesis of butyl α-D-glucopyranoside, *Tetrahedron* **49**, 10877-10882.

Fabre, J., Paul, F., Monsan, P., Blonski, C., Périé, J. (1994), Enzymic synthesis of amino acid ester of butyl α-D-glucopyranoside, *Tetrahedron Lett.* **35**, 3535-3536.

Fadel, A., Arzel, P. (1997a), Asymmetric construction of benzylic quaternary carbons by lipase-mediated enantioselective transesterification of prochiral α,α-disubstituted 1,3-propanediols, *Tetrahedron: Asymmetry* **8**, 283-291.

Fadel, A., Arzel, P. (1997b), Enzymatic asymmetrisation of prochiral α,α-disubstituted-malonates and -1,3-propanediols: formal asymmetric syntheses of (−)-aphanorphine and (+)-eptazocine, *Tetrahedron:Asymmetry* **8**, 371-374.

Fadnavis, N. W., Koteshwar, K. (1997), Remote control of stereoselectivity: lipase catalyzed enantioselective esterification of racemic α-lipoic acid, *Tetrahedron: Asymmetry* **8**, 337-339.

Falk-Heppner, M., Keller, M., Prinzbach, H. (1989), Ungesättigte 1,4-*cis*- und 1,4-*trans*-Diamino-tetradesoxycycloheptite - enantiomerenreine, polyfunktionalisierte Tropa-Derivate durch enzymatische Hydrolyse, *Angew. Chem.* **101**, 1281-1283.

Fallon, R. D., B.Stieglitz, Jr., I. T. (1997), A *Pseudomonas putida* capable of stereoselective hydrolysis of nitriles, *Appl. Microbiol. Biotechnol.* **47**, 156-161.

Farb, D., Jencks, W. D. (1980), Different forms of pig liver esterase, *Arch. Biochem. Biophys.* **203**, 214-226.

Farias, R. N., Torres, M., Canela, R. (1997), Spectrophotometric determination of the positional specificity of nonspecific and 1,3-specific lipases, *Anal. Biochem.* **252**, 186-189.

Fellous, R., Lizzani-Cuvelier, L., Loiseau, M. A., Sassy, E. (1994), Resolution of racemic ε-lactones, *Tetrahedron: Asymmetry* **5**, 343-346.

Fernández, S., Brieva, R., Rebolledo, F., Gotor, V. (1992), Lipase -catalyzed enantioselective acylation of *N*-protected or unprotected 2-aminoalkan-1-ols, *J. Chem. Soc., Perkin Trans. 1*, 2885-2889.

Fernández, S., Ferrero, M., Gotor, V., Okamura, W. H. (1995), Selective acylation of A-ring precursors of vitamin D using enzymes in organic solvents, *J. Org. Chem.* **60**, 6057-6061.

Ferraboschi, P., Brembilla, D., Grisenti, P., Santaniello, E. (1991), Enzymatic resolution of 2-substituted oxiranemethanols, a class of synthetically useful building blocks bearing a quaternary center, *J. Org. Chem.* **56**, 5478-5480.

Ferraboschi, P., Casati, S., Grisenti, P., Santaniello, E. (1993), A chemoenzymatic approach to enantiomerically-pure (*R*)- and (*S*)-2,3-epoxy-2-(4-pentenyl)-propanol, a chiral building block for the synthesis of (*R*)- and (*S*)-frontalin, *Tetrahedron: Asymmetry* **4**, 9-12.

Ferraboschi, P., Casati, S., Grisenti, P., Santaniello, E. (1994a), Selective enzymatic transformations of itaconic acid derivatives - an access to potentially useful building blocks, *Tetrahedron* **50**, 3251-3258.

Ferraboschi, P., Grisenti, P., Casati, S., Santaniello, E. (1994b), A new synthesis of (*R*)- and (*S*)-mevalonolactone from the enzymatic resolution of (*R,S*)-2-(3-methyl-2-butenyl)-oxiranemethanol, *Synlett*, 754-756.

Ferraboschi, P., Grisenti, P., Manzocchi, A., Santaniello, E. (1994c), Regio- and enantioselectivity of *Pseudomonas cepacia* lipase in the transesterification of 2-substituted-1,4-butanediols, *Tetrahedron: Asymmetry* **5**, 691-698.

Ferraboschi, P., Casati, S., Manzocchi, A., Santaniello, E. (1995a), Studies on the regio- and enantioselectivity of the lipase-catalyzed transesterification of 1'- and 2'-naphthyl alcohols in organic solvent, *Tetrahedron: Asymmetry* **6**, 1521-1524.

Ferraboschi, P., Casati, S., Verza, E., Santaniello, E. (1995b), Regio- and enantioselective properties of the lipase-catalyzed irreversible transesterification of some 2-substituted-1,4- butanediols in organic solvents, *Tetrahedron: Asymmetry* **6**, 1027-1030.

Ferraboschi, P., Molatore, A., Verza, E., Santaniello, E. (1996), The first example of lipase-catalyzed resolution of a stereogenic center in steroid side chains by transesterification in organic solvent, *Tetrahedron: Asymmetry* **7**, 1551-1554.

Ferreira-Dias, S., Fonseca, M. M. R. (1993), Enzymatic glycerolysis of olive oil: a reactional system with major analytic problems, *Biotechnol. Tech.* **7**, 447-452.

Ferreira-Dias, S., de Fonseca, M. M. R. (1995), Production of monoglycerides by glycerolysis of olive oil with immobilized lipases: effect of the water activity, *Bioprocess Eng.* **12**, 327-337.

Ferrero, M., Fernández, S., Gotor, V. (1997), Selective alkoxycarbonylation of A-ring precursors of vitamin D using enzymes in organic solvents. Chemoenzymatic synthesis of 1α,25-dihydroxyvitamin D3 C-5 A-ring carbamate derivatives, *J. Org. Chem.* **62**, 4358-4363.

Fiandanese, V., Hassan, O., Naso, F., Scilimati, A. (1993), A highly efficient kinetic resolution of ω-trimethylsilyl polyunsaturated secondary alcohols by lipase catalyzed transesterification, *Synlett*, 491-493.

Fiaud, J. C., Gil, R., Legros, J. Y., Aribi-Zouioueche, L., Koenig, W. A. (1992), Kinetic resolution of 3-*t*-butyl and 3-phenyl cyclobutylidinethanols through lipase-catalyzed acylation with succinic anhydride, *Tetrahedron Lett.* **33**, 6967-6970.

Fiechter, A. (1992), Biosurfactants; moving towards industrial application, *Trends Biotechnol.* **6**, 208-217.

Fink, A. L., Hay, G. W. (1969), Enzymic deacylation of esterified mono- and disaccharides. IV. Products of esterase-catalyzed deacetylations, *Can. J. Biochem.* **47**, 353-359.

Fischer, A., Bommarius, A. S., Drauz, K., Wandrey, C. (1994), A novel approach to enzymic peptide synthesis using highly solubilizing N^{α}-protecting groups of amino acids, *Biocatalysis* **8**, 289-307.

Fitzpatrick, P. A., Klibanov, A. M. (1991), How can the solvent affect enantioselectivity?, *J. Am. Chem. Soc.* **113**, 3166-3171.

Fitzpatrick, P. A., Steinmetz, A. C. U., Ringe, D., Klibanov, A. M. (1993), Enzyme crystal structure in a neat organic solvent, *Proc. Natl. Acad. Sci. USA* **90**, 8653-8657.

Fletcher, P. D. I., Robinson, B. H., Freedman, R. B., Olfield, C. (1985), Activity of lipase in water in oil microemulsions, *J. Chem. Soc. Faraday Trans. 1*, **81**, 2667-2679.

Fletcher, P. D. I., Freedman, R. B., Robinson, B. H., Rees, G. D., Schomäcker, R. (1987), Lipase-catalyzed ester synthesis in oil-continuous microemulsions, *Biochim. Biophys. Acta* **912**, 278-282.

Fouque, E., Rousseau, G. (1989), Enzymatic resolution of medium-ring lactones - synthesis of (S)-(+)-phoracantholide-I, *Synthesis*, 661-666.

Fowler, P. W., Macfarlane, E. L. A., Roberts, S. M. (1991), Highly diastereoselective inter-esterification reactions involving a racemic acetate and a racemic carboxylic acid catalysed by lipase enzymes, *J. Chem. Soc., Chem. Commun.*, 453-455.

Franssen, M. C. R., Jongejan, H., Kooijman, H., Spek, A. L., Mondril, N. L. F. L. C., de Santos, P. M. A. C. B., de Groot, A. (1996), Resolution of a tetrahydrofuran ester by *Candida rugosa* lipase (CRL) and an examination of CRL's stereochemical preference in organic media, *Tetrahedron: Asymmetry* **7**, 497-510.

Fregapane, G., Sarney, D. B., Vulfson, E. N. (1991), Enzymic solvent-free synthesis of sugar acetal fatty acid esters, *Enzyme Microb. Technol.* **13**, 796-800.

Fregapane, G., Sarney, D. B., Vulfson, E. N. (1994), Facile chemo-enzymatic synthesis of monosaccharide fatty acid esters, *Biocatalysis* **11**, 9-18.

Frykman, H., Öhrner, N., Norin, T., Hult, K. (1993), S-Ethyl thiooctanoate as acyl donor in lipase-catalyzed resolution of secondary alcohols, *Tetrahedron Lett.* **34**, 1367-1370.

Fuganti, C., Grasselli, P., Servi, S., Lazzarini, A., Casati, P. (1988), Substrate specificity and enantioselectivity of penicillinacylase catalyzed hydrolysis of phenacetyl esters of synthetically useful carbinols, *Tetrahedron* **44**, 2575-2582.

Fuganti, C., Rosell, C. M., Servi, S., Tagliani, A., Terreni, M. (1992), Enantioselective recognition of the phenylacetyl moiety in the penicillin G acylase catalyzed hydrolysis of phenylacetate esters, *Tetrahedron: Asymmetry* **3**, 383-386.

Fuganti, C., Grasselli, P., Mendozza, M., Servi, S., Zucchi, G. (1997), Microbially-aided preparation of (S)-2-methoxycyclohexanone key intermediate in the synthesis of sanfetrinen, *Tetrahedron* **53**, 2617-2624.

Fukusaki, E., Senda, S., Nakazono, Y., Omata, T. (1991), Lipase-catalyzed kinetic resolution of methyl 4-hydroxy-5- tetradecynoate and its application to a facile synthesis of Japanese beetle pheromone, *Tetrahedron* **47**, 6223-6230.

Fukusaki, E., Senda, S., Nakazono, Y., Omata, T. (1992a), Synthesis of the enantiomers of (Z-5-(1-octenyl)oxycyclopentan-2-one, a sey pheromone of the cupreous chafer beetle, *Anomala cuprea* Hope, *Biosci. Biotechnol. Biochem.* **56**, 1160-1161.

Fukusaki, E., Senda, S., Nakazono, Y., Yuasa, H., Omata, T. (1992b), Lipase-catalyzed kinetic resolution of 2,3-epoxy-8-methyl-1-nonanol, the key intermediate in the synthesis of the gypsy moth pheromone, *J. Ferment. Bioeng.* **73**, 280-283.

Fukusaki, E., Senda, S., Nakazono, Y., Yuasa, H., Omata, T. (1992c), Preparation of carboxyalkyl acrylate by lipase-catalyzed regioselective hydrolysis of corresponding methyl ester, *Bioorg. Med. Chem. Lett.* **2**, 411-414.

Fukusaki, E.-I., Satoda, S., Senda, S., Omata, T. (1998), Lipase-catalyzed kinetic resolution of 4-hydroxy-5-dodecynonitrile and its application to facile synthesis of a cupreous chafer beetle sex pheromone, *J. Ferment. Bioeng.* **86**, 508-509.

Fülling, G., Sih, C. J. (1987), Enzymatic second-order asymmetric hydrolysis of ketorolac esters: in situ racemization, *J. Am. Chem. Soc.* **109**, 2845-2846.

Furui, M., Furutani, T., Shibatani, T., Nakamoto, Y., Mori, T. H. (1996), A membrane bioreactor combined with crystallizer for production of optically active (2R,3S)-3-(4-methoxy-phenyl)glycidic acid methyl ester, *J. Ferment. Bioeng.* **81**, 21-25.

Furukawa, M., Kodera, Y., Uemura, T., Hiroto, M., Matsushima, A., Kuno, H., Matsushita, H., Inada, Y. (1994), Alcoholysis of ε-decalactone with polyethylene glycol-modified lipase in 1,1,1-trichloroethane, *Biochem. Biophys. Res. Commun.*, **199**, 41-45.

Gais, H. J., Elferink, V. H. M. (1995), Hydrolysis and formation of C–O-bonds, in.: *Enzyme Catalysis in Organic Synthesis* (Drauz, K. , Waldmann, H.; Eds.), **Vol. 1**, pp. 165-271 (ref. on pp 343-348). Weinheim: Wiley-VCH.

Gais, H.-J., Lukas, K. L. (1984), Enantioselective and enantioconvergent syntheses of building-blocks for the total synthesis of cyclopentanoid natural products, *Angew. Chem. Int. Ed. Engl.* **23**, 142-143.

Gais, H. J., von der Weiden, I. (1996), Preparation of enantiomerically pure α-hydroxymethyl S-tert-butyl sulfones by *Candida antarctica* lipase catalyzed resolution, *Tetrahedron: Asymmetry* **7**, 1253-1256.

Gais, H.-J., Lukas, K. L., Ball, W. A., Braun, S., Lindner, H. J. (1986), Enzyme-catalyzed asymmetric synthesis. 4. Synthesis of homochiral building-blocks for the enantioselective total synthesis of cyclopentanoids with an esterase-catalyzed asymmetric reaction as key step, *Liebigs Ann. Chem.*, 687-716.

Gais, H.-J., Zeissler, A., Maidonis, P. (1988), Diastereoselective D-galactopyranosyl transfer to *meso* diols catalyzed by β–galactosidases, *Tetrahedron Lett.* **29**, 5743-5744.

Gais, H.-J., Bülow, G., Zatorski, A., Jentsch, M., Maidonis, P., Hemmerle, H. (1989), Enzyme-catalyzed asymmetric synthesis. Enantioselectivity of pig liver esterase catalyzed hydrolyses of 4-substituted *meso*-cyclopentane 1,2-diesters, *J. Org. Chem.* **54**, 5115-5122.

Gais, H. J., Hemmerle, H., Kossek, S. (1992), Enzyme-catalyzed asymmetric synthesis. 10. Pseudomonas cepacia lipase mediated synthesis of enantiomerically pure (2R,3S)- and (2S,3R)-2,3-O-cyclohexylideneerythritol monoacetate from 2,3-O-cyclohexylideneerythritol, *Synthesis*, 169-173.

Gala, D., de Benedetto, D. J., Clark, J. E., Murphy, B. L., Schumacher, D. P., Steinman, M. (1996), Preparations of antifungal Sch 42427/SM 9164: preparative chromatographic resolution, and total asymmetric synthesis by enzymatic preparation of chiral α-hydroxy arylketones, *Tetrahedron Lett.* **37**, 611-614.

Gamalevich, G. D., Serebryakov, E. P., Vlasyuk, A. L. (1998), Lipase-mediated stereodivergent synthesis of both enantiomers of 4-(2,6-dimethylheptyl)benzoic acid, *Mendeleev Commun.*, 8-9.

Gancet, C. (1990), Catalysis by *Rhizopus arrhizus* mycelial lipase, *Ann. N. Y. Acad. Sci.* **613**, 600-604.

Gandhi, N. N. (1997), Applications of lipase, *J. Am. Oil Chem. Soc.* **74**, 621-634.

Garbay-Jaureguiberry, C., McCort-Tranchepain, I., Barbe, B., Ficheux, D., Roques, B. P. (1992), Improved synthesis of [*p*-phosphono and *p*-sulfo]methylphenylalanine. Resolution of [*p*-phosphono-, *p*-sulfo-, *p*-carboxy- and *p*-*N*-hydroxycarboxamido-]methylphenylalanine, *Tetrahedron: Asymmetry* **3**, 637-650.

Garcia, M. J., Azerad, R. (1997), Production of ring-substituted D-phenylglycines by microbial or enzymatic hydrolysis/deracemization of the corresponding DL-hydantoins, *Tetrahedron: Asymmetry* **8**, 85-92.

Garcia, M. J., Rebolledo, F., Gotor, V. (1992), Enzymic synthesis of 3-hydroxybutyramides and their conversion to optically active 1,3-amino alcohols, *Tetrahedron: Asymmetry* **3**, 1519-1522.

Garcia, M. J., Rebolledo, F., Gotor, V. (1993), Practical enzymic route to optically active 3-hydroxyamides. Synthesis of 1,3-amino alcohols, *Tetrahedron: Asymmetry* **4**, 2199-2210.

Garcia-Alles, L. F., Moris, F., Gotor, V. (1993), Chemo-enzymatic synthesis of 2'-deoxynucleosides urethanes, *Tetrahedron Lett.* **34**, 6337-6338.

Garcia-Granados, A., Parra, A., Simeó, Y., Extremera, A. L. (1998), Chemical, enzymatic and microbiological synthesis of 8,12-eudesmanolides: synthesis of sivasinolide and yomogin analogues, *Tetrahedron* **54**, 14421-14436.

Gardossi, L., Bianchi, D., Klibanov, A. M. (1991), Selective acylation of peptides catalyzed by lipases in organic solvents, *J. Am. Chem. Soc.* **113**, 6328-6329.

Garzon-Aburbeh, A., Poupaert, J. H., Claesen, M., Dumont, P., Atassi, G. (1983), 1,3-Dipalmitoylglycerol ester of chlorambucil as lymphotropic, orally administrable antineoplastic agent, *J. Med. Chem.* **26**, 1200-1203.

Garzon-Aburbeh, A., Poupaert, J. H., Claesen, M., Dumont, P. (1986), A lymphotrophic prodrug of L-Dopa: Synthesis, pharmacological properties and pharmacokinetic behaviour of 1,3-dihexadecanoyl-2-[(S)-2-amino-3-(3,4-dihydroxy-phenyl)propanoyl] propane-1,2,3,-triol, *J. Med. Chem.* **29**, 687-691.

Gaspar, J., Guerrero, A. (1995), Lipase-catalyzed enantioselective synthesis of naphthyl trifluoromethyl carbinols and their corresponding non-fluorinated counterparts, *Tetrahedron: Asymmetry* **6**, 231-238.

Gatfield, I. L. (1984), The enzymic synthesis of esters in nonaqueous systems, *Ann. N. Y. Acad. Sci.* **434**, 569-572.

Gaucher, A., Ollivier, J., Marguerite, J., Paugam, R., Salaun, J. (1994), Total asymmetric syntheses of (1*S*,2*S*)-norcoronamic acid, and of (1*R*,2*R*)- and (1*S*,2*S*)-coronamic acids from the diastereoselective cyclization of 2-(N-benzylideneamino)-4-chlorobutyronitriles, *Can. J. Chem.* **72**, 1312-1327.

Gavagnan, J. E., Fager, S. K., Fallon, R. D., Folsom, P. W., Herkes, F. E., Eisenberg, A., Hann, E. C., DiCosimo, R. (1998), Chemoenzymatic production of lactams from aliphatic α,ω-dinitriles, *J. Org. Chem.* **63**, 4792-4801.

Gelo-Pujic, M., Guibe-Jampel, e., Loupy, A., Galema, S. A., Mathe, D. (1996), Lipase-catalyzed esterification of some α-D-glucopyranosides in dry media under focused microwave irradiation, *J. Chem. Soc., Perkin Trans. 1*, 2777-2780.

Geresh, S., Gilboa, Y. (1990), Enzymatic synthesis of alkyds, *Biotechnol. Bioeng.* **36**, 270-274.

Geresh, S., Gilboa, Y. (1991), Enzymic syntheses of alkyds. II. lipase-catalyzed polytransesterification of dichloroethyl fumarate with aliphatic and aromatic diols, *Biotechnol. Bioeng.* **37**, 883-888.

Ghogare, A., Kumar, G. S. (1989), Oxime esters as novel irreversible acyl transfer agents for lipase catalysis in organic media, *J. Chem. Soc., Chem. Commun.*, 1533-1535.

Ghogare, A., Kumar, G. S. (1990), Novel route to chiral polymers involving biocatalytic transesterification of *O*-acryloyl oximes, *J. Chem. Soc., Chem. Commun.*, 134-135.

Ghosh, A. K., Chen, Y. (1995), Synthesis and optical resolution of high affinity P2-ligands for HIV-1- protease inhibitors, *Tetrahedron Lett.* **36**, 505-508.

Ghosh, D., Wawrzak, Z., Pletnev, V. Z., Li, N. Y., Kaiser, R., Pangborn, W., Jornvall, H., Erman, M., Duax, W. L. (1995), Structure of uncomplexed and linoleate-bound *Candida cylindracea* cholesterol esterase, *Structure* **3**, 279-288.

Ghosh, A. K., Kincaid, J. F., Haske, M. G. (1997), A convenient enzymatic route to optically active 1-aminoindan-2-ol: versatile ligands for HIV-1 protease inhibitors and asymmetric syntheses, *Synthesis*, 541-544.

Gil, G., Ferre, E., Meou, A., Petit, J. L., Triantaphylides, C. (1987), Lipase-catalyzed ester formation in organic solvents. Partial resolution of primary allenic alcohols, *Tetrahedron Lett.* **28**, 1647-1648.

Gilbert, E. J. (1993), *Pseudomonas* lipases: Biochemical properties and molecular cloning, *Enzyme Microb. Technol.* **15**, 634-645.

Gill, I., Valivety, R. (1997a), Polyunsaturated fatty acids, Part 1: Occurence, biological activities and applications, *Trends Biotechnol.* **15**, 401-409.

Gill, I., Valivety, R. (1997b), Polyunsaturated fatty acids, Part 2: Biotransformations and biotechnological applications, *Trends Biotechnol.* **15**, 470-478.

Giver, L., Gershenson, A., Freskgard, P.-O., Arnold, F. H. (1998), Directed evolution of a thermostable esterase, *Proc. Natl. Acad. Sci. USA* **95**, 12809-12813.

Gjellesvik, D. R., Lorens, J. B., Male, R. (1994), Pancreatic carboxylester lipase from Atlantic salmon (*Salmo salar*) cDNA sequence and computer-assisted modelling of tertiary structure, *Eur. J. Biochem.* **226**, 603-612.

Goderis, H. L., Ampe, G., Feyten, M. P., Fouwe, B. L., Guffens, W. M., Cauwenbergh, S. M. V., Tobback, P. P. (1987), Lipase-catalyzed ester exchange reactions in organic media with controlled humidity, *Biotechnol. Bioeng.* **30**, 258-266.

Goergens, U., Schneider, M. P. (1991a), Enzymatic preparation of enantiomerically pure and selectively protected 1,2- and 1,3-diols, *J. Chem. Soc., Chem. Commun.*, 1066-1068.

Goergens, U., Schneider, M. P. (1991b), A facile chemoenzymatic route to enantiomerically-pure oxiranes: building blocks for biologically active compounds, *J. Chem. Soc., Chem. Commun.*, 1064-1066.

Goj, O., Burchardt, A., Haufe, G. (1997), A versatile approach to optically active primary 2-fluoro-2-phenylalkanols through lipase-catalyzed transformations, *Tetrahedron: Asymmetry* **8**, 399-408.

Goldberg, M., Parvaresh, F., Thomas, D., Legoy, M. D. (1988), Enzymatic ester synthesis with continuous measurement of water activity, *Biochim. Biophys. Acta* **957**, 359-362.

Goldberg, M., Thomas, D., Legoy, M. D. (1990), Water activity as a key parameter of synthesis reactions: the example of lipase in biphasic (liquid/solid) media, *Enzyme Microb. Technol.* **12**, 976-981.

Gopalan, V., van der Jagt, D. J., Libell, D. P., Glew, R. H. (1992), Transglucosylation as a probe of the mechanism of action of mammalian cytosolic β-glucosidase, *J. Biol. Chem.* **267**, 9629-9638.

Goto, M., Kamiya, N., Miyata, M., Nakashio, F. (1994), Enzymatic esterification by surfactant-coated lipase in organic media, *Biotechnol. Prog.* **10**, 263-268.

Goto, M., Goto, M., Kamiya, N., Nakashio, F. (1995), Enzymic interesterification of triglyceride with surfactant-coated lipase in organic media, *Biotechnol. Bioeng.* **45**, 27-32.

Gotor, V., Morís, F. (1992), Regioselective acylation of 2'-deoxynucleosides through an enzymatic reaction with oxime esters, *Synthesis*, 626-628.

Gotor, V., Pulido, R. (1991), An improved procedure for regioselective acylation of carbohydrates: novel enzymatic acylation of α-D-glucopyranose and methyl α-D-glucopyranoside, *J. Chem. Soc., Perkin Trans. 1*, 491-492.

Gotor, V., Astorga, C., Rebollodo, F. (1990), An improved method for the preparation of acylhydrazines: the first example of an enzymatic hydrazinolysis reaction, *Synlett*, 387-388.

Gotor, V., Menendez, E., Mouloungui, Z., Gaset, A. (1993), Synthesis of optically active amides from β-furyl and β-phenyl esters by way of enzymic aminolysis, *J. Chem. Soc., Perkin Trans. 1*, 2453-2456.

Gou, D.-M., Liu, Y.-C., Chen, C.-S. (1992), An efficient chemoenzymic access to optically active *myo*-inositol polyphosphates, *Carbohydr. Res.* **234**, 51-64.

Gou, D.-M., Liu, Y.-C., Chen, C.-S. (1993), A practical chemoenzymatic synthesis of the taxol C-13 side chain N-benzoyl-(2R,2S)-3-phenylisoserine, *J. Org. Chem.* **58**, 1287-1289.

Gradley, M. L., Knowles, C. J. (1994), Asymmetric hydrolysis of chiral nitriles by *Rhodococcus rhodochrous* NCIMB 11216 nitrilase, *Biotechnol. Lett.* **16**, 41-46.

Gradley, M. L., Deverson, C. J. F., Knowles, C. J. (1994), Asymmetric hydrolysis of R-(−), S(+)-2-methylbutyronitrile by *Rhodococcus rhodochrous* NCIMB 11216, *Arch. Microbiol.* **161**, 246-251.

Graham, L. D., Haggett, K. D., Jennings, P. A., Le Brocque, D. S., Whittaker, R. G., Schober, P. A. (1993), Random mutagenesis of the substrate-binding site of a serine protease can generate enzymes with increased activities and altered primary specificities, *Biochemistry* **32**, 6250-6258.

Greener, A., Callahan, M., Jerpseth, B. (1996), An efficient random mutagenesis technique using *E. coli* mutator strain, *Methods Mol. Biol.* **57**, 375-385.

Greenstein, J. P., Winitz, M. (1961), Enzymes involved in the determination, characterization, and preparation of amino acids, in.: *Chemistry of the amino acids*, **Vol. 2**, pp. 1753-1816. New York: Wiley.

Griffin, M., Trudgill, P. W. (1976), Purification and properties of cyclopentanone oxygenase of *Pseudomonas* NCIB 9872, *Eur. J. Biochem.* **63**, 199-209.

Griffith, D. A., Danishefsky, S. J. (1991), Total synthesis of allosamidin - an application of the sulfonamidoglycosylation of glycals, *J. Am. Chem. Soc.* **113**, 5863-5864.

Griffith, O. H., Volwerk, J. J., Kuppe, A. (1991), Phosphatidylinositol-specific phospholipases C from *Bacillus cereus* and *Bacillus thuringiensis*, *Methods Enzymol.* **197**, 493-502.

Grochulski, P., Li, Y., Schrag, J. D., Bouthillier, F., Smith, P., Harrison, D., Rubin, B., Cygler, M. (1993), Insight into interfacial activation from an 'open' structure of *Candida rugosa* lipase, *J. Biol. Chem* **268**, 12843-12847.

Grochulski, P., Bouthillier, F., Kazlauskas, R. J., Serreqi, A. N., Schrag, J. D., Ziomek, E., Cygler, M. (1994), *Candida rugosa* lipase-inhibitor complexes simulating the acylation and deacylation transition states, *Biochemistry* **33**, 3494-3500.

Groeger, U., Drauz, K., Klenk, H. (1990), Transformation of amino acids. 4. Isolation of an L-stereospecific N-acyl-L-proline acylase and its use as a catalyst in organic synthesis, *Angew. Chem., Int. Ed. Engl.* **29**, 417-418.

Groeger, U., Drauz, K., Klenk, H. (1992), Amino acid transformation. 8. Enzymic resolution of N-alkyl amino acids, *Angew. Chem., Int. Ed. Engl.* **31**, 195-197.

Grogan, G., Roberts, S. M., Willetts, A. J. (1996), Novel aliphatic epoxide hydrolase activities from dematiaceous fungi, *FEMS Microbiol. Lett.* **141**, 239-243.

Grogan, G., Rippe, C., Willetts, A. (1997), Biohydrolysis of substituted styrene oxides by *Beauveria densa* CMC 3240, *J. Mol. Catal. B: Enzymatic* **3**, 253-257.

Gros, P., Betzel, C., Dauter, Z., Wilson, K. S., Hol, W. G. J. (1989), Molecular dynamics refinement of a thermitase-eglin-c complex at 1.98 A resolution and comparison of two crystal forms that differ in calcium content, *J. Mol. Biol.* **210**, 347-367.

Gruber-Khadjawi, M., Hönig, H., Illaszewicz, C. (1996), Chemoenzymatic methods for the preparation of optically active cyclic polyazido alcohols from easily available chiral starting materials, *Tetrahedron: Asymmetry* **7**, 807-814.

Gu, J., Li, Z. (1992), Enzyme reaction in aqueous-organic system. Kinetic resolution of racemic oxiranecarboxylic esters with lipase, *Chin. Chem. Lett.* **3**, 173-174.

Gu, Q. M., Sih, C. J. (1992), Improving the enantioselectivity of the *Candida cylindracea* lipase via chemical modification, *Biocatalysis* **6**, 115-126.

Gu, Q.-M., Chen, C.-S., Sih, C. J. (1986), A facile enzymatic resolution process for the preparation of (+)-*S*-2-(6-methoxy-2-naphthyl)propionic acid (naproxen), *Tetrahedron Lett.* **27**, 1763-1766.

Gu, R. L., Lee, I. S., Sih, C. J. (1992), Chemo-enzymic asymmetric synthesis of amino acids. Enantioselective hydrolyses of 2-phenyl-oxazolin-5-ones, *Tetrahedron Lett.* **33**, 1953-1956.

Guanti, G., Riva, R. (1995), Synthesis of optically active *N*-benzyl-2,4-bis(hydroxymethyl) substituted azetidines by lipase catalyzed acetylations, *Tetrahedron: Asymmetry* **6**, 2921-2924.

Guanti, G., Banfi, L., Narisano, E., Riva, R., Thea, S. (1986), Enzymes in asymmetric synthesis: Effect of reaction media on the PLE catalysed hydrolysis of diesters, *Tetrahedron Lett.* **27**, 4639-4642.

Guanti, G., Banfi, L., Narisano, E. (1989), Enzymes as selective reagents in organic synthesis: enantioselective preparation of asymmetrized tris(hydroxymethyl)methane, *Tetrahedron Lett.* **30**, 2697-2698.

Guanti, G., Banfi, L., Narisano, E. (1990a), Asymmetrized tris(hydroxymethyl)methane as a highly stereodivergent chiral building block: preparation of all four stereoisomers of protected 2-hydroxymethyl-1,3-butanediol, *Tetrahedron Lett.* **31**, 6421-6424.

Guanti, G., Banfi, L., Narisano, E. (1990b), Enzymes in organic synthesis: remarkable influence of a π system on the enantioselectivity in PPL catalyzed monohydrolysis of 2-substituted 1,3-diacetoxypropanes, *Tetrahedron: Asymmetry* **1**, 721-724.

Guanti, G., Narisano, E., Podgorski, T., Thea, S., Williams, A. (1990c), Enzyme catalyzed monohydrolysis of 2-aryl-1,3-propanediol diacetates. A study of structural effects of the aryl moiety on the enantioselectivity, *Tetrahedron* **46**, 7081-7092.

Guanti, G., Banfi, L., Narisano, E. (1992), Chemoenzymic preparation of asymmetrized tris(hydroxymethyl)methane (THYM) and of asymmetrized bis(hydroxymethyl)acetaldehyde (BHYMA) as new highly versatile chiral building blocks, *J. Org. Chem.* **57**, 1540-1554.

Guanti, G., Banfi, L., Brusco, S., Narisano, E. (1994a), Enzymatic preparation of homochiral 2-(*N*-carbobenzyloxypiperid-4-yl)-1,3-propanediol monoacetate - a facile entry to both enantiomers of 3-hydroxymethylquinuclidine, *Tetrahedron: Asymmetry* **5**, 537-540.

Guanti, G., Banfi, L., Riva, R. (1994b), Enzymatic asymmetrization of some prochiral and *meso* diols through monoacetylation with pig pancreatic lipase (PPL), *Tetrahedron: Asymmetry* **5**, 9-12.

Guevel, R., Paquette, L. A. (1993), An enzyme-based synthesis of (*S*)-(-)-3-methyl-2- [(phenylsulfonyl)methyl]butyl phenyl sulfide and the stereochemical course of its alkylation, *Tetrahedron: Asymmetry* **4**, 947-956.

Guibé-Jampel, E., Rousseau, G., Salaün, J. (1987), Enantioselective hydrolysis of racemic diesters by porcine pancreatic lipase, *J. Chem. Soc., Chem. Commun.*, 1080-1081.

Gunstone, F. D. (1998), Movements towards tailor-made fats, *Prog. Lipid Res.* **37**, 277-305.

Guo, Z.-W. (1993), Novel plots of data from combined multistep enzymatic resolutions of enantiomers, *J. Org. Chem.* **58**, 5748-5752.

Guo, Z. W., Sih, C. J. (1988), Enzymic synthesis of macrocyclic lactones, *J. Am. Chem. Soc.* **110**, 1999-2001.

Guo, Z. W., Sih, C. J. (1989), Enantioselective inhibition: strategy for improving the enantioselectivity of biocatalytic systems, *J. Am. Chem. Soc.* **111**, 6836-6841.

Guo, Z. W., Ngooi, T. K., Scilimati, A., Fülling, G., Sih, C. J. (1988), Macrocyclic lactones via biocatalysis in nonaqueous media, *Tetrahedron Lett.* **29**, 5583-5586.

Guo, Z. W., Wu, S. H., Chen, C. S., Girdaukas, G., Sih, C. J. (1990), Sequential biocatalytic kinetic resolutions, *J. Am. Chem. Soc.* **112**, 4942-4945.

Gupta, A. K., Kazlauskas, R. J. (1993), Substrate modification to increase the enantioselectivity of hydrolases. A route to optically-active cyclic allylic alcohols, *Tetrahedron: Asymmetry* **4**, 879-888.

Gustafsson, J., Sandstroem, J., Sterner, O. (1995), Enantioselective esterification of 5-hydroxy-2-methyl-6- (diethoxymethyl)-1-cyclohexenecarbaldehyde. Synthesis of both enantiomers of 6-methylbicyclo[4.1.0]hept-2-ene-1,2-dicarbaldehyde, *Tetrahedron: Asymmetry* **6**, 595-602.

Gutman, A. L. (1990), Enzymatic synthesis of chiral lactones and polyesters, *Mat. Res. Soc. Symp. Proc.* **174**, 217-222.

Gutman, A. L., Bravdo, T. (1989), Enzyme-catalyzed enantioconvergent lactonization of γ-hydroxy diesters in organic solvents, *J. Org. Chem.* **54**, 4263-4265.

Gutman, A. L., Zuobi, K., Boltansky, A. (1987), Enzymic lactonization of γ-hydroxy esters in organic solvents. Synthesis of optically pure γ-methylbutyrolactones and γ-phenylbutyrolactone, *Tetrahedron Lett.* **28**, 3861-3864.

Gutman, A. L., Meyer, E., Kalerin, E., Polyak, F., Sterling, J. (1992), Enzymatic resolution of racemic amines in a continuous reactor in organic solvents, *Biotechnol. Bioeng.* **40**, 760-767.

Ha, H. J., Yi, D. G., Yoon, K. N., Song, C. E. (1996), Lipase-catalyzed preparation of optically active γ-substituted methyl-γ-butyrolactones, *Bull. Korean Chem. Soc.* **17**, 544-547.

Ha, H.-J., Yoon, K.-N., Lee, S.-Y., Park, Y.-S., Lim, M.-S., Yim, Y.-G. (1998), Lipase PS (*Pseudomonas cepacia*) mediated resolution of γ-substituted γ-((acetoxy)methyl)-γ-butyrolactones: complete stereochemical reversion by substituents, *J. Org. Chem.* **63**, 8062-8066.

Haas, M. J., Joerger, R. D. (1995), Lipases of the genera *Rhizopus* and *Rhizomucor*: versatile catalysts in nature and the laboratory, in.: *Food Biotechnology: Microorganisms.* (Khachatourians, G. G. , Hui, H. Y.; Eds.), **Vol.** pp. 549-588. Weinheim: Wiley-VCH.

Haase, B., Schneider, M. P. (1993), Enzyme assisted synthesis of enantiomerically pure δ-lactones, *Tetrahedron: Asymm.* **4**, 1017-1026.

Halling, P. J. (1990), High affinity binding of water by proteins is similar in air and organic solvents, *Biochim. Biophys. Acta* **1040**, 225-228.

Halling, P. J. (1992), Salt hydrates for water activity control with biocatalysts in organic media, *Biotechnol. Lett.* **6**, 271-276.

Halling, P. J. (1994), Thermodynamic predictions for biocatalysis in non-conventional media: theory, tests and recommendations for experimental design and analysis, *Enz.yme Microb. Technol.* **16**, 178-206.

Halling, P. J. (1996), Predicting the behaviour of lipases in low-water media, *Biochem. Soc. Trans.* **25**, 170-174.

Hamada, H., Shiromoto, M., Funahashi, M., Itoh, T., Nakamura, K. (1996), Efficient synthesis of optically pure 1,1,1-trifluoro-2-alkanols through lipase-catalyzed acylation in organic media, *J. Org. Chem.* **61**, 2332-2336, correction, p. 9635.

Hamaguchi, S., Asada, M., Hasegawa, J., Watanabe, K. (1985), Biological resolution of racemic 2-oxazolidinones. Part V. Stereospecific hydrolysis of 2-oxazolidinone esters and separation of products with an immobilized lipase column, *Agric. Biol. Chem.* **49**, 1661-1667.

Hammond, D. A., Karel, A., Klibanov, A. M., Krukonis, V. J. (1985), Enzymatic reactions in supercritical gases, *Appl. Biochem. Biotechnol.* **11**, 393-400.

Han, D., Rhee, J. S. (1986), Characteristics of lipase-catalyzed hydrolysis of olive oil in AOT-isooctane reversed micelles, *Biotechnol. Bioeng.* **28**, 1250-1255.

Han, D., Kwon, D. Y., Rhee, J. S. (1987), Determination of lipase activity in AOT-isooctane reversed micelles, *Agric. Biol. Chem.* **51**, 615-618.

Handa, S., Earlam, G. J., Geary, P. J., Hawes, J. E., Phillips, G. T., Pryce, R. J., Ryback, G., Shears, J. H. (1994), The enantioselective synthesis of an important intermediate to the antiviral (-)-Carbovir, *J. Chem. Soc. Perkin Trans. 1*, 1885-1886.

Haraldsson, G. G. (1992), The application of lipases in organic synthesis, Supplement B: The Chemistry of Acid Derivatives (Patai, S.; Ed.), **Vol. 2**, pp. 1395-1473. Chichester: John Wiley & Sons.

Haraldsson, G. G., Thorarensen, A. (1994), The generation of glyceryl ether lipids highly enriched with eicosapentaenoic acid and docosahexaenoic acid by lipase, *Tetrahedron Lett.* **35**, 7681-7684.

Haraldsson, G. G., Gudmundsson, B. Ö., Almarsson, Ö. (1993), The preparation of homogeneous triglycerides of eicosapentaenoic acid and docosahexaenoic acid by lipase, *Tetrahedron Lett.* **34**, 5791-5794.

Haraldsson, G. G., Gudmundsson, B. Ö., Almarsson, Ö. (1995), The synthesis of homogeneous triglycerides of eicosapentaenoic acid and docosahexaenoic acid by lipase, *Tetrahedron* **51**, 941-952.

Harayama, S. (1998), Artificial evolution by DNA shuffling, *Trends Biotechnol.* **16**, 76-82.

Harris, K. J., Gu, Q. M., Shih, Y. E., Girdaukas, G., Sih, C. J. (1991), Enzymatic preparation of (3*S*,6*R*) and (3*R*,6*S*)-3-hydroxy-6- acetoxycyclohex-1-ene, *Tetrahedron Lett.* **32**, 3941-3944.

Hashimoto, H., Katayama, C., Goto, M., Okinaga, T., Kitahata, S. (1995), Enzymatic synthesis of α-linked galactooligosaccharides using the reverse reaction of a cell-bound α-galactosidase from *Candida guilliermondii* H-404, *Biosci. Biotech. Biochem.* **59**, 179-183.

Hatakeyama, S., Kawamura, M., Takano, S., Irie, H. (1994), Enantiospecific synthesis of (+)-paeonilactone C and (+)-paeoniflorigenone from *R*-(−)-carvone, *Tetrahedron Lett.* **35**, 7993-7996.

Hayes, D. G., Gulari, E. (1991), 1-Monoglyceride production from lipase-catalyzed esterification of glycerol and fatty acid in reverse micelles, *Biotechnol. Bioeng.* **38**, 507-517.

Hayes, D. G., Gulari, E. (1992), Formation of polyol-fatty acid esters by lipases in reverse micellar media, *Biotechnol. Bioeng.* **40**, 110-118.

Hayes, D. G., Gulari, E. (1994), Improvement of enzyme activity and stability for reverse micellar-encapsulated lipases in the presence of short-chain and polar alcohols, *Biocatalysis* **11**, 223-231.

Hedström, G., Backlund, M., Slotte, J. P. (1993), Enantioselective synthesis of ibuprofen esters in AOT/isooctane microemulsions by *Candida cylindracea* lipase, *Biotechnol. Bioeng.* **42**, 618-624.

Heim, J., Schmidt-Dannert, C., Atomi, H., Schmid, R. D. (1998), Functional expression of a mammalian acetylcholinesterase in *Pichia pastoris*: comparison to acetylcholinesterase, expressed and reconstituted from *Escherichia coli*, *Biochim. Biophys. Acta* **1396**, 306-319.

Hein, G. E., Niemann, C. (1962a), Steric course and specificity of α-chymotrypsin-catalyzed reactions. I., *J. Am. Chem. Soc.* **84**, 4487-4494.

Hein, G. E., Niemann, C. (1962b), Steric course and specificity of α-chymotrypsin-catalyzed reactions. II., *J. Am. Chem. Soc.* **84**, 4495-4503.

Heisler, A., Rabiller, C., Hublin, L. (1991), Lipase catalysed isomerisation of 1,2-(2,3)-diglyceride into 1,3-diglyceride. The crucial role of water, *Biotechnol. Lett.* **13**, 327-332.

Heiss, L., Gais, H.-J. (1995), Polyethylene glycol monomethyl ether-modified pig liver esterase: preparation, characterization and catalysis of enantioselective hydrolysis in water and acylation in organic solvents, *Tetrahedron Lett.* **36**, 3833-3836.

Hemmerle, H., Gais, H.-J. (1987), Asymmetric hydrolysis and esterfication catalyzed by esterases from porcine pancreas in the synthesis of both enantiomers of cyclopentanoid building blocks, *Tetrahedron Lett.* **28**, 3471-3474.

Henegar, K. E., Ashford, S. W., Baughman, T. A., Sih, J. C., Gu, R. L. (1997), Practical asymmetric syntehsis of (S)-4-ethyl-7,8-dihydro-4-hydroxy-1H-pyrano[3,4-f]indolizine-3,6,10(4H)-trione, a key intermediate for the synthesis of irinotecan and other camptothecin analogs, *J. Org. Chem.* **62**, 6588-6597.

Henkel, B., Kunath, A., Schick, H. (1992), Enantioselective lactonization of methyl 3,5-dihydroxyalkanoates. An access to (3R,5S,6E)-3-hydroxy-7-phenyl-6-hepten-5-olide by enzyme-catalyzed kinetic resolution in organic solvents, *Liebigs Ann. Chem.*, 809-811.

Henkel, B., Kunath, A., Schick, H. (1993), Enzyme-catalyzed lactonization of methyl (±)-(E)-3,5-dihydroxy-7-phenyl-6-heptenoates. A comparison of the behaviour of *syn*- and *anti*-compounds, *Tetrahedron: Asymmetry* **4**, 153-156.

Henkel, B., Kunath, A., Schick, H. (1994), Synthesis of (3R,5S)-3-hydroxy-7-phenyl-6-heptyn-5-olide by an enantioselective enzyme-catalyzed lactonization of a racemic 3,5-dihydroxy ester, *Tetrahedron: Asymmetry* **5**, 17-18.

Henkel, B., Kunath, A., Schick, H. (1995), Chemo-enzymatic synthesis of (5R,6S)-6-acetoxyhexadecan-5-olide - the major component of the mosquito oviposition attractant pheromone, *Liebigs Ann.*, 921-923.

Hennen, W. J., Sweers, H. M., Wang, Y.-F., Wong, C.-H. (1988), Enzymes in carbohydrate synthesis: Lipase-catalyzed selective acylation and deacylation of furanose and pyranose derivatives, *J. Org. Chem.* **53**, 4939-4945.

Hermoso, J., Pignol, D., Kerfelec, B., Crenon, I., Chapus, C., Fontecillacamps, J. C. (1996), Lipase activation by nonionic detergents - the crystal structure of the porcine lipase-colipase-tetraethylene glycol monooctyl ether complex, *J. Biol. Chem.* **271**, 18007-18016.

Hernáiz, M. J., Sánchez-Montero, J. M., Sinisterra, J. V. (1994), Comparison of the enzymatic activity of commercial and semipurified lipases of *Candida cylindracea* in the hydrolysis of the esters of (R,S)-2-aryl propionic acids, *Tetrahedron* **50**, 10749-10760.

Hernáiz, M. J., Sánchez-Montero, J. M., Sinisterra, J. V. (1996), Improved stability of the lipase from *Canida ruogsa* in different purification degrees by chemical modification, *Biotechnol. Techn.* **10**, 917-922.

Herradón, B. (1994), Influence of the nature of the solvent on the enantioselectivity of lipase-catalyzed transesterification: a comparison between PPL and PCL, *J. Org. Chem.* **59**, 2891-2893.

Hess, R., Bornscheuer, U., Capewell, A., Scheper, T. (1995), Lipase-catalyzed synthesis of monostearoylglycerol in organic solvents using 1,2-*O*-isopropylideneglycerol, *Enzyme Microb. Technol.* **17**, 725-728.

Heymann, E., Junge, W. (1979), Characterization of the isoenzymes of pig-liver esterase. 1. Chemical studies, *Eur. J. Biochem.* **95**, 509-518.

Heymann, E., Peter, K. (1993), A note on the identity of porcine liver carboxylesterase and prolyl-beta-naphthylamidase, *Biol. Chem. Hoppe Seyler* **374**, 1033-1036.

Hills, M. J., Kiewitt, I., Mukherjee, K. D. (1989), Enzymatic fractionation of evening primrose oil by rape lipase: enrichment of γ-linolenic acid, *Biotechnol. Lett.* **11**, 629-632.

Hills, M. J., Kiewitt, I., Mukherjee, K. D. (1990a), Enzymatic fractionation of fatty acids: enrichment of γ-linoleic acid and dosahexaenoic acid by selective esterification catalyzed by lipases, *J. Am. Oil Chem. Soc.* **67**, 561-564.

Hills, M. J., Kiewitt, I., Mukherjee, K. D. (1990b), Lipase from *Brassica napus* L. discriminates against *cis*-4 and *cis*-6 unsaturated fatty acids and secondary and tertiary alcohols, *Biochim. Biophys. Acta* **1042**, 237-240.

Hiratake, J., Yamamoto, K., Yamamoto, Y., Oda, J. (1989), Highly regioselective ring-opening of α-substituted cyclic acid anhydrides catalyzed by lipase, *Tetrahedron Lett.* **30**, 1555-1556.

Hirche, F., Schierhorn, A., Scherer, G., Ulbrich-Hofmann, R. (1997), Enzymatic introduction of N-heterocyclic and As-containing head groups into glycerophospholipids, *Tetrahedron Lett.* **38**, 1369-1370.

Hirose, Y., Kariya, K., Sasaki, I., Kurono, Y., Ebiike, H., Achiwa, K. (1992), Drastic solvent effect on lipase-catalyzed enantioselective hydrolysis of prochiral 1,4-dihydropyridines, *Tetrahedron Lett.* **33**, 7157-7160.

Hirose, Y., Kariya, K., Nakanishi, Y., Kurono, Y., Achiwa, K. (1995), Inversion of enantioselectivity in hydrolysis of 1,4-dihydropyridines by point mutation of lipase PS, *Tetrahedron Lett.* **36**, 1063-1066.

Hobson, A. H., Buckley, C. M., Aamand, J. L., Joergensen, S. T., Diderichsen, B., McConnell, D. J. (1993), Activation of a bacterial lipase by its chaperone, *Proc. Natl. Acad. Sci. USA* **90**, 5682-5686.

Hoegh, I., Patkar, S., Halkier, T., Hansen, M. T. (1995), Two lipases from *Candida antarctica* - cloning and expression in *Aspergillus oryzae*, *Can. J. Botan.* **73**, S869-S875.

Hoenke, C., Kluewer, P., Hugger, U., Krieger, R., Prinzbach, H. (1993), Biocatalysis en route to diamino-di(tri)deoxycyclohexitols and diamino-tetradeoxylcycloheptitols, *Tetrahedron Lett.* **34**, 4761-4764.

Hof, R. P., Kellog, R. M. (1994), Lipase catalyzed resolutions of some α,α- disubstituted 1,2-diols in organic solvents; near absolute regio and chiral recognition, *Tetrahedron: Asymmetry* **5**, 565-568.

Hof, R. P., Kellog, R. M. (1996a), Lipase AKG mediated resolutions of α,α-disubstituted 1,2-diols in organic solvents - remarkably high regio- and enantioselectivity, *J. Chem. Soc., Perkin Trans. 1*, 2051-2060.

Hof, R. P., Kellog, R. M. (1996b), Synthesis and lipase-catalyzed resolution of 5-(hydroxymethyl)-1,3-dioxolan-4-ones: masked glycerol analog as potential building blocks for pharmaceuticals, *J. Org. Chem.* **61**, 3423-3427.

Hoff, B. H., Anthonsen, H. W., Anthonsen, T. (1996), The enantiomer ratio strongly depends on the alkyl part of the acyl donor in transesterification with lipase B from *Candida antarctica*, *Tetrahedron: Asymmetry* **7**, 3187-3192.

Höft, E., Hamann, H.-J., Kunath, A., Adam, W., Hoch, U., Saha-Möller, C. R., Schreier, P. (1995), Enzyme-catalyzed kinetic resolution of racemic secondary hydroperoxides, *Tetrahedron: Asymmetry* **6**, 603-608.

Högberg, H.-E., Edlund, H., Berglund, P., Hedenström, E. (1993), Water activity influences enantioselectivity in a lipase-catalyzed resolution by esterification in an organic solvent, *Tetrahedron: Asymmetry* **4**, 2123-2126.

Holdgrün, X. K., Sih, C. J. (1991), A chemoenzymic synthesis of optically active dihydropyridines, *Tetrahedron Lett.* **32**, 3465-3468.

Holla, E. W. (1989), Enzymic synthesis of selectively protected glycals, *Angew. Chem. Intl. Ed. Engl.* **28**, 220-221.

Holla, E. W., Sinnwell, V., Klaffke, W. (1992), Two syntheses of 3-azido-3-deoxy-D-mannose, *Synlett*, 413-415.

Holmberg, K. (1994), Organic and bioorganic reactions in microemulsions, *Adv. Coll. Interface Sci.* **51**, 137-174.

Holmberg, K., Österberg, E. (1988), Enzymatic preparation of monoglycerides in microemulsion, *J. Am. Oil Chem. Soc.* **65**, 1544-1548.

Holmberg, E., Szmulik, P., Norin, T., Hult, K. (1989a), Hydrolysis and esterification with lipase from *Candida cylindracea*. Influence of the reaction conditions and acid moiety on the enantiomeric excess, *Biocatalysis* **2**, 217-223.

Holmberg, K., Lassen, B., Stark, M. B. (1989b), Enzymatic glycerolysis of a triglyceride in aqueous and nonaqueous microemulsions, *J. Am. Oil Chem. Soc.* **66**, 1796-1800.

Holmberg, E., Holmquist, M., Hedenström, E., Berglund, P., Norin, T., Hult, K. et al. (1992), Reaction conditions for the resolution of 2-methylalkanoic acids in esterification and hydrolysis with lipase from *Candida cylindracea*, *Appl. Microbiol. Biotechnol.* **35**, 572-578.

Holmquist, M., Haeffner, F., Norin, T., Hult, K. (1996), A structural basis for enantioselective inhibition of *Candida rugosa* lipase by long-chain aliphatic alcohols, *Protein Sci.* **5**, 83-88.

Holzwarth, H. C., Pleiss, J., Schmid, R. D. (1997), Computer-aided modelling of stereoselective triglyceride hydrolysis catalyzed by *Rhizopus oryzae* lipase, *J. Mol. Catal. B: Enzymatic* **3**, 73-82.

Hom, S. S. M., Scott, E. M., Atchison, R. E., Picataggio, S., Mielenz, J. R. (1991), Characterization and over-expression of a cloned *Pseudomonas* lipase gene, in.: *Lipases* (Schmid, R. D., Alberghina, L. , Verger, R.; Eds.), *GBF Monographs* **Vol. 16**, pp. 267-270. Weinheim: Wiley-VCH.

Honda, T., Ogino, T. (1998), Enantiodivergent synthesis of the key intermediate for a marine natural furanoterpene by chemoenzymatic process, *Tetrahedron: Asymmetry* **9**, 2663-2669.

Hongo, H., Iwasa, K., Kabuto, C., Matsuzaki, H., Nakano, H. (1997), Preparation of optically active photopyridone by lipase-catalyzed asymmetric resolution, *J. Chem. Soc., Perkin Trans. 1*, 1747-1754.

Hönig, H., Seufer-Wasserthal, P. (1990), A general method for the separation of enantiomeric *trans*-2-substituted cyclohexanols, *Synthesis*, 1137-1140.

Hönig, H., Shi, N., Polanz, G. (1994), Enzymatic resolutions of heterocyclic alcohols, *Biocatalysis* **9**, 61-69.

Hoq, M. M., Tagami, H., Yamane, T., Shimizu, S. (1985), Some characteristics of continuous glyceride synthesis by lipase in a microporous hydrophobic membrane bioreactor, *Agric. Biol. Chem.* **49**, 335-342.

Hoshino, O., Fuchino, H., Kikuchi, M. (1994), Lipase-catalyzed resolution of acetates of racemic phenolic aporphines and homoaporphines in organic solvent, *Heterocycles* **39**, 553-560.

Hosokawa, M., Takahashi, K., Kikuchi, Y., Hatano, M. (1995a), Preparation of therapeutic phospholipids through porcine pancreatic phospholipase A_2-mediated esterification and Lipozyme-mediated hydrolysis, *J. Am. Oil Chem. Soc.* **72**, 1287-1291.

Hosokawa, M., Takahashi, K., Miyazaki, N., Okamura, K., Hatano, M. (1995b), Application of water mimics on preparation of eicosapentaenoic and docosahexaenoic acids containing glycerolipids, *J. Am. Oil Chem. Soc.* **72**, 421-425.

Hosoya, N., Hatayama, A., Irie, R., Sasaki, H., Katsuki, T. (1994), Rational design of Mn-Salen epoxidation catalysts: Preliminary results, *Tetrahedron* **50**, 4311-4322.

Hou, C. T. (1993), Screening of microbial esterases for asymmetric hydrolysis of 2-ethylhexyl butyrate, *J. Ind. Microbiol.* **11**, 73-81.

Houng, J.-Y., Hsieng, C. L., Chen, S. T. (1996a), Lipase-catalyzed kinetic resolution of ethyl D,L-2-amino-4-phenylbutyrate by hydrolysis, *Biotechnol. Techn.* **10**, 353-358.

Houng, J.-Y., Wu, M.-L., Chen, S.-T. (1996b), Kinetic resolution of amino acid esters catalyzed by lipases, *Chirality* **8**, 418-422.

Hoye, T. R., Ye, Z. X., Yao, L. J., North, J. T. (1996), Synthesis of the C-2-symmetric, macrocyclic alkaloid, (+)-xestospongin A its C(9)-epimer, (-)-xestospongin C - impact of substrate rigidity and reaction conditions on the efficiency of the macrocyclic dimerzation reaction, *J. Am. Chem. Soc.* **118**, 12074-12081.

Hoyle, A. J., Bunch, A. W., Knowles, C. J. (1998), The nitrilases of *Rhodococcus rhodochrous* NCIMB 11216, *Enzyme Microb. Technol.* **23**, 475-482.

Hsu, S.-H., Wu, S.-S., Wang, Y.-F., Wong, C.-H. (1990), Lipase-catalyzed irreversible transesterification using enol esters: XAD-8 immobilized lipoprotein lipase-catalyzed resolution of secondary alcohols, *Tetrahedron Lett.* **31**, 6403-6406.

Huang, K.-H., Akoh, C. C. (1994), Lipase-catalyzed incorporation of n-3 polyunsaturated fatty acids into vegetable oils, *J. Am. Oil Chem. Soc.* **71**, 1277-1280.

Huang, W., Jia, J., Cummings, J., Nelson, M., Schneider, G., Lindqvist, Y. (1997), Crystal structure of nitrile hydratase reveals a novel iron centre in a novel fold, *Structure* **15**, 691-699.

Huber, R. E., Gaunt, M. T., Hurlburt, K. L. (1984), Binding and reactivity at the "glucose" site of galactosyl-β-galactosidase (*Escherichia coli*), *Arch. Biochem. Biophys.* **234**, 151-160.

Huge-Jensen, B., Galluzzo, D. R., Jensen, R. G. (1987), Partial purification and characterization of free and immobilized lipases from *Mucor miehei*, *Lipids* **22**, 559-565.

Huge-Jensen, B., Andreasen, F., Christensen, T., Christensen, M., Thim, L., Boel, E. (1989), *Rhizomucor miehei* triglyceride lipase is processed and secreted from transformed *Aspergillus oryzae*, *Lipids* **24**, 781-785.

Hughes, D. L., Bergan, J. J., Amato, J. S., Reider, P. J., Grabowski, E. J. J. (1989), Synthesis of chiral dithioacetals: a chemoenzymic synthesis of a novel LTD4 antagonist, *J. Org. Chem.* **54**, 1787-1788.

Hughes, D. L., Bergan, J. J., Amato, J. S., Bhupathy, M., Leazer, J. L., McNamara, J. M., Sidler, D. R., Reider, P. J., Grabowski, E. J. J. (1990), Lipase-catalyzed asymmetric hydrolysis of esters having remote chiral/prochiral centers, *J. Org. Chem.* **55**, 6252-6259.

Hughes, D. L., Song, Z., Smith, G. B., Bergan, J. J., Dezeny, G. C., Reider, P. J., Grabowski, E. J. J. (1993), A practical chemoenzymic synthesis of an LTD4 antagonist, *Tetrahedron: Asymmetry* **4**, 865-874.

Hui, D. Y., Kissel, J. A. (1990), Sequence identity between human pancreatic cholesterol esterase and bile salt-stimulated milk lipase, *FEBS Lett.* **276**, 131-134.

Hulshof, L. A., Roskam, J. H. (1989), Phenylglycidate stereoisomers,conversion products thereof with e.g. 2-nitrophenol and preparation of diltiazem,

Hultin, P. G., Jones, J. B. (1992), Dilemma regarding the active site model for porcine pancreatic lipase, *Tetrahedron Lett.* **33**, 1399-1402.

Hyatt, J. A., Skelton, C. (1997), A kinetic resolution route to the (*S*)-chromanmethanol intermediate for synthesis of the natural tocols, *Tetrahedron: Asymmetry* **8**, 523-526.

Iacazio, G., Roberts, S. M. (1993), Investigation of the regioselectivity of some esterifications involving methyl 4,6-*O*-benzylidene D-glycopyranosides and *Pseudomonas fluorescens* lipase, *J. Chem. Soc., Perkin Trans. 1*, 1099-1101.

Igarashi, Y., Otsutomo, S., Harada, M., Nakano, S., Watanabe, S. (1997), Lipase mediated resolution of indene bromohydrin, *Synthesis*, 549-552.

Iglesias, L. E., Baldessari, A., Gros, E. G. (1996), Lipase-catalyzed chemospecific *O*-acylation of 3-mercapto-1-propanol and 4-mercapto-1-butanol, *Bioorg. Med. Chem. Lett.* **6**, 853-856.

Iglesias, L. E., Sánchez, V. M., Rebolledo, F., Gotor, V. (1997), Candida antarctica B lipase catalysed resolution of (±)-1-(heteroaryl) ethylamines, *Tetrahedron: Asymmetry* **8**, 2675-2677.

Ihara, M., Suzuki, M., Fukumoto, K., Kabuto, C. (1990), Asymmetric total synthesis of atisine via intramolecular double Michael reaction, *J. Am. Chem. Soc.* **112**, 1164-1171.

Iimori, T., Azumaya, I., Hayashi, Y., Ikegami , S. (1997), A practical preparation of optically active *endo*-bicyclo[3.3.0]octen-2-ols, *Chem. Pharm. Bull.* **45**, 207-208.

Iizumi, T., Nakamura, K., Shimada, Y., Sugihara, A., Tominaga, Y., Fukase, T. (1991), Cloning, nucleotide sequence and expression in *Escherichia coli* of a lipase and its activator genes from *Pseudomonas* sp. KWI-56, *Agric. Biol. Chem.* **55**, 2349-2357.

Ikeda, I., Klibanov, A. M. (1993), Lipase-catalyzed acylation of sugars solublized in hydrophobic solvents by complexation, *Biotechnol. Bioeng.* **42**, 788-791.

Ikushima, Y., Saito, N., Yokohama, T., Hatakeda, K., Ito, S., Arai, M., Blanch, H. W. (1993), Solvent effects on an enzymatic ester synthesis in supercritical carbon dioxide, *Chem. Lett.* 109-112.

Ikushima, Y., Saito, N., Arai, M., Blanch, H. W. (1995), Activation of a lipase triggered by interactions with supercritical carbon dioxide in the near critical region, *J. Phys. Chem.* **99**, 8941-8944.

Imperiali, B., Prins, T. J., Fisher, S. L. (1993), Chemoenzymatic synthesis of 2-amino-3-(2,2'-bipyridinyl)propanoic acids, *J. Org. Chem.* **58**, 1613-1616.

Inagaki, M., Hiratake, J., Nishioka, T., Oda, J. (1989), Lipase-catalyzed stereoselective acylation of [1,1'-binaphthyl]-2,2'-diol and deacylation of its esters in organic solvent, *Agric. Biol. Chem.* **53**, 1879-1884.

Inagaki, M., Hiratake, J., Nishioka, T., Oda, J. (1991), Lipase-catalyzed kinetic resolution with *in situ* racemization: one-pot synthesis of optically active cyanohydrin acetates from aldehydes, *J. Am. Chem. Soc.* **113**, 9360-9361.

Inagaki, M., Hiratake, J., Nishioka, T., Oda, J. (1992), One-pot synthesis of optically active cyanohydrin acetates from aldehydes via lipase -catalyzed kinetic resolution coupled with in situ formation and racemization of cyanohydrins, *J. Org. Chem.* **57**, 5643-5649.

Ingvorsen, K., Yde, B., Godtfredsen, S. E., Tsuchiya, R. T. (1988), Microbial hydrolysis of organic nitriles and amides, in.: *Ciba Geigy Symp.: Cyanide Compounds in Biology*, **Vol. 140**, pp. 16-31. New York: John Wiley & Sons.

Isono, Y., Nabetani, H., Nakajima, M. (1995), Interesterification of triglyceride and fatty acid in a microaqueous reaction system using lipase-surfactant complex, *Biosci. Biotechnol. Biochem.* **59**, 1632-1635.

Isowa, Y., Ohmori, M., Ichikawa, T., Mori, K., Nonaka, Y., Kihara, K., Oyama, K., Satoh, H., Nishimura, S. (1979), The thermolysin-catalyzed condensation reactions of *N*-substituted aspartic acid and glutamic acids with phenylalanine alkyl esters., *Tetrahedron Lett.* 2611-2612.

Itoh, T., Ohta, T. (1990), A simple method for the preparation of optically active α-hydroxystannanes by the enantioselective hydrolysis using a lipase, *Tetrahedron Lett.* **31**, 6407-6408.

Itoh, T., Ohta, T., Sano, M. (1990), An efficient preparation of the optically active γ–hydroxy stannanes using lipase-catalyzed hydrolysis, *Tetrahedron Lett.* **31**, 6387-6390.

Itoh, T., Takagi, Y., Nishiyama, S. (1991), Enhanced enantioselectivity of an enzymatic reaction by the sulfur functional group. A simple preparation of optically active ß-hydroxy nitriles using a lipase, *J. Org. Chem.* **56**, 1521-1524.

Itoh, T., Chika, J., Takagi, Y., Nishiyama, S. (1993a), An efficient enantioselective total synthesis of antitumor lignans: synthesis of enantiomerically pure 4-hydroxyalkanenitriles via an enzymic reaction, *J. Org. Chem.* **58**, 5717-5723.

Itoh, T., Hiyama, Y., Betchaku, A., Tsukube, H. (1993b), Enhanced reaction rate and enantioselectivity in lipase-catalyzed hydrolysis by addition of a crown ether, *Tetrahedron Lett.* **34**, 2617-2620.

Itoh, T., Ohara, H., Takagi, Y., Kanada, N., Uneyama, K. (1993c), Preparation of a new chiral building block for synthesizing broadly varied types of tertiary alcohols, *Tetrahedron Lett.* **34**, 4215-4218.

Itoh, T., Shiromoto, M., Inoue, H., Hamada, H., Nakamura, K. (1996a), Simple preparation of optically pure bis(trifluoromethyl)-alkanediols through lipase-catalyzed reaction, *Tetrahedron Lett.* **37**, 5001-5002.

Itoh, T., Takagi, Y., Murakami, T., Hiyama, Y., Tsukube, H. (1996b), Crown ethers as regulators of enzymatic reactions: enhanced reaction rate and enantioselectivity in lipase-catalyzed hydrolysis of 2-cyano-1-methylethyl acetate, *J. Org. Chem.* **61**, 2158-2163.

Itoh, T., Uzu, A., Kanda, N., Takagi, Y. (1996c), Preparation of 3-alkyl-4-hydroxy-2-butenyl acetate through highly regioselective lipase catalyzed hydrolysis of the corresponding diacetates, *Tetrahedron Lett.* **37**, 91-92.

Iwasaki, Y., Niwa, S., Nakano, H., Nagasawa, T., Yamane, T. (1994), Purification and properties of phosphatidylinositol-specific phospholipase C from *Streptomyces antibioticus*, *Biochim. Biophys. Acta* **1214**, 221-228.

Iwasaki, Y., Mishima, N., Mizumoto, K., Nakano, H., Yamane, T. (1995), Extracellular production of phospholipase D of *Streptomyces antibioticus* using recombinant *Escherichia coli*, *J. Ferment. Bioeng.* **79**, 417-421.

Izumi, T., Aratani, S. (1993), Lipase-catalyzed resolution of 1-hydroxymethyl-2-methylferrocene, *J. Chem. Technol. Biotechnol.* **57**, 33-36.

Izumi, T., Tamura, F., Sasaki, K. (1992), Enzymic kinetic resolution of [4](1,2)ferrocenophane derivatives, *Bull. Chem. Soc. Jpn.* **65**, 2784-2788.

Izumi, T., Hino, T., Ishihara, A. (1993), Enzymic kinetic resolution of [3](1,1')ferrocenophane derivatives, *J. Chem. Technol. Biotechnol.* **56**, 45-49.

Jackson, M. A., King, J. W. (1997), Lipase-catalyzed glycerolysis of soy bean oil in supercritical carbon dioxide, *J. Am. Oil Chem. Soc.* **74**, 103-106.

Jackson, M. A., King, J. W., List, G. R., Neff, W. E. (1997), Lipase-catalyzed randomization of fats and oils in flowing supercritical carbon dioxide, *J. Am. Oil Chem. Soc.* **74**, 635-639.

Jaeger, K.-E., Reetz, M. T. (1998), Microbial lipases form versatile tools for biotechnology, *Trends Biotechnol.* **16**, 396-403.

Jaeger, K.-E., Ransac, S., Dijkstra, B. W., Colson, C., van Heuvel, M., Misset, O. (1994), Bacterial lipases, *FEMS Microbiol. Rev.* **15**, 29-63.

Jaeger, K. E., Liebeton, K., Zonta, A., Schimossek, K., Reetz, M. T. (1996), Biotechnological application of *Pseudomonas aeruginosa* lipase: efficient kinetic resolution of amines and alcohols, *Appl. Microbiol. Biotechnol.* **46**, 99-105.

Jandacek, R. J., Whiteside, J. A., Holcombe, B. N., Volpenhein, R. A., Taulbee, J. D. (1987), The rapid hydrolysis and efficient absorption of triacylglycerides with octanoic acid in the 1 and 3 positions and long-chain fatty acid in the 2 position, *Am. J. Clin. Nutr.* **45**, 940-945.

Janes, L. E., Kazlauskas, R. J. (1997a), Empirical rules for the enantiopreference of lipase from *Aspergillus niger* toward secondary alcohols and carboxylic acids, especially α-amino acids, *Tetrahedron: Asymmetry* **8**, 3719-3733.

Janes, L. E., Kazlauskas, R. J. (1997b), Quick E. A fast spectrophotometric method to measure the enantioselectivity of hydrolases, *J. Org. Chem.* **62**, 4560-4561.

Janes, L. E., Löwendahl, A. C., Kazlauskas, R. J. (1998), Quantitative screening of hydrolase librarires using pH indicators: Identifying active and enantioselective hydrolases, *Chem. Eur. J.* **4**, 2324-2331.

Janssen, A. E. M., Halling, P. J. (1994), Specificities of enzymes 'corrected for solvation' depend on the choice of the standard state, *J. Am. Chem. Soc.* **116**, 9827-9830.

Janssen, A. E. M., Lefferts, A. G., van't Riet, K. (1990), Enzymatic synthesis of carbohydrate esters in aqueous media, *Biotechnol. Lett.* **12**, 711-716.

Janssen, A. E. M., Klabbers, C., Franssen, M. C. R., van't Riet, K. (1991a), Enzymic synthesis of carbohydrate esters in 2-pyrrolidone, *Enzyme Microb. Technol.* **13**, 565-572.

Janssen, A. J. M., Klunder, A. J. H., Zwanenburg, B. (1991b), Resolution of secondary alcohols by enzyme-catalyzed transesterification in alkyl carboxylates as the solvent, *Tetrahedron* **47**, 7645-7662.

Janssen, A. E. M., van der Padt, A., van't Riet, K. (1993a), Solvent effects on lipase-catalyzed esterification of glycerol and fatty acids, *Biotechnol. Bioeng.* **42**, 953-962.

Janssen, A. E. M., van der Padt, A., van Sonsbeek, H. M., van't Riet, K. (1993b), The effect of organic solvents on the equilibrium position of enzymic acylglycerol synthesis, *Biotechnol. Bioeng.* **41**, 95-103.

Janssens, R. J. J., van der Lugt, J. P., Oostrom, W. H. M. (1992), The integration of biocatalysis and downstream processing in supercritical carbon dioxide, in.: *High Press. Biotechnol.* (Balny, C., Hayashi, R., Heremans, K. , Masson, P.; Eds.), **Vol. 224**, pp. 447-449.

Jensen, R. G. (1974), Characteristics of the lipase from the mold *Geotrichum candidum*: a review, *Lipids* **9**, 149-157.

Jerina, D. M., Daly, J. W. (1974), Arene oxides: a new aspect of drug metabolism, *Science* **185**, 573-582.

Jeromin, G. E., Welsch, V. (1995), Diketene a new esterification reagent in the enzyme-aided synthesis of chiral alcohols and chiral acetoacetic acid esters, *Tetrahedron Lett.* **36**, 6663-6664.

Jiménez, O., Bosch, M. P., Guerrero, A. (1997), Lipase-catalyzed enantioselective synthesis of methyl (*R*)- and (*S*)-2-tetradecyloxiranecarboxylate through sequential kinetic resolution, *J. Org. Chem.* **62**, 3496-3499.

Joerger, R. D., Haas, M. J. (1994), Alteration of chain length selectivity of a *Rhizopus delemar* lipase through site-directed mutagenesis, *Lipids* **29**, 377-384.

Johnson, C. J., Bis, S. J. (1992), Enzymic asymmetrization of *meso*-2-cycloalken-1,4-diols and their diacetates in organic and aqueous media, *Tetrahedron Lett.* **33**, 7287-7290.

Johnson, C. R., Bis, S. J. (1995), Enzymatic asymmetrization of 6-amino-2-cycloheptene-1,4-diol derivatives: synthesis of tropane alkaloids (+)- and (−)-calystegine A(3), *J. Org. Chem.* **60**, 615-623.

Johnson, C. R., Penning, T. D. (1986), Triply convergent synthesis of (-)-prostaglandin-E$_2$, *J. Am. Chem. Soc.* **108**, 5655-5656.

Johnson, C. R., Sakaguchi, H. (1992), Enantioselective transesterifications using immobilized, recombinant *Candida antarctica* lipase B: resolution of 2-iodo-2-cycloalken-1-ols, *Synlett*, 813-816.

Johnson, C. R., Senanayake, C. H. (1989), Polyhydroxylated seven-membered chiral building blocks. Asymmetric synthesis of compactin analogues, *J. Org. Chem.* **54**, 735-736.

Johnson, C. R., Plé, P. A., Adams, J. P. (1991), Enantioselective synthesis of (+)- and (−)-conduritol C from benzene via microbial oxidation and enzymic asymmetrization, *J. Chem. Soc., Chem. Commun.*, 1006-1007.

Johnson, C. R., Golebiowski, A., Steensma, D. H., Scialdone, M. A. (1993), Enantio- and diastereoselective transformations of cycloheptatriene to sugars and related products, *J. Org. Chem.* **58**, 7185-7194.

Johnson, C. R., Harikrishnan, L. S., Golebiowski, A. (1994), Enantioselective synthesis of 7-cycloocten-1,3,5,6-tetraol derivatives by enzymic asymmetrization, *Tetrahedron Lett.* **35**, 7735-7738.

Johnson, C. R., Xu, Y., Nicolaou, K. C., Yang, Z., Guy, R. K., Dong, J. G., Berova, N. (1995), Enzymic resolution of a key stereochemical intermediate for the synthesis of (−)-taxol, *Tetrahedron Lett.* **36**, 3291-3294.

Johnston, B. R., Morgan, B., Oehlschlager, A. C., Ramaswamy, S. (1991), A convenient synthesis of both enantiomers of seudenol and their conversion to 1-methyl-2-cyclohexen-1-ol, *Tetrahedron: Asymmetry* **2**, 377-380.

Jones, J. B. (1993), Probing the specificity of synthetically useful enzymes, *Can. J. Chem.* **71**, 1273-1282.

Jones, J. B., Beck, J. F. (1976), Asymmetric syntheses and resolutions using enzymes, in.: *Applications of Biochemical Systems in Organic Chemistry* (Jones, J. B., Sih, C. J., Perlman, D.; Eds.), **Vol. X**, pp. 107-401. New York: Wiley.

Jones, J. B., Marr, P. W. (1973), Potential of α-chymotrysin for the resolution of alicyclic acids and esters, *Tetrahedron Lett.* 3165-3168.

Jones, J. B., Hinks, R. S., Hultin, P. G. (1985), Enzymes in organic synthesis. 33. Stereoselective pig liver esterase-catalyzed hydrolyses of meso cyclopentyl-, tetrahydrofuranyl-, and tetrahydrothiophenyl-1,3-diesters, *Can. J. Chem.* **63**, 452-456.

Jones, B. C. N. M., Simons, C., Nishimura, H., Zemlicka, J. (1995), Resolution of (±)-cytallene - a highly active anti-HIV agent with axial dissymmetry, *Nucleos. Nucleot.* **14**, 431-434.

Jorgensen, S., Skov, K. W., Diderichsen, B. (1991), Cloning, sequence, and expression of a lipase gene from *Pseudomonas cepacia*: lipase production in heterologous hosts requires two *Pseudomonas* genes, *J. Bacteriol.* **173**, 559-567.

Jouglet, B., Rousseau, G. (1993), Enzymatic resolution of *N*-hydroxylmethyl γ-butyrolactams. An access to optically active γ-butyrolactams, *Tetrahedron Lett.* **34**, 2307-2310.

Joyce, G. F. (1992), Directed molecular evolution, *Scientific Amer.* 90-97.

Junge, W., Heymann, E. (1979), Characterization of the isoenzymes of pig liver esterase. 2. Kinetic studies, *Eur. J. Biochem.* **95**, 519-525.

Justiz, O. H., Fernandez-Lafuente, R., Guisan, J. M., Negri, P., Pagani, G., Pregnolato, M., Terreni, M. (1997), One-pot chemoenzymic synthesis of 3'-functionalized cephalosporines (Cefazolin) by three consecutive biotransformations in fully aqueous medium., *J. Org. Chem.* **62**, 9099-9106.

Kaga, H., Siegmund, B., Neufellner, E., Faber, K., Paltauf, F. (1994), Stabilization of *Candida* lipase against acetaldehyde by adsorption onto Celite, *Biotechnol. Tech.* **8**, 369-374.

Kagan, H. B., Fiaud, J. C. (1988), Kinetic resolution, *Top. Stereochem.* **18**, 249-330.

Kakeya, H., Sakai, N., Sano, A., Yokoyama, M., Sugai, T., Ohta, H. (1991a), Microbial hydrolysis of 3-substituted glutaronitriles, *Chem. Lett.* 1823-1824.

Kakeya, H., Sakai, N., Sugai, T., Ohta, H. (1991b), Microbial hydrolysis as a potent method for the preparation of optically active nitriles, amides and carboxylic acids, *Tetrahedron Lett.* **32**, 1343-1346.

Kakeya, H., Sakai, N., Sugai, T., Ohta, H. (1991c), Preparation of optically active α-hydroxy acid derivatives by microbial hydrolysis of cyanohydrins and its application to the synthesis of (*R*)-4-dodecanolide, *Agric. Biol. Chem.* **55**, 1877-1881.

Kalaritis, P., Regenye, R. W., Partridge, J. J., Coffen, D. L. (1990), Kinetic resolution of 2-substituted esters catalyzed by a lipase ex. *Pseudomonas fluorescens, J. Org. Chem.* **55**, 812-815.

Kamat, S., Burrera, J., Beckman, E. J., Russell, A. J. (1992), Biocatalysis synthesis of acrylates in organic solvents and supercritical fluids: I. Optimization of enzyme environment, *Biotechnol. Bioeng.* **40**, 158-166.

Kamat, S., Critchley, G., Beckman, E. J., Russell, A. J. (1995), Biocatalytic synthesis of acrylates in organic solvents and supercritical fluids: III. Does carbon dioxide covalently modify enzymes?, *Biotechnol. Bioeng.* **46**, 610-620.

Kamat, S., Iwaskezyz, B., Beckman, E. J., Russell, A. J. (1993), Biocatalysis synthesis of acrylates in organic solvents and supercritical fluids: II. Tuning enzyme activity by changing pressure, *Proc. Natl. Acad. Sci. USA* **90**, 2940-2944.

Kaminska, J., Gornicka, I., Sikora, M., Gora, J. (1996), Preparation of homochiral (*S*)- And (*R*)-1-(2-furyl)ethanols by lipase-catalyzed transesterification, *Tetrahedron: Asymmetry* **7**, 907-910.

Kamiya, N., Goto, M., Nakashio, F. (1995), Surfactant-coated lipase suitable for the enzymic resolution of menthol as a biocatalyst in organic media, *Biotechnol. Prog.* **11**, 270-275.

Kamphuis, J., Boesten, W. H. J., Kapten, B., Hermes, H. F. M., Sonke, T., Broxterman, Q. B., van den Tweel, W. J. J., Schoemaker, H. E. (1992), The production and uses of optically pure natural and unnatural amino acids, in.: *Chirality in Industry* (Collins, A. N., Sheldrake, G. N. , Crosby, J.; Eds.), pp. 187-208. New York: Wiley.

Kanerva, L. T., Sundholm, O. (1993a), Enzymic acylation in the resolution of methyl *threo*-2-hydroxy-3-(4-methoxyphenyl)-3-(2-X-phenylthio)propionates in organic solvents, *J. Chem. Soc., Perkin Trans. 1*, 2407-2410.

Kanerva, L. T., Sundholm, O. (1993b), Lipase catalysis in the resolution of racemic intermediates of diltiazem synthesis in organic solvents, *J. Chem. Soc., Perkin Trans. 1*, 1385-1389.

Kanerva, L. T., Vänttinen, E. (1997), Optimized double kinetic resolution for the preparation of (*S*)-solketal, *Tetrahedron: Asymmetry* **8**, 923-933.

Kanerva, L. T., Csomos, P., Sundholm, O., Bernath, G., Fulop, F. (1996), Approach to highly enantiopure β-amino acid esters by using lipase catalysis in organic media, *Tetrahedron: Asymmetry* **7**, 1705-1716.

Kang, S.-K., Jeon, J.-H., Yamaguchi, T., Kim, J.-S., Ko, B.-S. (1995), Enzymic synthesis of (S)-(–)-1-(2-thienyl)propyl acetate, *Tetrahedron: Asymmetry* **6**, 2139-2142.

Kaptein, B., Boesten, W. H. J., Broxterman, Q. B., Peters, P. J. H., Schoemaker, H. E., Kamphuis, J. (1993), Enzymatic resolution of α,α-disubstituted α-amino-acid estes and amides, *Tetrahedron: Asymmetry* **4**, 1113-1116.

Kataoka, M., Shimizu, K., Sakamoto, K., Yamada, H., Shimizu, S. (1995), Lactonohydrolase-catalyzed optical resolution of pantoyl lactone: selection of a potent enzyme producer and optimization of culture and reaction conditions for practical resolution, *Appl. Microbiol. Biotechnol.* **44**, 333-338.

Kato, K., Katayama, M., Gautam, R. K., Fujii, S., Kimoto, H. (1994), Lipase-catalyzed optical resolution of 2,2,2-trifluoro-1-(1-naphthyl)ethanol, *Biosci. Biotechnol. Biochem.* **58**, 1353-1354.

Kato, K., Katayama, M., Gautam, R. K., Fujii, S., Fukaya, H., Kimoto, H. (1995a), A facile preparation of enantiomers of ethyl 4,4,4-trifluoro-3- (indole-3-)butyrate, a novel plant growth regulator, *J. Ferment. Bioeng.* **79**, 171-173.

Kato, K., Katayama, M., Gautam, R. K., Fujii, S., Kimoto, H. (1995b), Preparation and lipase-catalyzed optical resolution of 2,2,2-trifluoro-1-(naphthyl)ethanols, *Biosci. Biotechnol. Biochem.* **59**, 271-276.

Kato, K., Katayama, M., Fujii, S., Kimoto, H. (1996), Effective preparation of optically active 4,4,4-trifluoro-3-(indole-3-) butyric acid, a novel plant growth regulator, using lipase from *Pseudomonas fluorescens*, *J. Ferment. Bioeng.* **82**, 355-360.

Katsuki, T., Martin, V. S. (1996), Asymmetric epoxidation of allylic alcohols: The Katsuki-Sharpless epoxidation reaction, *Org. React. (N.Y.)* **48**, 1-299.

Kawakami, K., Yoshida, S. (1995), Sol-gel entrapment of lipase using a mixture of tetramethoxysilane and methyltrimethoxysilane as the alkoxide precursor: esterification activity in organic media, *Biotechnol. Tech.* **9**, 701-704.

Kawanami, Y., Moriya, H., Goto, Y. (1994), Lipase-catalyzed kinetic resolution of *trans*-2,5-disubstituted pyrrolidine derivatives, *Chem. Lett.* 1161-1162.

Kawanami, Y., Moriya, H., Goto, Y., Tsukao, K., Hashimoto, M. (1996), Lipase-catalyzed kinetic resolution of *trans*-2,5-disubstituted pyrrolidine derivatives, *Tetrahedron* **52**, 565-570.

Kawashiro, K., Sugahara, H., Tsukioka, T., Sugiyama, S., Hayashi, H. (1996), Effect of ester moiety of substrates on enantioselectivity of protease catalysis in organic media, *Biotechnol. Lett.* **18**, 1381-1386.

Kazlauskas, R. J. (1989), Resolution of binaphthols and spirobiindanols using cholesterol esterase, *J. Am. Chem. Soc.* **111**, 4953-4959.

Kazlauskas, R. J. (1991), (S)-(-)- and (R)-(+)-1,1'-bi-2-naphthol, *Org. Synth.* **70**, 60-67.

Kazlauskas, R. J., Bornscheuer, U. T. (1998), Biotransformations with lipases, in.: *Biotechnology-Series* (Rehm, H. J., Reed, G., Pühler, A., Stadler, P. J. W. , Kelly, D. R.; Eds.), **Vol. 8a**, pp. 37-191. Weinheim: Wiley-VCH.

Kazlauskas, R. J., Weissfloch, A. N. E. (1997), A structure-based rationalization of the enantiopreference of subtilisin toward secondary alcohols and isosteric primary amines, *J. Mol. Catal. B: Enzymatic* **3**, in press.

Kazlauskas, R. J., Weissfloch, A. N. E., Rappaport, A. T., Cuccia, L. A. (1991), A rule to predict which enantiomer of a secondary alcohol reacts faster in reactions catalyzed by cholesterol esterase, lipase from *Pseudomonas cepacia*, and lipase from *Candida rugosa*, *J. Org. Chem.* **56**, 2656-2665.

Ke, T., Klibanov, A. M. (1998), Insights into the solvent dependence of chymotryptic prochiral selectivity, *J. Am. Chem. Soc.* **120**, 4259-4263.

Ke, T., Wescott, C. R., Klibanov, A. M. (1996), Prediction of the solvent dependence of enzymatic prochiral selectivity by means of structure-based thermodynamic calculations, *J. Am. Chem. Soc.* **118**, 3366-3374.

Keil, O., Schneider, M. P., Rasor, J. P. (1995), New hydantoinases from thermophilic microorganisms. Synthesis of enantiomerically pure D-amino acids, *Tetrahedron: Asymmetry* **6**, 1257-1260.

Kelly, D. R., Wan, P. W. H., Tang, J. (1998), Flavin monoxygenases - uses as catalysts for Baeyer-Villiger ring expansion and heteroatom oxidation, in.: *Biotechnology-Series* (Rehm, H. J., Reed, G., Pühler, A., Stadler, P. J. W. , Kelly, D. R.; Eds.), **Vol. 8a**, pp. 535-587. Weinheim: Wiley-VCH.

Kelly, N. M., Reid, R. G., Willis, C. L., Winton, P. L. (1996), Chemo-enzymatic synthesis of isotopically labelled L-valine, L-isoleucine and *allo*-isoleucine, *Tetrahedron Lett.* **37**, 1517-1520.

Kennedy, J. P. (1991), Structured lipids: fats of the future, *Food Technol.* **45**, 76-83.

Kerscher, V., Kreiser, W. (1987), Enantiomerenreine Glycerin-Derivate durch enzymatische Hydrolyse prochiraler Ester, *Tetrahedron Lett.* **28**, 531-534.

Khalaf, N., Govardhan, C. P., Lalonde, J. J., Persichetti, R. A., Wang, Y. F., Margolin, A. L. (1996), Cross-linked enzyme crystals as highly active catalysts in organic solvents, *J. Am. Chem. Soc.* **118**, 5494-5495.

Khalameyzer, V., Fischer, I., Bornscheuer, U. T., Altenbuchner, J. (1999), Screening, nucleotide sequence and biochemical characterization of an esterase from *Pseudomonas fluorescens* with high activity toward lactones, *Appl. Environm. Microbiol.* **65**, 477-482.

Khan, S. A., Halling, P. J., Bell, G. (1990), Measurement and control of water activity with an aluminum oxide sensor in organic two-phase reaction mixtures for enzymic catalysis, *Enzyme Microb. Technol.* **12**, 453-458.

Khlebnikov, V., Mori, K., Terashima, K., Tanaka, Y., Sato, M. (1995), Lipase-catalyzed resolution of sterically crowded 1,2-diols, *Chem. Pharm. Bull.* **43**, 1659-1662.

Khmelnitsky, Y. L., Welch, S. H., Clark, D. S., Dordick, J. S. (1994), Salts dramatically enhance activity of enzymes suspended in organic solvents, *J. Am. Chem. Soc* **116**, 2647-2648.

Khmelnitsky, Y. L., Budde, C., Arnold, J. M., Usyantinsky, A., Clark, D. S., Dordick, J. S. (1997), Synthesis of water-soluble paclitaxel derivatives by enzymatic acylation, *J. Am. Chem. Soc* **119**, 11554-11555.

Khushi, T., O'Toole, K. J., Sime, J. T. (1993), Biotransformation of phosphonate esters, *Tetrahedron Lett.*. **34**, 2375-2378.

Kiefer, M., Vogel, R., Helmchen, G., Nuber, B. (1994), Resolution of (1,1'-binaphthalene)-2,2'-dithiol by enzyme-catalyzed hydrolysis of a racemic diacyl derivative, *Tetrahedron* **50**, 7109-7114.

Kielbasinski, P., Zurawinski, R., Pietrusiewicz, K. M., Zablocka, K., Mikolajczyk, M. (1994), Enzymatic heteroatom chemistry. 1. Enzymatic resolution of racemic phosphinoylacetates having a stereogenic phosphorus atom, *Tetrahedron Lett.* **35**, 7081-7084.

Kikkawa, S., Takahashi, K., Katada, T., Inada, Y. (1989), Esterification of chiral secondary alcohols with fatty acid in organic solvents by polyethylene glycol-modified lipase, *Biochem. Int.* **19**, 1125-1131.

Kikuchi, M., Koshiyama, I., Fukushima, D. (1983), A new enzyme, proline acylase (*N*-acyl-L-proline amidohydrolase) from *Pseudomonas* species, *Biochim. Biophys. Acta* **744**, 180-188.

Kim, M. J., Cho, H. (1992), *Pseudomonas* lipases as catalysts in organic synthesis: specificity of lipoprotein lipase, *J. Chem. Soc., Chem. Commun.*, 1411-1413.

Kim, M. J., Choi, Y. K. (1992), Lipase-catalyzed enantioselective transesterification of *O*-trityl 1,2-diols. Practical synthesis of (*R*)-tritylglycidol, *J. Org. Chem.* **57**, 1605-1607.

Kim, S. B., Choi, C. Y. (1995), Effect of solid salt hydrates on the asymmetric esterfication of 2-chloropropionic acid: control of water activity in organic solvent, *Biotechnol. Lett.* **17**, 1075-1076.

Kim, T., Chung, K. (1989), Some characteristics of palm kernel olein hydrolysis by *Rhizopus arrhizus* lipase in reversed micelle of AOT in isooctane, and additive effects, *Enzyme Microb. Technol.* **11**, 528-532.

Kim, S. M., Rhee, J. S. (1991), Production of medium-chain glycerides by immobilized lipase in a solvent-free system, *J. Am. Oil Chem. Soc.* **68**, 499-503.

Kim, H., Ziani-Cherif, C., Oh, J., Cha, J. K. (1995a), New [4+3] cycloaddition approach to cis-2,8-disubstiuted oxocanes, *J. Org. Chem.* **60**, 792-793.

Kim, M.-J., Choi, G.-B., Kim, J.-Y., Kim, H.-J. (1995b), Lipase-catalyzed transesterification as a practical route to homochiral acyclic anti-1,2-diols. A new synthesis of (+)- and (−)-endo-brevicomin, *Tetrahedron Lett.* **36**, 6253-6256.

Kim, K. K., Song, H. K., Shin, D. H., Hwang, K. Y., Choe, S., Yoo, O. J., Suh, S. W. (1997a), Crystal structure of carboxylesterase from Pseudomonas fluorescens, an alpha/beta hydrolase with broad substrate specificity, *Structure* **5**, 1571-1584.

Kim, K. K., Song, H. K., Shin, D. H., Hwang, K. Y., Suh, S. W. (1997b), The crystal structure of a triacylglycerol lipase from *Pseudomonas cepacia* reveals a highly open conformation in the absence of a bound inhibitor, *Structure* **5**, 173-185.

Kingerywood, J., Johnson, J. S. (1996), Resolution of anti-3-oxotricyclo[2.2.1.0]heptane-7-carboxylic acid by *Candida antartica* lipase A, *Tetrahedron Lett.* **37**, 3975-3976.

Kinney, A. J. (1996), Designer oils for better nutrition, *Nature Biotechnol.* **14**, 946.

Kirk, O., Christensen, M. W., Damhus, T., Godtfredsen, S. E. (1994), Enzyme catalyzed degradation and formation of peroxycarboxylic acids, *Biocatalysis* **11**, 65-77.

Kita, Y., Takebe, Y., Murata, K., Naka, T., Akai, S. (1996), 1-Ethoxyvinyl acetate as a novel, highly reactive, and reliable acyl donor for enzymatic resolution of alcohols, *Tetrahedron Lett.* **37**, 7369-7372.

Kitaguchi, H., Fitzpatrick, P. A., Huber, J. E., Klibanov, A. M. (1989), Enzymic resolution of racemic amines: crucial role of the solvent, *J. Am. Chem. Soc.* **111**, 3094-3095.

Kitayama, T. (1996), Asymmetric syntheses of pheromones for *Bactrocera nigrotibialis*, *Andrena wilkella*, and *Andrena haemorrhoa* F from a chiral nitro alcohol, *Tetrahedron* **52**, 6139-6148.

Klein, R. R., King, G., Moreau, R. A., Haas, M. J. (1997), Altered acyl chain length specificity of *Rhizopus delemar* lipase through mutagenesis and molecular modeling, *Lipids* **32**, 123-130.

Klempier, N., Faber, K., Griengl, H. (1989), Chemoenzymic large-scale preparation of homochiral bicyclo[3.2.0]hept-2-en-6-one, *Biotechnol. Lett.* **11**, 685-688.

Klempier, N., Geymayer, P., Stadler, P., Faber, K., Griengl, H. (1990), Biocatalytic preparation of bicyclo[3.2.0]heptane derivatives, *Tetrahedron: Asymmetry* **1**, 111-118.

Klempier, N., de Raadt, A., Faber, K., Griengl, H. (1991), Selective transformation of nitriles into amides and carboxylic acids by an immobilized nitrilase, *Tetrahedron Lett.* **32**, 341-344.

Klibanov, A. M. (1989), Enzymatic catalysis in anhydrous organic solvents, *Trends Biochem. Sci.* **14**, 141-144.

Klibanov, A. M. (1990), Asymmetric transformations catalyzed by enzymes in organic solvents, *Acc. Chem. Res.* **23**, 114-120.

Klibanov, A. M. (1997), Why are enzymes less active in organic solvents than in water?, *Trends Biotechnol.* **15**, 97-101.

Kloosterman, M., Mosmuller, E. W. J., Schoemaker, H. E., Meijer, E. M. (1987), Application of lipases in the removal of protective groups on glycerides and glycosides, *Tetrahedron Lett.* **28**, 2989-2992.

Kloosterman, M., Elferink, V. H. M., van Iersel, J., Roskam, J. H., Meijer, E. M., Hulshof, L. A., Sheldon, R. A. (1988), Lipases in the preparation of β-blockers, *Trends Biotechnol.* **6**, 251-256.

Kloosterman, M., de Nijs, M. P., Weijnen, J. G. J., Schoemaker, H. E., Meijer, E. M. (1989), Regioselective hydrolysis of carbohydrate secondary acyl esters by lipases, *J. Carbohydr. Chem.* **8**, 333-341.

Klotz-Berendes, B., Kleemiß, W., Jegelka, U., Schäfer, H. J., Kotila, S. (1997), Enzymatic synthesis of optically active mono-alkylated malonic monoesters, *Tetrahedron: Asymmetry* **8**, 1821-1823.

Knani, D., Kohn, D. H. (1993), Enzymatic polymerization in organic media. II. Enzyme-catalyzed synthesis of lateral-substituted aliphatic polyesters and copolyesters, *J. Polym. Sci. A: Polymer Chem.* **31**, 2887-2897.

Knani, D., Gutman, A. L., Kohn, D. H. (1993), Enzymatic polyesterification in organic enzyme-catalyzed synthesis of linear polyesters. I. condensation polymerization of hydroxyesters. II. ring-opening polymerization of ε-caprolactone, *J. Polym. Sci. A: Polymer Chem.* **31**, 1221-1232.

Knez, Z., Habulin, M. (1992), Lipase catalyzed transesterification in supercritical carbon dioxide, in.: *Biocatalysis in Non-Conventional Media* (Tramper, J.; Eds.), **Vol. 8**, pp. 401-406. Amsterdam: Elsevier.

Knezovic, S., Sunjic, V., Levai, A. (1993), Enantioselective hydrolysis of some 3-(2-nitrophenoxy)butanoates catalyzed by *Pseudomonas fluorescens* and *Pseudomonas* sp. lipase, *Tetrahedron: Asymmetry* **4**, 313-320.

Kobayashi, M., Shimizu, S. (1994), Versatile nitrilases: Nitrile-hydrolysing enzymes, *FEMS Microbiol. Lett.* **120**, 217-224.

Kobayashi, M., Shimizu, S. (1998), Metalloenzyme nitrile hydratase: structure, regulation, and application to biotechnology, *Nature Biotechnol.* **16**, 733-736.

Kobayashi, S., Uyama, H. (1993), Enzymatic polymerization of cyclic acid anhydrides and glycols by a lipase catalyst, *Makromol. Chem., Rapid Commun.*, **14**, 841-844.

Kobayashi, M., Nagasawa, T., Yamada, H. (1988), Regiospecific hydrolysis of dinitrile compounds by nitrilase from *Rhodococcus rhodochrous* J1, *Appl. Microbiol. Biotechnol.* **29**, 231-233.

Kobayashi, M., Yanaka, N., Nagasawa, T., Yamada, H. (1990a), Monohydrolysis of an aliphatic dinitrile compound by nitrilase from *Rhodococcus rhodochrous* K22, *Tetrahedron* **46**, 5587-5590.

Kobayashi, S., Kamiyama, K., Ohno, M. (1990b), Chiral synthon obtained with pig liver esterase: Introduction of chiral centers into cyclohexene skeleton, *Chem. Pharm. Bull.* **38**, 350-354.

Kobayashi, M., Nagasawa, T., Yamada, H. (1992), Enzymatic syntheses of acrylamide: a success story not yet over, *Trends Biotechnol.* **10**, 402-408.

Kobayashi, M., Izui, H., Nagasawa, T., Yamada, H. (1993), Nitrilase in biosynthesis of the plant hormone indole-3-acetic acid from indole-3-acetonitrile: Cloning of the *Alcaligenes* gene and site-directed mutagenesis of cysteine residues, *Proc. Natl. Acad. Sci. USA* **90**, 247-251.

Kobayashi, S., Shoda, S., Uyama, H. (1994), Enzymatic polymerization, *J. Syn. Org. Chem. Jpn.* **52**, 754-764.

Kobayashi, M., Suzuki, T., Fujita, T., Masuda, M., Shimizu, S. (1995), Occurence of enzymes involved in biosynthesis of indole-3-acetic acid from indole-3-acetonitrile in plant-associated bacteria, *Agrobacterium* and *Rhizobium*, *Proc. Natl. Acad. Sci. USA* **92**, 714-718.

Kodera, Y., Furukawa, M., Yokoi, M., Kuno, H., Matsushita, H., Inada, Y. (1993), Lactone synthesis from 16-hydroxyhexadecanoic acid ethyl ester in organic solvents catalyzed with polyethylene glycol-modified lipase, *J. Biotechnol.* **31**, 219-224.

Kodera, Y., Nishimura, H., Matsushima, A., Hiroto, M., Inada, Y. (1994), Lipase made active in hydrophobic media by coupling with polyethylene glycol, *J. Am. Oil Chem. Soc.* **71**, 335-338.

Koga, T., Nagao, A., Terao, J., Sawada, K., Mukai, K. (1994), Synthesis of a phosphatidyl derivative of Vitamin E and its antioxidant activity in phospholipid bilayers, *Lipids* **29**, 83-89.

Koh, J. H., Jeong, H. M., Park, J. (1998), Efficient catalytic racemization of secondary alcohols, *Tetrahedron Lett.* **39**, 5545-5548.

Kohno, M., Funatsu, J., Mikami, B., Kugimiya, W., Matsuo, T., Morita, Y. (1996), The crystal structure of lipase II from *Rhizopus niveus* at 2.2 angstrom resolution, *J. Biochem.* **120**, 505-510.

Koizumi, Y., Mukai, K., Murakawa, K., Yamane, T. (1987), Scale-up of microporous hydrophobic membrane bioreactor with respect to continuous glycerolysis of fat by lipase, *J. Jpn. Oil Chem. Soc.* **36**, 561-564.

Konigsberger, K., Luna, H., Prasad, K., Repic, O., Blacklock, T. J. (1996), Separation of *cis/trans*-cyclohexanecarboxylates by enzymic hydrolysis: preference for diequatorial isomers, *Tetrahedron Lett.* **37**, 9029-9032.

Koshiro, S., Sonomoto, K., Tanaka, A., Fukui, S. (1985), Stereoselective esterification of *dl*-menthol by polyurethane-entrapped lipase in organic solvent, *J. Biotechnol.* **2**, 47-57.

Koskinen, A. M. P., Klibanov, A. M. (Eds.) (1996), *Enzymatic Reactions in Organic Media*, Glasgow: Chapman & Hall.

Kosugi, Y., Azuma, N. (1994), Synthesis of triacylglycerol from polyunsaturated fatty acid by immobilized lipase, *J. Am. Oil Chem. Soc.* **71**, 1397-1403.

Kötting, J., Eibl, H. (1994), Lipases and phospholipases in organic synthesis, in.: *Lipases: Their structure, biochemistry, and application* (Woolley, P. , Petersen, S. B.; Eds.), pp. 289-313. Cambridge: Cambridge University Press.

Koyama, N., Doi, Y. (1996), Miscibility, thermal properties, and enzymatic degradability of binary blends of poly[(*R*)-3-hydroxybutyric acid] with poly(ε-caprolactone-co-lactide), *Macromolecules* **29**, 5843-5851.

Krebsfänger, N., Schierholz, K., Bornscheuer, U. T. (1998), Enantioselectivity of a recombinant esterase from *Pseudomonas fluorescens* towards alcohols and carboxylic acids, *J. Biotechnol.* **60**, 105-111.

Krief, A., Surleraux, D., Ropson, N. (1993), Novel enantioselective synthesis of optically active (1*R*)-*cis*- and (1*R*)-*trans*-chrysanthemic acids, *Tetrahedron: Asymmetry* **4**, 289-292.

Kroutil, W., Mischitz, M., Plachota, P., Faber, K. (1996), Deracemization of (±)-*cis*-2,3-epoxy-heptane *via* enantioconvergent biocatalytic hydrolysis using *Nocardia* EH1-epoxide hydrolase, *Tetrahedron Lett.* **37**, 8379-8382.

Kroutil, W., Kleewein, A., Faber, K. (1997a), A computer program for analysis, simulation and optimization of asymmetric catalytic processes proceeding through two consecutive steps. Type 1: asymmetrization-kinetic resolutions, *Tetrahedron: Asymmetry* **8**, 3251-3261.

Kroutil, W., Kleewein, A., Faber, K. (1997b), A computer program for analysis, simulation and optimization of asymmetric catalytic processes proceeding through two consecutive steps. Type 2: sequential kinetic resolutions, *Tetrahedron: Asymmetry* **8**, 3263-3274.

Kroutil, W., Mischitz, M., Faber, K. (1997c), Deracemization of (±)-2,3-disubstituted oxiranes *via* biocatalytic hydrolysis using bacterial epoxide hydrolases: kinetics of an enantioconvergent process, *J. Chem. Soc., Perkin Trans. 1*, 3629-3636.

Kroutil, W., Osprian, I., Mischitz, M., Faber, K. (1997d), Chemoenzymatic synthesis of (*S*)-(–)-frontalin using bacterial epoxide hydrolases, *Synthesis*, 156-158.

Kroutil, W., Genzel, Y., Pietzsch, M., Syldatk, C., Faber, K. (1998a), Purification and characterization of a highly selective epoxide hydrolase from *Nocardia* sp. EH1, *J. Biotechnol.* **61**, 143-150.

Kroutil, W., Orru, R. V. A., Faber, K. (1998b), Stabilization of *Nocardia* EH1 epoxide hydrolase by immobilization, *Biotechnol. Lett.* **20**, 373-377.

Kruizinga, W. H., Bolster, J., Kellogg, R. M., Kamphuis, J., Boesten, W. H. J., Meijer, E. M., Schoemaker, H. E. (1988), Synthesis of optically pure α-alkylated α-amino acids and a single-step method for enantiomeric excess determination, *J. Org. Chem.* **53**, 1826-1827.

Kuboki, A., Okazaki, H., Sugai, T., Ohta, H. (1997), An expeditions route to *N*-glycolylneura-minic acid based on enzyme-catalyzed reaction, *Tetrahedron* **53**, 2387-2400.

Kuchner, O., Arnold, F. H. (1997), Directed evolution of enzyme catalysts, *Trends Biotechnol.* **15**, 523-530.

Kudo, I., Murakami, M., Hara, S., Inoue, K. (1993), Mammalian non-pancreatic phospholipases A2, *Biochim Biophys Acta* **1170**, 217-231.

Kuge, Y., Shioga, K., Sugaya, T., Tomioka, S. (1993), Enzymatic optical resolution of dibenzoxepins and its application to an optically active antiallergic agent with thromboxane A2 receptor antagonist activity, *Biosci. Biotechnol. Biochem.* **57**, 1157-1160.

Kullmann, W. (1987), *Enzymatic peptide synthesis*, Boca Raton: CRC-Press.

Kundu, N., Roy, S., Maenza, F. (1972), Esterase activity of chymotrypsin on oxygen-substituted tyrosine substrates, *Eur. J. Biochem* **28**, 311-315.

Kunz, H., Kowalczyk, D., Braun, P., Braum, G. (1994), Enzymatic hydrolysis of hydrophilic diethyleneglycol and polyethyleneglycol esters of peptides and glycopeptides by lipases, *Angew. Chem. Int. Ed. Engl.* **33**, 336-339.

Kvittingen, L., Sjursnes, B. J., Anthonsen, T., Halling, P. (1992), Use of salt hydrates to buffer optimal water level during lipase catalyzed synthesis in organic media: a practical procedure for organic chemists, *Tetrahedron* **48**, 2793-2802.

Kwon, S. J., Han, J. J., Rhee, J. S. (1995), Production and *in situ* preparation of mono- or diacyl-glycerol catalyzed by lipases in n-hexane, *Enzyme Microb. Technol.* **17**, 700-704.

Laane, C. (1987), Medium-engineering for bioorganic synthesis, *Biocatalysis* **1**, 17-22.

Laane, C., Boeren, S., Vos, K., Verger, C. (1987), Rules for optimization of biocatalysis in organic solvents, *Biotechnol. Bioeng.* **30**, 81-87.

Lacourciere, G. M., Armstrong, R. N. (1993), The catalytic mechanism of microsomal epoxide hydrolase involves an ester intermediate, *J. Am. Chem. Soc.* **115**, 10466-10467.

Ladner, W. E., Whitesides, G. M. (1984), Lipase-catalyzed hydrolysis as a route to esters of chiral epoxy alcohols, *J. Am. Chem. Soc.* **106**, 7250-7251.

Laib, T., Ouazzani, J., Zhu, J. (1998), Horse liver esterase catalyzed enantioselective hydrolysis of *N,O*-diacetyl-2-amino-1-arylethanol, *Tetrahedron: Asymmetry* **9**, 169-178.

Lallemand, J. Y., Leclaire, M., Levet, R., Aranda, G. (1993), Easy access to an optically pure precursor of forskolin, *Tetrahedron: Asymmetry* **4**, 1775-1778.

Lalonde, J. J. (1995), The preparation of homochiral drugs and peptides using cross-linked enzyme crystals, *Chimica oggi* **9**, 31-35.

Lalonde, J. J., Bergbreiter, D. E., Wong, C.-H. (1988), Enzymatic kinetic resolution of α-nitro α-methyl carboxylic acids, *J. Org. Chem.* **53**, 2323-2327.

Lalonde, J. J., Govardhan, C., Khalaf, N., Martinez, A. G., Visuri, K., Margolin, A. L. (1995), Cross-linked crystals of *Candida rugosa* lipase: highly efficient catalysts for the resolution of chiral esters, *J. Am. Chem. Soc.* **117**, 6845-6852.

Lam, L. K. P., Hui, R. A. H. F., Jones, J. B. (1986), Enzymes in organic synthesis. 35. Stereoselective pig liver esterase catalyzed hydrolyses of 3-substituted glutarate diesters. Optimization of enantiomeric excess via reaction conditions control, *J. Org. Chem.* **51**, 2047-2050.

Lam, L. K. P., Brown, C. M., de Jeso, B., Lym, L., Toone, E. J., Jones, J. B. (1988), Enzymes in organic synthesis. 42. Investigation of the effects of the isozymal composition of pig liver esterase on its stereoselectivity in preparative-scale ester hydrolyses of asymmetric synthetic value, *J. Am. Chem. Soc.* **110**, 4409-4411.

Lamare, S., Legoy, M. D. (1995), Working at controlled water activity in a continuous process: the gas/solid system as a solution, *Biotechnol. Bioeng.* **45**, 387-397.

Lambusta, D., Nicolosi, G., Patti, A., Piatelli, M. (1993), Enzyme-mediated regioprotection-deprotection of hydroxyl groups in (+)-catechin, *Synthesis*, 1155-1158.

Lambusta, D., Nicolosi, G., Patti, A., Piattelli, M. (1996), Lipase-mediated resolution of racemic 2-hydroxymethyl-1-methylthioferrocene, *Tetrahedron Lett.* **37**, 127-130.

Lampe, T. F. J., Hoffmann, H. M. R., Bornscheuer, U. T. (1996), Lipase mediated desymmetrization of *meso* 2,6-di(acetoxymethyl)tetrahydropyran-4-one derivatives. An innovative route to enantiopure 2,4,6-trifunctionalized *C*-glycosides, *Tetrahedron: Asymmetry* **7**, 2889-2900.

Lang, D., Hofmann, B., Haalck, L., Hecht, H. J., Spener, F., Schmid, R. D., Schomburg, D. (1996), Crystal structure of a bacterial lipase from *Chromobacterium viscosum* ATCC 6918 refined at 1.6 Å resolution, *J. Mol. Biol.* **259**, 704-717.

Langrand, G., Secchi, M., Buono, G., Baratti, J., Triantaphylides, C. (1985), Lipase-catalyzed ester formation in organic solvents. An easy preparative resolution of α-substituted cyclohexanols, *Tetrahedron Lett.* **26**, 1857-1860.

Langrand, G., Baratti, J., Buono, G., Triantaphylides, C. (1986), Lipase catalyzed reactions and strategy for alcohol resolution, *Tetrahedron Lett.* **27**, 29-32.

Lankiewicz, L., Kasprzykowski, F., Grzonka, Z., Kettmann, U., Hermann, P. (1989), Resolution of racemic amino acids with thermitase, *Bioorg. Chem.* **17**, 275-280.

Larsson, K. M., Adlercreutz, P., Mattiasson, B. (1990), Enzymatic catalysis in microemulsions: enzyme reuse and product recovery, *Biotechnol. Bioeng.* **36**, 135-141.

Larsson, A. L. E., Persson, B. A., Backvall, J.-E. (1997), Enzymic resolution of alcohols coupled with ruthenium-catalyzed racemization of the substrate alcohol, *Angew. Chem. Int. Ed. Engl.* **36**, 1211-1212.

Laughlin, L. T., Tzeng, H.-F., Lin, S., Armstrong, R. N. (1998), Mechanism of microsomal epoxide hydrolase. Semifunctional site-specific mutants affecting the alkylation half-reaction, *Biochemistry* **37**, 2897-2904.

Laumen, K. E. (1987), Esterhydrolasen – Anwendung in der organischen Synthese: chirale Bausteine aus Estern prochiraler und racemischer Alkohole. *Ph.D. thesis*, Bergische Universität Wuppertal, Germany.

Laumen, K., Breitgoff, D., Schneider, M. P. (1988), Enzymic preparation of enantiomerically pure secondary alcohols. Ester synthesis by irreversible acyl transfer using a highly selective ester hydrolase from *Pseudomonas* sp.; an attractive alternative to ester hydrolysis, *J. Chem. Soc., Chem. Commun.*, 1459-1461.

Laumen, K., Breitgoff, D., Seemayer, R., Schneider, M. P. (1989), Enantiomerically pure cyclohexanols and cyclohexane-1,2-diol derivatives, chiral auxiliaries and substitutes for (-)-8-phenylmenthol. A facile enzymic route, *J. Chem. Soc., Chem. Commun.*, 148-150.

Laumen, K., Ghisalba, O. (1994), Preparative-scale chemo-enzymic synthesis of optically pure D-*myo*-inositol 1-phosphate, *Biosci. Biotechnol. Biochem.* **58**, 2046-2049.

Laumen, K., Schneider, M. (1984), Enzymatic hydrolysis of prochiral *cis*-1,4-diacyl-2-cyclopentene diols - preparation of (1*S*,4*R*) and (1*R*,4*S*)-4-hydroxy-2-cylopentenyl derivatives, versatile building blocks for cyclopentanoic natural products, *Tetrahedron Lett.* **25**, 5875-5878.

Laumen, K., Schneider, M. P. (1986), A facile chemoenzymatic route to optically pure building blocks for cyclopentanoid natural products, *J. Chem. Soc. Chem. Commun.*, 1298-1299.

Laumen, K., Schneider, M. P. (1988), A highly selective ester hydrolase from *Pseudomonas* sp. for the enzymic preparation of enantiomerically pure secondary alcohols chiral auxiliaries in organic synthesis, *J. Chem. Soc., Chem. Commun.*, 598-600.

Lawson, W. B. (1967), The conformation of substrates during hydrolysis at the active site of chymotrypsin, *J. Biol. Chem.* **242**, 3397-3401.

Layh, N., Stolz, A., Förster, S., Effenberger, F., Knackmuss, H.-J. (1992), Enantioselective hydrolysis of O-acetylmandelonitrile to O-acetylmandelic acid by bacterial nitrilases, *Arch. Microbiol.* **158**, 405-411.

Layh, N., Stolz, A., Böhme, J., Effenberger, F., Knackmuss, H.-J. (1994), *J. Biotechnol.* **33**, 175-182.

Layh, N., Knackmuss, H.-J., Stolz, A. (1995), Enantioselective hydrolysis of ketoprofen amide by *Rhodococcus* sp. C3II and *Rhodococcus erythropolis* MP 50, *Biotechnol. Lett.* **17**, 187-192.

Layh, N., Hirrlinger, B., Stolz, A., Knackmuss, H.-J. (1997), Enrichment strategies for nitrile-hydrolysing bacteria, *Appl. Microbiol. Biotechnol.* **47**, 668-674.

Leanna, M. R., Morton, H. E. (1993), *N*-(Boc)-L-(2-Bromoallyl)-glycine: a versatile intermediate for the synthesis of optically active unnatural amino acids, *Tetrahedron Lett.* **34**, 4485-4488.

Lefker, B. A., Hada, W. A., McGarry, P. J. (1994), An efficient synthesis of enantiomerically enriched aryllactic esters, *Tetrahedron Lett.* **35**, 5205-5208.

LeGrand, D. M., Roberts, S. M. (1992), Enzyme-catalyzed hydrolysis of 3,5-*cis*-diacetoxy-4-*trans*-benzyloxymethylcyclopentene and the synthesis of aristeromycin precursors, *J. Chem. Soc., Perkin Trans. 1*, 1751-1752.

Lemke, K., Theil, F., Kunath, A., Schick, H. (1996), Lipase-catalyzed kinetic resolution of phenylethan-1,2-diol by sequential transesterification - the influence of the solvent, *Tetrahedron: Asymmetry* **7**, 971-974.

Lemke, K., Lemke, M., Theil, F. (1997), A three-dimensional predictive active site model for lipase from *Pseudomonas cepacia*, *J. Org. Chem.* **62**, 6268-6273.

Lemoult, S. C., Richardson, P. F., Roberts, S. M. (1995), Lipase-catalyzed Baeyer-Villiger reactions, *J. Chem. Soc., Perkin Trans. 1*, 89-91.

Levayer, F., Rabiller, C., Tellier, C. (1995), Enzyme catalysed resolution of 1,3-diarylpropan-1,3-diols, *Tetrahedron: Asymmetry* **6**, 1675-1682.

Ley, S. V., Mio, S., Meseguer, B. (1996), Dispiroketals in synthesis (part 20): preparation of chiral 2,2'-bis(halomethyl) and 2,2'bis(phenylthiomethyl)dihydropyrans, *Synlett*, 787.

Li, Y.-F., Hammerschmidt, F. (1993), Enzymes in organic chemistry, part 1: enantioselective hydrolysis of α-(acyloxy)phosphonates by esterolytic enzymes, *Tetrahedron: Asymmetry* **4**, 109-120.

Li, Z. Y., Ward, O. P. (1993), Lipase-catalyzed esterification of glycerol and *n*-3 polyunsaturated fatty acid concentrate in organic solvent, *J. Am. Oil Chem. Soc.* **70**, 745-748.

Li, Z. Y., Ward, O. P. (1994), Synthesis of monoglyceride containing omega-3 fatty acids by microbial lipase in organic solvent, *J. Ind. Microbiol.* **13**, 49-52.

Liang, S., Paquette, L. A. (1990), Biocatalytic-based synthesis of optically-pure (C-6)-functionalized 1-(tert-butyldimethylsilyloxy) 2-methyl-(*E*)-2-heptenes, *Tetrahedron: Asymmetry* **1**, 445-452.

Lie-Ken-Jie, M. S. F., Syed-Rahmatullah, M. S. K. (1995), Chemical and enzymic preparation of acylglycerols containing C18 furanoid fatty acids, *Lipids* **30**, 79-84.

Lin, Y. Y., Palmer, D. N., Jones, J. B. (1974), The specificity of the nucleophilic site of α-chymotrypsin and its potential for the resolution of alcohols. Enzyme-catalyzed hydrolyses of some (+)-, (–)-, and (±)-2-butyl, -2-octyl, and -α-phenethyl esters, *Can. J. Chem.* **52**, 469-476.

Linderman, R. J., Walker, E. A., Haney, C., Roe, R. M. (1995), Determination of the regiochemistry of insect epoxide hydrolase catalyzed epoxide hydration of juvenile hormone by ^{18}O-labelling studies, *Tetrahedron* **51**, 10845-10856.

Ling, L., Ozaki, S. (1993), Enzyme aided synthesis of D-*myo*-inositol 1,4,5-trisphosphate, *Tetrahedron Lett.* **34**, 2501-2504.

Ling, L., Watanabe, Y., Akiyama, T., Ozaki, S. (1992), A new efficient method of resolution of *myo*-inositol derivatives by enzyme catalyzed regio- and enantio-selective esterification in organic solvent, *Tetrahedron Lett.* **33**, 1911-1914.

Linker, T. (1997), The Jacobsen-Katsuki epoxidation and its controversial mechanism, *Angew. Chem. Int. Ed. Engl.* **36**, 2060-2062.

Linko, Y. Y., Seppälä, J. (1996), Producing high molecular weight biodegradable polyesters, *Chemtech* **26**, 25-31.

Linko, Y.-Y., Wang, Z.-L., Seppälä, J. (1995a), Lipase-catalyzed linear aliphatic polyester synthesis in organic solvent, *Enzyme Microb. Technol.* **17**, 506-511.

Linko, Y.-Y., Wang, Z.-L., Seppälä, J. (1995b), Lipase-catalyzed synthesis of poly(1,4-butyl sebacate) from sebacic acid or its derivatives with 1,4-butanediol, *J. Biotechnol.* **40**, 133-138.

Liu, Y.-C., Chen, C.-S. (1989), An efficient synthesis of optically active D-*myo*-inositol 1,4,5-triphosphate, *Tetrahedron Lett.* **30**, 1617-1620.

Liu, K. K. C., Nozaki, K., Wong, C. H. (1990), Problems of acyl migration in lipase-catalyzed enantioselective transformation of *meso*-1,3-diol systems, *Biocatalysis* **3**, 169-177.

Ljunger, G., Adlercreutz, P., Mattiasson, B. (1994), Lipase catalyzed acylation of glucose, *Biotechnol. Lett.* **16**, 1167-1172.

Lobell, M., Schneider, M. P. (1993), Lipase-catalyzed formation of lactones via irreversible intramolecular acyl transfer, *Tetrahedron: Asymmetry* **4**, 1027-1030.

Lokotsch, W., Fritsche, K., Syldatk, C. (1989), Resolution of D,L-menthol by interesterification with triacetin using the free and immobilized lipase of *Candida cylindracea, Appl. Microbiol. Biotechnol.* **31**, 467-472.

López, R., Montero, E., Sanchez, F., Canada, J., Fernandez-Mayoralas, A. (1994), Regioselective acetylations of alkyl β-D-xylopyranosides by use of lipase PS in organic solvents and application to the chemoenzymatic synthesis of oligosaccharides, *J. Org. Chem.* **59**, 7027-7032.

Lord, M. D., Negri, J. T., Paquette, L. A. (1995), Oxonium ion-initiated pinacolic ring expansion reactions - application to the enantioselective synthesis of the spirocyclic sesquiterpene ethers dactyloxene-B and -C, *J. Org. Chem.* **60**, 191-195.

Lotti, M., Grandori, R., Fusetti, F., Longhi, S., Brocca, S., Tramontano, A., Alberghina, L. (1993), Cloning and analysis of *Candida cylindracea* lipase sequences, *Gene* **124**, 45-55.

Lovey, R. G., Saksena, A. K., Girijavallabhan, V. M. (1994), PPL-catalyzed enzymatic asymmetrization of a 5-substituted prochiral 1,3-diol with remote chiral functionality - improvements toward synthesis of the eutomers of SCH 45012, *Tetrahedron Lett.* **35**, 6047-6050.

Lundell, K., Raijola, T., Kanerva, L. T. (1998), Enantioselectivity of *Pseudomonas cepacia* and *Candida rugosa* lipases for the resolution of secondary alcohols: The effect of *Candida rugosa* isoenzymes, *Enzyme Microb. Technol.* **22**, 86-93.

Lundh, H., Nordin, O., Hedenström, E., Högberg, H. E. (1995), Enzyme catalysed irreversible transesterifications with vinyl acetate - are they really irreversible?, *Tetrahedron: Asymmetry* **6**, 2237-2244.

Lutz, D., Huffer, M., Gerlach, D., Schreier, P. (1992), Carboxylester-lipase-mediated reactions, in.: *Flavor precursors: thermal and enzymatic conversions* (Teranishi, R., Takeoka, G. R. , Guentert, M.; Eds.), pp. 32-45. Washington DC: American Chemical Society.

MacBeath, G., Kast, P., Hilvert, D. (1998), Redesigning enzyme topology by directed evolution, *Science* **279**, 1958-1961.

MacDonald, R. T., Pulapura, S. K., Svirkin, Y. Y., Gross, R. A., Kaplan, D. L., Akkara, J., Swift, G., Wolk, S. (1995), Enzyme-catalyzed ε-caprolactone ring-opening polymerization, *Macromolecules* **28**, 73-78.

Macfarlane, E. L. A., Roberts, S. M., Turner, N. J. (1990), Enzyme-catalyzed inter-esterification procedure for the preparation of esters of a chiral secondary alcohol in high enantiomeric purity, *J. Chem. Soc., Chem. Commun.*, 569-571.

MacKeith, R. A., McCague, R., Olivo, H. F., Palmer, C. F., Roberts, S. M. (1993), Conversion of (−)-4-hydroxy-2-oxabicyclo[3.3.0]oct-7-en-3-one into the anti-HIV agent carbovir, *J. Chem. Soc., Perkin Trans. 1*, 313-314.

MacKeith, R. A., McCague, R., Olivo, H. F., Roberts, S. M., Taylor, S. J. C., Xiong, H. (1994), Enzyme-catalyzed kinetic resolution of 4-endo-hydroxy-2- oxabicyclo[3.3.0]oct-7-en-3-one and employment of the pure enantiomers for the synthesis of antiviral and hypocholestemic agents, *Bioorg. Med. Chem.* **2**, 387-394.

Macrae, A. R. (1983), Lipase-catalyzed interesterification of oils and fats, *J. Am. Oil Chem. Soc.* **60**, 291-294.

Macrae, A. R., Hammond, R. C. (1985), Present and future applications of lipases, *Biotechnol. Genet. Eng. Rev.* **3**, 193-217.

Maelicke, A. (1991), Acetylcholine esterase: the structure, *Trends Biochem. Sci.* **16**, 355-356.

Majeric, M., Sunjic, V. (1996), Preparation of (*S*)-2-ethylhexyl-*p*-methoxycinnamate by lipase catalyzed sequential kinetic resolution, *Tetrahedron: Asymmetry* **7**, 815-824.

Makita, A., Nihira, T., Yamada, Y. (1987), Lipase-catalyzed synthesis of macrocyclic lactones in organic solvents, *Tetrahedron Lett.* **28**, 805-808.

Maleczka, R. E. J., Paquette, L. A. (1991), Adaptation of oxyanionic sigmatropy to the convergent enantioselective synthesis of ambergris-type odorants, *J. Org. Chem.* **56**, 6538-6546.

Malézieux, B., Jaouen, G., Salaün, J., Howell, J. A. S., Palin, M. G., et al. (1992), Enzymatic generation of planar chirality in the (arene)tricarbonyl-chromium series, *Tetrahedron: Asymmetry* **3**, 375-376.

Marangoni, A. G., Rousseau, D. (1995), Engineering triacylglycerols: The role of interesterification, *Trends Food Sci. Technol.* **6**, 329-335.

Margolin, A. L. (1993a), Enzymes in the synthesis of chiral drugs, *Enzyme Microb. Technol.* **15**, 266-280.

Margolin, A. L. (1993b), Synthesis of optically pure mechanism-based inhibitors of γ-aminobutyic acid aminotransferase (GABA-T) via enzyme-catalyzed resolution, *Tetrahedron Lett.* **34**, 1239-1242.

Margolin, A. L. (1996), Novel crystalline catalysts, *Trends Biotechnol.* **14**, 223-230.

Margolin, A. L., Crenne, J.-Y., Klibanov, A. M. (1987), Stereoselective oligomerizations catalyzed by lipases in organic solvents, *Tetrahedron Lett.* **28**, 1607-1610.

Margolin, A. L., Delinck, D. L., Whalon, M. R. (1990), Enzyme-catalyzed regioselective acylation of castanospermine, *J. Am. Chem. Soc.* **112**, 2849-2854.

Marr, R., Gamse, T., Schilling, T., Klingsbichel, E., Schwab, H., Michor, H. (1996), Enzymatic catalysis in supercritical carbon dioxide: comparison of different lipases and a novel esterase, *Biotechnol. Lett.* **18**, 79-84.

Marshalko, S. J., Schweitzer, B. I., Beardsley, G. P. (1995), Chiral chemical synthesis of DNA containing (*S*)-9-(1,3-dihydroxy-2-propoxymethyl)-guanine (dhpg) and effects on thermal stability, duplex structure, and thermodynamics of duplex formation, *Biochemistry* **34**, 9235-9248.

Martin, S. F., Hergenrother, P. J. (1998), Enzymatic synthesis of a modified phospholipid and its evalutation as a substrate for *B. cereus* phospholipase C, *Bioorg. Med. Chem. Lett.* **8**, 593-596.

Martinek, K., Levashov, A. V., Khmelnitski, Y. L., Klyachko, N., Berezin, I. V. (1982), Colloidal solution of water in organic solvents: A microheterogeneous medium for enzymatic reactions, *Science* **218**, 889-891.

Martinez, C., Geus, P. D., Lauwereys, M., Matthyssens, G., Cambillau, C. (1992), Fusarium solani cutinase is a lipolytic enzyme with a catalytic serine accessible to the solvent, *Nature* **356**, 615-618.

Martinez, C., Nicolas, A., Tilbeurgh, H. van, Egloff, M.-P., Cudrey, C., Verger, R., Cambillau, C. (1994), Cutinase, a lipolytic enzyme with a preformed oxyanion hole, *Biochemistry* **33**, 83-89.

Martínková, L., Stolz, A., Knackmuss, H.-J. (1996), Enantioselectivity of the nitrile hydratase from *Rhodococcus equi* A4 towards substituted (*R,S*)-2-arylpropionitriles, *Biotechnol. Lett.* **18**, 1073-1076.

Martins, J. F., Sampaio, T. C., Carvalho, I. B., da Ponte, M. N., Barreiros, S. (1992), Lipase catalyzed esterification of glycidol in chloroform and in supercritical carbon dioxide, in.: *High Press. Biotechnol.* (Balny, C., Hayashi, R., Heremans, K. , Masson, P.; Eds.), **Vol. 224**, pp. 411-415.

Martres, M., Gil, G., Meon, A. (1994), Preparation of optically active aziridine carboxylates by lipase-catalyzed alcoholysis, *Tetrahedron Lett.* **35**, 8787-8790.

Marty, A., Chulalaksananukul, W., Condoret, J. S., Willemot, R. M., Durand, G. (1990), Comparison of lipase-catalysed esterification in supercritical carbon dioxide and in n-hexane, *Biotechnol. Lett.* **12**, 11-16.

Marty, A., Chulalaksananukul, W., Willemot, R. M., Condorét, J. S. (1992), Kinetics of lipase-catalyzed esterification in supercritical carbon dioxide, *Biotechnol. Bioeng.* **39**, 273-280.

Mathew, C. D., Nagasawa, T., Kobayashi, M., Yamada, H. (1988), Nitrilase-catalyzed production of nicotinic acid from 3-cyanopyridine in *Rhodococcus rhodochrous* J1, *Appl. Environm. Microbiol.* **54**, 1030-1032.

Matsumae, H., Furui, M., Shibatani, T. (1993), Lipase-catalyzed asymmetric hydrolysis of 3-phenylglycidic acid ester, the key intermediate in the synthesis of diltiazem hydrochloride, *J. Ferment. Bioeng.* **75**, 93-98.

Matsumae, H., Furui, M., Shibatani, T., Tosa, T. (1994), Production of optically active 3-phenylglycidic acid ester by the lipase from *Serratia marcescens* on a hollow-fiber membrane reactor, *J. Ferment. Bioeng.* **78**, 59-63.

Matsumoto, K. (1992), Production of 6-APA, 7-ACA, and 7-ADCA by immobilized penicillin and cephalosporin amidases, in.: *Industrial Applications of Immobilized Biocatalysts* (Tanaka, A., Tosa, T. , Kobayashi, T.; Eds.), New York: Marcel Dekker.

Matsumoto, K., Fuwa, S., Kitajima, H. (1995), Enzyme-mediated enantioselective hydrolysis of cyclic carbonates, *Tetrahedron Lett.* **36**, 6499-6502.

Matsumoto, K., Fuwa, S., Shimojo, S., Kitajima, H. (1996), Preparation of optically active diol derivatives by the enzymatic hydrolysis of cyclic carbonates, *Bull. Chem. Soc. Jpn.* **69**, 2977-2987.

Matsuo, N., Ohno, N. (1985), Preparation of optically active 1-acetoxy-2-aryloxypropionitriles and its application to a facile synthesis of (*S*)-(-)-Propranolol, *Tetrahedron Lett.* **26**, 5533-5534.

Matsuo, T., Sawamura, N., Hashimoto, Y., Hashida, W. (1981), The enzyme and method for enzymatic transesterification of lipid, *European Patent* EP 0 035 883 (Fuji Oil Co.) (Chem. Abstr. 96:4958).

Matsushima, M., Inoue, H., Ichinose, M., Tsukada, S., Miki, K., Kurokawa, K., Takahashi, T., Takahashi, K. (1991), The nucleotide and deduced amino acid sequence of porcine liver proline-β-naphthylamidase. Evidence for the idendity with carboxylesterase, *FEBS Lett.* **293**, 37-41.

Matta, M. S., Rohde, M. F. (1972), α-Chymotrypsin and rigid substrates. Reactivity of some *p*-nitrophenyl 1,2,3,4-tetrahydro-2-naphthoates and indan-2-carboxylates, *J. Am. Chem. Soc.* **94**, 8573-8578.

Matthews, B. W. (1988), Structural basis of the action of thermolysin and related zinc peptidases, *Acc. Chem. Res.* **21**, 333-340.

Matthews, B. W., Sigler, P. B., Henderson, R., Blow, D. M. (1967), Three-dimensional structure of tosyl-α-chymotrypsin., *Nature* **214**, 652-656.

Mattson, A., Öhrner, N., Norin, T., Hult, K. (1993), Resolution of diols with C_2 symmetry by lipase-catalyzed transesterification, *Tetrahedron: Asymmetry* **4**, 925-930.

Mattson, A., Orrenius, C., Oehrner, N., Unelius, C. R., Hult, K., Norin, T. (1996), Kinetic resolution of chiral auxiliaries with C_2-symmetry by lipase-catalyzed alcoholysis and aminolysis, *Acta Chem. Scand.* **50**, 918-921.

Mazdiyasni, H., Konopacki, D. B., Dickman, D. A., Zydowsky, T. M. (1993), Enzyme-catalyzed synthesis of optically-pure β-sulfonamidopropionic acids. Useful starting materials for P-3 site modified renin inhibitors, *Tetrahedron Lett.* **34**, 435-438.

Mazur, A. W., Hiler, G. D. I., El-Nokaly, M. (1991), Preparation of 2-monoglycerides, *ACS Symp. Ser.* **448**, 51-61.

McCague, R., Taylor, S. J. C. (1997), Four case studies in the development of biotransformation-based processes, in.: *Chirality in Industry II* (Collins, A. N., Sheldrake, G. N. , Crosby, J.; Eds.), pp. 183-206. Chichester: John Wiley & Sons.

McNeill, G. P., Sonnet, P. E. (1995), Low-calorie triglyceride synthesis by lipase-catalyzed esterification of monoglycerides, *J. Am. Oil Chem. Soc.* **72**, 1301-1307.

McNeill, G. P., Yamane, T. (1991), Further improvements in the yield of monoglycerides during enzymic glycerolysis of fats and oils, *J. Am. Oil Chem. Soc.* **68**, 6-10.

McNeill, G. P., Shimizu, S., Yamane, T. (1990), Solid-phase enzymatic glycerolysis of beef tallow resulting in a high yield of monoglycerides, *J. Am. Oil Chem. Soc.* **67**, 779-783.

McNeill, G. P., Shimizu, S., Yamane, T. (1991), High-yield enzymic glycerolysis of fats and oils, *J. Am. Oil Chem. Soc.* **68**, 1-5.

McNeill, G. P., Ackman, R. G., Moore, S. R. (1996), Lipase-catalyzed enrichment of long-chain polyunsaturated fatty acids, *J. Am. Oil Chem. Soc.* **73**, 1403-1407.

McPhalen, C. A., Schnebli, H. P., James, M. N. G. (1985), Crystal and molecular structure of the inhibitor eglin from leeches in complex with subtilisin Carlsberg, *FEBS Lett.* **188**, 55-58.

Megremis, C. J. (1991), Medium chain triglycerides: A nonconventional fat, *Food Technol.* **45**, 108-114.

Meng, D., Sorensen, E. J., Bertinato, P., Danishefsky, S. J. (1996), Studies toward a synthesis of epothilone A: Use of hydropyran templates for the management of acyclic stereochemical relationships, *J. Org. Chem.* **61**, 7998-7999.

Merlo, V., Reece, F. J., Roberts, S. M., Gregson, M., Storer, R. (1993), Synthesis of optically active 5'-noraristeromycin: enzyme-catalyzed kinetic resolution of 9-(4-hydroxycyclopent-2-enyl)purines, *J. Chem. Soc., Perkin Trans. 1*, 1717-1718.

Mertoli, P., Nicolosi, G., Patti, A., Piattelli, M. (1996), Convenient lipase-assisted preparation of both enantiomers of suprofen, a non-steroidal anti-inflammatory drug, *Chirality* **8**, 377-380.

Meth-Cohn, O., Wang, M. X. (1997a), An in-depth study of the biotransformation of nitriles into amides and/or acids using *Rhodococcus rhodochrous* AJ270, *J. Chem. Soc., Perkin Trans. 1*, 1099-1104.

Meth-Cohn, O., Wang, M. X. (1997b), Rationalisation of the regioselective hydrolysis of aliphatic dinitriles with *Rhodococcus rhodochrous* AJ270, *Chem. Commun.*, 1041-1042.

Michor, H., Marr, R., Gamse, T., Schilling, T., Klingsbichel, E., Schwab, H. (1996), Enzymic catalysis in supercritical carbon dioxide: comparison of different lipases and a novel esterase, *Biotechnol. Lett.* **18**, 79-84.

Miller, C., Austin, H., Posorske, L., Gonzlez, J. (1988), Characteristics of an immobilized lipase for the commercial synthesis of esters, *J. Am. Oil Chem. Soc.* **65**, 927-931.

Miller, D. A., Blanch, H. W., Prausnitz, J. M. (1990), Enzymic interesterification of triglycerides in supercritical carbon dioxide, *Ann. N. Y. Acad. Sci.* **613**, 534-537.

Millqvist, A., Adlercreutz, P., Mattiasson, B. (1994), Lipase-catalyzed alcoholysis of triglycerides for the preparation of 2-monoglycerides, *Enzyme Microb. Technol.* **16**, 1042-1047.

Millqvist-Fureby, A., Adlercreutz, P., Mattiasson, B. (1996a), Glyceride synthesis in a solvent-free system, *J. Am. Oil Chem. Soc.* **71**, 1489-1495.

Millqvist-Fureby, A., Virto, C., Adlercreutz, P., Mattiasson, B. (1996b), Acyl group migrations in 2-monoolein, *Biocatal. Biotransform.* **14**, 89-111.

Millqvist-Fureby, A., Tian, P., Adlercreutz, P., Mattiasson, B. (1997), Preparation of diglycerides by lipase-catalyzed alcoholysis of triglycerides, *Enzyme Microb. Technol.* **20**, 198-206.

Milton, J., Brand, S., Jones, M. F., Rayner, C. M. (1995), Enzymatic resolution of α-acetoxysulfides - a new approach to the synthesis of homochiral *S,O*-acetals, *Tetrahedron: Asymmetry* **6**, 1903-1906.

Mingarro, I., Abad, C., Braco, L. (1995), Interfacial activation-based molecular bioimprinting of lipolytic enzymes, *Proc. Natl. Acad. Sci. U. S. A.* **92**, 3308-3312.

Mingarro, I., Gonzalez-Navarro, H., Braco, L. (1996), Trapping of different lipase conformers in water-restricted environments, *Biochemistry* **35**, 9935-9944.

Minning, S., Schmidt-Dannert, C., Schmid, R. D. (1998), Functional expression of *Rhizopus oryzase* lipase in *Pichia pastoris*: High-level production and some properties, *J. Biotechnol.* **66**, 147-156.

Misawa, E., ChanKwoChion, C. K. C., Archer, I. V., Woodland, M. P., Zhou, N.-Y., Carter, S. F., Widdowson, D. A., Leak, D. J. (1998), Characterization of a catabolic epoxide hydrolase from *Corynebacterium* sp., *Eur. J. Biochem.* **253**, 173-183.

Mischitz, M., Faber, K. (1994), Asymmetric opening of an epoxide by azide catalyzed by an immobilized enzyme preparation from *Rhodococcus* sp., *Tetrahedron Lett.* **35**, 81-84.

Mischitz, M., Faber, K. (1996), Chemo-enzymatic synthesis of (2*R*,5*S*)- and (2*R*,5*R*)-5-(1-hydroxy-1-methylethyl)-2-methyl-2-vinyl-tetrahydrofuran ('linalool oxide'): Preparative application of a highly selective epoxide hydrolase, *Syntlett* 978-979.

Mischitz, M., Faber, K., Willetts, A. (1995a), Isolation of a highly enantioselective epoxide hydrolase from *Rhodococcus* sp. NCIMB11216, *Biotechnol. Lett.* **17**, 893-898.

Mischitz, M., Kroutil, W., Wandel, U., Faber, K. (1995b), Asymmetric microbial hydrolysis of epoxides, *Tetrahedron: Asymmetry* **6**, 1261-1272.

Mischitz, M., Mirtl, C., Saf, R., Faber, K. (1996), Regioselectivity of *Rhodococcus* sp. NCIMB11216 epoxide hydrolase: Applicability of *E*-values for description of enantioselectivity depends on substrate structure, *Tetrahedron: Asymmetry* **7**, 2041-2046.

Misset, O., Gerritse, G., Jaeger, K.-E., Winkler, U., Colson, C., Schanck, K., Lesuisse, E., Dartois, V., Blaauw, M., Ransac, S., Dijkstra, B. W. (1994), The structure-function relationship of the lipases from *Pseudomonas aeruginosa* and *Bacillus subtilis*, *Protein Eng.* **7**, 523-529.

Mitrochkine, A., Gil, G., Reglier, M. (1995a), Synthesis of enantiomerically pure *cis*- and *trans*-2-amino-1-indanol, *Tetrahedron: Asymmetry* **6**, 1535-1538.

Mitrochkine, A., Mycke, F., Martres, M., Gil, G., Heumann, A., Reglier, M. (1995b), Synthesis of enantiomerically pure (1*S*,2*R*)-epoxyindan and *cis*-(1*R*,2*S*)-2-amino-1-indanol, *Tetrahedron: Asymmetry* **6**, 59-62.

Miyaoka, H., Sagawa, S., Nagaoka, H., Yamada, Y. (1995), Efficient synthesis of enantiomerically pure 5,5-dimethyl-4-hydroxy-2-cyclopentenone, *Tetrahedron: Asymmetry* **6**, 587-594.

Miyazawa, T., Kurita, S., Ueji, S., Yamada, T., Kuwata, S. (1992), Resolution of racemic carboxylic acids via the lipase-catalyzed irreversible transesterification using vinyl esters: effects of alcohols as nucleophiles and organic solvents on enantioselectivity, *Biotechnol. Lett.* **14**, 941-946.

Mizuguchi, E., Nagai, H., Uchida, H., Achiwa, K. (1994), Lipase-catalyzed enantioselective reaction-based on remote recognition of the stereogenic carbon-atom away from the reaction site, *J. Synth. Org. Chem. Jpn.* **52**, 638-648.

Mohamed, H. M. A., Bloomer, S., Hammadi, K. (1993), Modification of fats by lipase interesterification. I: Changes in glyceride structure, *Fat Sci. Technol.* **95**, 428-431.

Mohar, B., Stimac, A., Kobe, J. (1994), Chiral building blocks for carbocyclic *N*- and *C*-ribonucleosides through biocatalytic asymmetrization of *meso*-cyclopentane-1,3-dimethanols, *Tetrahedron: Asymmetry* **5**, 863-878.

Mohr, P., Waespe-Sarcevic, N., Tamm, C., Gawronska, K., Gawronski, J. K. (1983), A study of stereoselective hydrolysis of symmetrical diesters with pig liver esterase, *Helv. Chim. Acta* **66**, 2501-2511.

Molinari, F., Brenna, O., Valenti, M., Aragozzini, F. (1996), Isolation of a novel carboxylesterase from *Bacillus coagulans* with high enantioselectivity toward racemic esters of 1,2-O-isopropylideneglycerol, *Enzyme Microb. Technol.* **19**, 551-556.

Mölm, D., Risch, N. (1995), Resolution of racemic 1-azaadamantane derivatives by pig liver esterase catalysis, *Liebigs Ann.*, 1901-1902.

Moore, J. C., Arnold, F. H. (1996), Directed evolution of a para-nitrobenzyl esterase for aqueous-organic solvents, *Nature Biotechnol.* **14**, 458-467.

Moore, S. R., McNeill, G. P. (1996), Production of triglycerides enriched in long-chain n-3 polyunsaturated fatty acids from fish oil, *J. Am. Oil Chem. Soc.* **73**, 1409-1414.

Moorlag, H., Kellogg, R. M., Kloosterman, M., Kaptein, B., Kamphuis, J., Schoemaker, H. E. (1990), Pig-liver-esterase-catalyzed hydrolyses of racemic α-substituted α-hydroxy esters, *J. Org. Chem.* **55**, 5878-5881.

Moravcová, J., Vanclová, Z., Capková, J., Kefurt, K., Stanek, J. (1997), Enzymic hydrolysis of methyl 2,3-di-O-acetyl-5-deoxy-α and β-D-xylofuranosides - an active-site model of pig liver esterase, *J. Carbohydr. Chem.* **16**, 1011-1028.

Moree, W. J., Sears, P., Kawashiro, K., Witte, K., Wong, C. H. (1997), Exploitation of subtilisin BPN' as catalyst for the synthesis of peptides containing noncoded amino acids, peptide mimetics and peptide conjugates, *J. Am. Chem. Soc.* **119**, 3942-3947.

Morgan, B., Oehlschlager, A. C., Stokes, T. M. (1991), Enzyme reactions in apolar solvents. The resolution of branched and unbranched 2-alkanols by porcine pancreatic lipase, *Tetrahedron* **47**, 1611-1620.

Morgan, B., Oehlschlager, A. C., Stokes, T. M. (1992), Enzyme reactions in apolar solvents. 5. The effect of adjacent unsaturation on the PPL-catalyzed kinetic resolution of secondary alcohols, *J. Org. Chem.* **57**, 3231-3236.

Morgan, B., Dodds, D. R., Zaks, A., Andrews, D. R., Klesse, R. (1997a), Enzymatic desymmetrization of prochiral 2-substituted-1,3-propanediols: a practical chemoenzymatic synthesis of a key precursor of SCH5108, a broad spectrum orally active antifungal agent, *J. Org. Chem.* **62**, 7736-7743.

Morgan, B., Stockwell, B. R., Dodds, D. R., Andrews, D. R., Sudhakar, A. R., Nielsen, C. M., Mergelsberg, I., Zumbach, A. (1997b), Chemoenzymatic approaches to SCH 56592, a new azole antifungal, *J. Am. Oil Chem. Soc.* **74**, 1361-1370.

Morgan, J., Pinhey, J. T., Sherry, C. J. (1997c), Reaction of organolead triacetates with 4-ethoxycarbonyl-2-methyl-4,5-dihydro-1,3-oxazol-5-one. The synthesis of α-aryl-and α-vinyl-*N*-acetylglycines and their ethyl esters and their enzymic resolution, *J. Chem. Soc., Perkin Trans. 1*, 613-619.

Mori, K. (1995), Biochemical methods in enantioselective synthesis of bioactive natural products, *Synlett*, 1097-1109.

Morihara, K., Oka, T. (1983), Enzymic semisynthesis of human insulin by transpeptidation method with *Achromobacter* protease: comparison with the coupling method, *Pept. Chem.* **20**, 231-236.

Mori, K., Hazra, B. G., Pfeiffer, R. J., Gupta, A. K., Lindgren, B. S. (1987), Synthesis and bioactivity of optically-active forms of 1-methyl-2-cyclohexen-1-ol, an aggregation pheromone of *Dendroctonus pseudotsugae*, *Tetrahedron* **43**, 2249-2254.

Mori, K., Puapoomchareon, P. (1991), Preparation of optically pure 2,4,4-trimethyl-2-cyclohexen-1-ol, a new and versatile chiral building block in terpene synthesis, *Liebigs Ann. Chem.*, 1053-1056.

Morimoto, T., Murakami, N., Nagatsu, A., Sakakibara, J. (1994), Enzymatic regioselective acylation of 3-*O*-β-D-galactopyranosyl-*sn*-glycerol by *Achromobacter* sp lipase, *Chem. Pharm. Bull.* **42**, 751-753.

Morís, F., Gotor, V. (1992a), Lipase-mediated alkoxycarbonylation of nucleosides with oxime carbonates, *Tetrahedron* **48**, 9869-9876.

Morís, F., Gotor, V. (1992b), A novel and convenient route to 3'-carbonates from unprotected 2'-deoxynucleosides through an enzymatic reaction, *J. Org. Chem.* **57**, 2490-2492.

Morís, F., Gotor, V. (1993a), Enzymic acylation and alkoxycarbonylation of α-, *xylo-*, anhydro-, and *arabino*-nucleosides, *Tetrahedron* **49**, 10089-10098.

Morís, F., Gotor, V. (1993b), A useful and versatile procedure for the acylation of nucleosides through an enzymic reaction, *J. Org. Chem.* **58**, 653-660.

Morís, F., Gotor, V. (1994), Selective aminoacylation of nucleosides through an enzymatic reaction with oxime aminoacyl esters, *Tetrahedron* **50**, 6927-6934.

Morisseau, C., Nellaiah, H., Archelas, A., Furstoss, R., Baratti, J. C. (1997), Asymmetric hydrolysis of racemic *para*-nitrostyrene oxide using an epoxide hydrolase preparation from *Aspergillus niger*, *Enzyme Microb. Technol.* **20**, 446-452.

Morrone, R., Nicolosi, G., Patti, A., Piattelli, M. (1995), Resolution of racemic flurbiprofen by lipase-mediated esterification in organic solvent, *Tetrahedron: Asymmetry* **6**, 1773-1778.

Moussou, P., Archelas, A., Baratti, J., Furstoss, R. (1998a), Determination of the regioselectivity during epoxide hydrolase oxirane ring opening: A new method from racemic epoxides, *J. Mol. Catal. B: Enzymatic* **5**, 213-217.

Moussou, P., Archelas, A., Baratti, J., Furstoss, R. (1998b), Microbiological transformations. Part 39: Determination of the regioselectivity occurring during oxirane ring opening by epoxide hydrolases : a theoretical analysis and a new method for its determination, *Tetrahedron: Asymmetry* **9**, 1539-1547.

Moussou, P., Archelas, A., Furstoss, R. (1998c), Microbiological transformations. 40. Use of fungal epoxide hydrolases for the synthesis of enantiopure alkyl epoxides, *Tetrahedron* **54**, 1563-1572.

Mozhaev, V. V., Budde, C. L., Rich, J. O., Usyatinsky, A. Y., Michels, P. C., Khmelnitsky, Y. L., Clark, D. S., Dordick, J. S. (1998), Regioselective enzymatic acylation as a tool for producing solution-phase combinatorial libraries, *Tetrahedron* **54**, 3971-3982.

Muchmore, D. C. (1993), Enantiomeric enrichment of (*R,S*)-3-quinuclidinol, *US Patent* US 5 215 918 (Chem. Abstr. 119: 135 038).

Mukherjee, K. D. (1990), Lipase-catalyzed reactions for modification of fats and other lipids, *Biocatalysis* **3**, 277-293.

Multzsch, R., Lokotsch, W., Steffen, B., Lang, S., Metzger, J. O., Schäfer, H. J., Warwel, S., Wagner, F. (1994), Enzymatic production and physicochemical characterization of uncommon wax esters and monoglycerides, *J. Am. Oil Chem. Soc.* **71**, 721-725.

Mulvihill, M. J., Gage, J. L., Miller, M. J. (1998), Enzymatic resolution of aminocyclopentenols as precursors to D- and L-carbocyclic nucleosides, *J. Org. Chem.* **63**, 3357-3363.

Mulzer, J., Greifenberg, S., Beckstett, A., Gottwald, M. (1992), Selective acetate hydrolysis of diastereomers with porcine pancreatic lipase (PPL) as an access to useful chiral building blocks, *Liebigs Ann. Chem.*, 1131-1135.

Mustranta, A. (1992), Use of lipases in the resolution of racemic ibuprofen, *Appl. Microbiol. Biotechnol.* **38**, 61-66.

Myrnes, B., Barstad, H., Olsen, R. L., Elveoll, E. O. (1995), Solvent-free enzymic glycerolysis of marine oils, *J. Am. Oil Chem. Soc.* **72**, 1339-1344.

Na, A., Eriksson, C., Eriksson, S.-G., Österberg, E., Holmberg, K. (1990), Synthesis of phosphatidylcholine with (n-3) fatty acids by phospholipase A2 in microemulsion, *J. Am. Oil Chem. Soc.* **67**, 766-770.

Naemura, K., Miyabe, H., Shingai, Y. (1991), Synthesis and enantiomer recognition of crown ethers containing cyclohexane-1,2-diol derivatives as the chiral center and enzymatic resolution of the chiral subunits, *J. Chem. Soc., Perkin Trans. 1*, 957-959.

Naemura, K., Takahashi, N., Tanaka, S. (1992), Resolution of the diols of bicyclo[2.2.1]heptane, bicyclo[2.2.2]octane and bicyclo[3.2.1]octane by enzymatic hydrolysis, and their absolute-configurations, *J. Chem. Soc., Perkin Trans. 1*, 2337-2343.

Naemura, K., Fukuda, R., Takahashi, N., Konishi, M., Hirose, Y., Tobe, Y. (1993a), Enzyme-catalyzed asymmetric acylation and hydrolysis of *cis*-2,5-disubstituted tetrahydrofuran derivatives - contribution to the development of models for reactions catalyzed by porcine liver esterase and porcine pancreatic lipase, *Tetrahedron: Asymmetry* **4**, 911-918.

Naemura, K., Ida, H., Fukuda, R. (1993b), Lipase YS-catalyzed enantioselective transesterification of alcohols of bicarbocyclic compounds, *Bull. Chem. Soc. Jpn.* **66**, 573-577.

Naemura, K., Fukuda, R., Konishi, M., Hirose, K., Tobe, Y. (1994), Lipase YS catalysed acylation of alcohols: a predictive active site model for lipase YS to identify which enantiomer of a primary or secondary alcohol reacts faster in this acylation, *J. Chem. Soc. Perkin Trans. 1*, 1253-1256.

Naemura, K., Fukuda, R., Murata, M., Konishi, M., Hirose, K., Tobe, Y. (1995), Lipase YS-catalyzed enantioselective acylation of alcohols: a predictive active site model for lipase YS to identify which enantiomer of an alcohol reacts faster in this acylation, *Tetrahedron: Asymmetry* **6**, 2385-2394.

Naemura, K., Murata, M., Tanaka, R., Yano, M., Hirose, K., Tobe, Y. (1996), Enantioselective acylation of primary and secondary alcohols catalyzed by lipase QL from *Alcaligenes* sp.: a predictive active site model for lipase QL to identify which enantiomer of an alcohol reacts faster in this acylation, *Tetrahedron: Asymmetry* **7**, 3285-3294.

Nagai, H., Shiozawa, T., Achiwa, K., Terao, Y. (1993), Convenient syntheses of optically active β-lactams by enzymatic resolution, *Chem. Pharm. Bull.* **41**, 1933-1938.

Nagao, A., Kito, M. (1990), Lipase-catalyzed synthesis of fatty acid esters useful in the food industry, *Biocatalysis* **3**, 295-305.

Nagao, Y., Kume, M., Wakabayashi, R. C., Nakamura, T., Ochiai, M. (1989), Efficient preparation of new chiral synthons useful for (+)-carbacyclin synthesis by utilizing enzymatic hydrolysis of prochiral σ-symmetric diesters, *Chem. Lett.* 239-242.

Nagasawa, T., Yamada, H. (1995), Microbial production of commodity chemicals, *Pure Appl. Chem.* **67**, 1241-1256.

Nagasawa, T., Nanba, H., Ryuno, K., Takeuchi, K., Yamada, H. (1987), Nitrile hydratase of *Pseudomonas chlororaphis* B23. Purification and characterization., *Eur. J. Biochem.* **162**, 691-698.

Nagasawa, T., Mathew, C. D., Mauger, J., Yamada, H. (1988), Nitrile hydratase-catalyzed production of nicotinamide from 3-cyanopyridine in *Rhodococcus rhodochrous* J1, *Appl. Environm. Microbiol.* **54**, 1766-1769.

Nagasawa, T., Nakamura, T., Yamada, H. (1990), Production of acrylic acid and methacrylic acid using *Rhodococcus rhodochrous* J1 nitrilase, *Appl. Microbiol. Biotechnol.* **34**, 322-324.

Nagasawa, T., Takeuchi, K., Yamada, H. (1991), Characterization of a new cobalt-containg nitrile hydratase purified from urea-induced cells of *Rhodococcus rhodochrous* J1, *Eur. J. Biochem.* **196**, 581-589.

Nagashima, S., Nakasako, M., Dohmae, N., Tsujimura, M., Takio, K., Odaka, M., Yohda, M., Kamiya, N., Endo, I. (1998), Novel non-heme iron center of nitrile hydratase with a claw setting of oxygen atoms, *Nature Struct. Biol.* **5**, 347-351.

Nagata, M. (1996), Enzymatic degradation of aliphatic polyesters copolymerized with various diamines, *Macromol. Rapid Commun.*, **17**, 583-587.

Nagumo, S., Arai, T., Akita, H. (1996), Enzymatic hydrolysis of *meso-(syn-syn)*-1,3,5-tricetoxy-2,4-dimethylpentane and acetylation of *meso-(syn-syn)*-3-benzyloxy-2,4-dimethylpentane-1,5-diol by lipase, *Chem. Pharm. Bull.* **44**, 1391-1394.

Nair, M. S., Anilkumar, A. T. (1996), Versatile chiral intermediates for terpenoid synthesis using lipase-catalyzed acylation, *Tetrahedron: Asymmetry* **7**, 511-514.

Nakada, M. (1995), The first total synthesis of antitumor macrolide, Rhizoxin, *Mem. Sch. Sci. Eng. Waseda Univ.* **59**, 167-195.

Nakamura, K. (1990), Kinetics of lipase-catalyzed esterification in supercritical CO_2, *Trends Biotechnol.* **8**, 288-291.

Nakamura, K., Chi, Y. M., Yamada, Y., Yano, T. (1986), Lipase activity and stability in supercritical carbon dioxide, *Chem. Eng. Commun.*, **45**, 207-212.

Nakamura, K., Inoue, Y., Kitayama, T., Ohno, A. (1990a), Stereoselective preparation of (*R*)-4-nitro-2-butanol and (*R*)-5-nitro-2-pentanol mediated by a lipase, *Agric. Biol. Chem.* **54**, 1569-1570.

Nakamura, K., Ishihara, K., Ohno, A., Uemura, M., Nishimura, H., Hayashi, Y. (1990b), Kinetic resolution of (η^6-arene)chromium complexes by a lipase, *Tetrahedron Lett.* **31**, 3603-3604.

Nakamura, K., Takebe, Y., Kitayama, T., Ohno, A. (1991), Effect of solvent structure on enantioselectivity of lipase-catalyzed transesterification, *Tetrahedron Lett.* **32**, 4941-4944.

Nakanishi, Y., Watanabe, H., Washizu, K., Narahashi, Y., Kurono, Y. (1991), Cloning, sequencing and regulation of the lipase gene from *Pseudomonas* sp. M-12-33, in.: *Lipases* (Schmid, R. D., Alberghina, L. , Verger, R.; Eds.), *GBF-Monographs*, **Vol. 16**, pp. 263-266. Weinheim: Wiley-VCH.

Nakano, H., Okuyama, Y., Iwasa, K., Hongo, H. (1996), Lipase-catalyzed resolution of 2-azabicyclo[2.2.1]hept-5-en-3-ones, *Tetrahedron: Asymmetry* **7**, 2381-2386.

Nakano, H., Ide, Y., Tsuda, T., Yang, J., Ishii, S., Yamane, T. (1999), Improvement in the organic solvent stability of *Pseudomonas* lipase by random mutation, *Ann. N. Y. Acad. Sci.*, **864**, 431-434.

Natoli, M., Nicolosi, G., Piattelli, M. (1990), Enzyme-catalyzed alcoholysis of flavone acetates in organic solvent, *Tetrahedron Lett.* **31**, 7371-7374.

Natoli, M., Nicolosi, G., Piattelli, M. (1992), Regioselective alcoholysis of flavonoid acetates with lipase in an organic solvent, *J. Org. Chem.* **57**, 5776-5778.

Neidhart, D. J., Petsko, G. A. (1988), The refined crystal structure of subtilisin Carlsberg at 2.5 Å resolution, *Protein Eng.* **2**, 271-276.

Nellaiah, H., Morisseau, C., Archelas, A., Furstoss, R., Baratti, J. C. (1996), Enantioselective hydrolysis of *p*-nitrostyrene oxide by an epoxide hydrolase preparation from *Aspergillus niger*, *Biotechnol. Bioeng.* **49**, 70-77.

Nettekoven, M., Psiorz, M., Waldmann, H. (1995), Synthesis of enantiomerically pure 4-alkylsubstituted tryptophan derivatives by a combination of organometallic reactions with enantioselective enzymatic transformations, *Tetrahedron Lett.* **36**, 1425-1428.

New, R. R. C. (1993), Biological and biotechnological applications of phospholipids, in.: *Phospholipid Handbook* (Cevc, G.; Eds.), pp. 855-878. New York: Marcel Dekker.

Ng-Youn-Chen, M. C., Serreqi, A. N., Huang, Q., Kazlauskas, R. J. (1994), Kinetic resolution of pipecolic acid using partially-purified lipase from *Aspergillus niger*, *J. Org. Chem.* **59**, 2075-2081.

Nicolosi, G., Morrone, R., Patti, A., Piattelli, M. (1992), Lipase mediated desymmetrization of 1,2-bis(hydroxymethyl)ferrocene in organic medium: production of both enantiomers of 2-acetoxymethyl-1-hydroxymethyl ferrocene, *Tetrahedron: Asymmetry* **3**, 753-758.

Nicolosi, G., Piattelli, M., Sanfilippo, C. (1993), Lipase-catalyzed regioselective protection of hydroxyl groups in aromatic dihydroxyaldehydes and ketones, *Tetrahedron* **49**, 3143-3148.

Nicolosi, G., Patti, A., Piattelli, M. (1994a), Lipase -mediated separation of the stereoisomers of 1-(1-hydroxyethyl)-2-(hydroxymethyl)ferrocene, *J. Org. Chem.* **59**, 251-254.

Nicolosi, G., Patti, A., Piattelli, M., Sanfilippo, C. (1994b), Desymmetrization of *meso*-hydrobenzoin via stereoselective enzymatic esterification, *Tetrahedron: Asymmetry* **5**, 283-288.

Nicolosi, G., Patti, A., Piattelli, M., Sanfilippo, C. (1995a), Desymmetrization of cis-1,2-dihydroxycycloalkanes by stereoselective lipase mediated esterification, *Tetrahedron: Asymmetry* **6**, 519-524.

Nicolosi, G., Patti, A., Piattelli, M., Sanfilippo, C. (1995b), Enzymatic access to homochiral 1-acetoxy-2-hydroxycyclohexane-3,5-diene through lipase-assisted acetylation in organic solvent, *Tetrahedron Lett.* **36**, 6545-6546.

Nicotra, F., Riva, S., Secundo, F., Zucchelli, L. (1989), An interesting example of complementary regioselective acylation of secondary hydroxyl groups by different lipases, *Tetrahedron Lett.* **30**, 1703-1704.

Nielsen, T. (1985), Industrial application possibilities for lipase, *Fat Sci. Technol.* **87**, 15-19.

Nilsson, K. G. I. (1987), A simple strategy for changing the regioselectivity of glycosidase-catalyzed formation of disaccharides, *Carbohydr. Res.* **167**, 95-103.

Nilsson, J., Blackberg, L., Carlsson, P., Enerback, S., Hernell, O., Bjursell, G. (1990), cDNA cloning of human-milk bile-salt-stimulated lipase and evidence for its identity to pancreatic carboxylic ester hydrolase, *Eur. J. Biochem.* **192**, 543-550.

Nishio, T., Kamimura, M., Murata, M., Terao, Y., Achiwa, K. (1989), Production of optically active esters and alcohols from racemic alcohols by lipase-catalyzed stereoselective transesterification in nonaqueous reaction system, *J. Biochem.* **105**, 510-512.

Nishizawa, M., Gomi, H., Kishimoto, F. (1993), Purification and some properties of carboxylesterase from Arthobacter globiformis; Stereoselective hydrolysis of ethyl chrysanthemate, *Biosci. Biotech. Biochem.* **57**, 594-598.

Nishizawa, M., Shimizu, M., Ohkawa, H., Kanaoka, M. (1995), Stereoselective production of (+)-*trans*-chrysanthemic acid by a microbial esterase: cloning, nucleotide sequence, and overexpression of the esterase gene of *Arthrobacter globiformis* in *Escherichia coli*, *Appl. Environ. Microbiol.* **61**, 3208-3215.

Nishizawa, K., Ohgami, Y., Matsuo, N., Kisida, H., Hirohara, H. (1997), Studies on hydrolysis of chiral, achiral and racemic alcohol esters with *Pseudomonas cepacia* lipase - mechanism of stereospecificity of the enzyme, *J. Chem. Soc., Perkin Trans. 2*, 1293-1298.

Nobes, G. A. R., Kazlauskas, R. J., Marchessault, R. H. (1996), Lipase-catalyzed ring-opening polyerization of lactones: A novel route to poly(hydroxyalkanoate)s, *Macromolecules* **29**, 4829-4833.

Noble, M. E. M., Cleasby, A., Johnson, L. N., Egmond, M. R., Frenken, L. G. J. (1993), The crystal structure of triacylglycerol lipase from *Pseudomonas glumae* reveals a partially reductant catalytic aspartate, *FEBS Lett.* **331**, 123-128.

Noble, M. E. M., Cleasby, A., Johnson, L. N., Egmond, M. R., Frenken, L. G. J. (1994), Analysis of the structure of *Pseudomonas glumae* lipase, *Protein Eng.* **7**, 559-562.

Node, M., Inoue, T., Araki, M., Nakamura, D., Nishide, K. (1995), An asymmetric synthesis of C_2-symmetric dimethyl 3,7-dioxo-cis-bicyclo[3.3.0]octane-2,6-dicarboxylate by lipase-catalyzed demethoxycarbonylation, *Tetrahedron Lett.* **36**, 2255-2256.

Noguchi, T., Hoshino, M., Tsujimura, M., Odaka, M., Inoue, Y., Endo, I. (1996), Resonance Raman evidence that photodissociation of nitric oxide from the non-heme iron center activates nitrile hydratase from *Rhodococcus* sp. N-771, *Biochemistry* **35**, 16777-16781.

Norin, M., Hult, K., Mattson, A., Norin, T. (1993), Molecular modelling of chymotrypsin-substrate interactions: calculation of enantioselectivity, *Biocatalysis* **7**, 131-147.

Noritomi, H., Almarsson, Ö., Barletta, G. L., Klibanov, A. M. (1996), The influence of the mode of enzyme preparation on enzymatic enantioselectivity in organic solvents and its temperature dependence, *Biotechnol. Bioeng.* **51**, 95-99.

Nozaki, K., Uemura, A., Yamashita, J., Yasumoto, M. (1990), Enzymatic regioselective acylation of the 3' hydroxyl groups of the 2'-deoxy5-fluorouridine (FUdR) and 2'-deoxy5-trifluoromethyluridine (CF_3UdR), *Tetrahedron Lett.* **31**, 7327-7328.

Oberhauser, T., Bodenteich, M., Faber, K., Penn, G., Griengl, H. (1987), Enzymic resolution of norbornane-type esters, *Tetrahedron* **43**, 3931-3944.

Oddon, G., Uguen, D. (1997), Toward a total synthei of an aglycone of spiramycine, two complementary accesses to a C-5/C-9 fragment, *Tetrahedron Lett.* **38**, 4411-4414.

Oesch, F. (1973), Mammalian epoxide hydrases: inducible enzymes catalysing the inactivation of carcinogenic and cytotoxic metabolites derived from aromatic and olefinic compounds, *Xenobiotica* **3**, 305-340.

Oesch, F. (1974), Purification and specificity of a human microsomal epoxide hydratase, *Biochem. J.* **139**, 77-88.

Oetting, J., Holzkamp, J., Meyer, H. H., Pahl, A. (1997), Total synthesis of the piperidinol alkaloid (−)-(2*R*,3*R*,6*S*)-cassine, *Tetrahedron: Asymmetry* **8**, 477-484.

Ogawa, J., Shimizu, S. (1997), Diversity and versatility of microbial hydantoin-transforming enzymes, *J. Mol. Catal. B: Enzymatic* **2**, 163-176.

Oguntimein, G. B., Erdmann, H., Schmid, R. D. (1993), Lipase catalysed synthesis of sugar ester in organic solvents, *Biotech. Lett.* **15**, 175-180.

O'Hagan, D., Rzepa, H. S. (1994), Stereoelectronic influence of fluorine in enzyme resolutions of α-fluoroesters, *J. Chem. Soc., Perkin Trans. 2*, 3-4.

O'Hagan, D., Zaidi, N. A. (1992), Hydrolytic resolution of tertiary acetylenic acetate esters with the lipase from *Candida cylindracea*, *J. Chem. Soc. Perkin Trans. 1*, 947-948.

O'Hagan, D., Zaidi, N. A. (1993), Polymerization of 10-hydroxydecanoic acid with lipase from *Candida cylindracea*, *J. Chem. Soc., Perkin Trans. 1*, 2389-2390.

O'Hagan, D., Zaidi, N. A. (1994a), Enzyme-catalyzed condensation polymerization of hydroxyundecanoic acid with lipase from *Candida cylindracea*, *Polymer* **35**, 3576-3578.

O'Hagan, D., Zaidi, N. A. (1994b), The resolution of tertiary α-acetylene acetate esters by the lipase from *Candida cylindracea*, *Tetrahedron: Asymmetry* **5**, 1111-1118.

Ohno, M., Otsuka, M. (1990), Chiral synthons by ester hydrolysis catalysed by pig liver esterase, *Org. React.* **37**, 1-55.

Ohno, M., Kobayashi, S., Iimori, T., Wang, Y. F., Izawa, T. (1981), Synthesis of (*S*)- and (*R*)-4-[(methoxycarbonyl)methyl]-2-azetidinone by chemicoenzymatic approach, *J. Am. Chem. Soc.* **103**, 2405-2406.

Öhrner, K., Mattson, A., Norin, T., Hult, K. (1990), Enantiotopic selectivity of pig liver esterase isoenzymes, *Biocatalysis* **4**, 81-88.

Öhrner, N., Martinelle, M., Mattson, A., Norin, T., Hult, K. (1992), Displacement of the equilibrium in lipase catalyzed transesterification in ethyl octanoate by continuous evaporation of ethanol, *Biotechnol. Lett.* **14**, 263-268.

Öhrner, N., Orrenius, C., Mattson, A., Norin, T., Hult, K. (1996), Kinetic resolutions of amine and thiol analogues of secondary alcohols catalyzed by the *Candida antarctica* lipase B, *Enzyme Microb. Technol.* **19**, 328-331.

Ohsawa, K., Shiozawa, T., Achiwa, K., Terao, Y. (1993), Synthesis of optically active nucleoside analogs containing 2,3-dideoxyapiose in the presence of a catalytic amount of trimethylsilyl iodide, *Chem. Pharm. Bull.* **41**, 1906-1909.

Ohta, H. (1996), Stereochemistry of enzymatic hydrolysis of nitriles, *Chimia* **50**, 434-436.

Ohta, H., Kimura, Y., Sugano, Y., Sugai, T. (1989), Enzymatic kinetic resolution of cyanohydrin acetates and its application to the synthesis of (*S*)-(−)-frontalin, *Tetrahedron* **45**, 5469-5476.

Ohtani, T., Kikuchi, K., Kamezawa, M., Hamatani, H., Tachibana, H., Totani, T., Naoshima, Y. (1996), Chemoenzymatic synthesis of (+)-(4E,15E)-docosa-4,15-dien-1-yn-3-ol, a component of the marine sponge *Cribrochalina vasculum*, *J. Chem. Soc., Perkin Trans. 1*, 961-962.

Ohtani, T., Nakatsukasa, H., Kamezawa, M., Tachibana, H., Naoshima, Y. (1998), Enantioselectivity of *Candida antarctica* lipase for some synthetic substrates including aliphatic secondary alcohols, *J. Mol. Catal. B: Enzymatic* **4**, 53-60.

Okahata, Y., Ijiro, K. (1988), A lipid-coated lipase as a new catalyst for triglyceride synthesis in organic solvents, *J. Chem. Soc., Chem. Commun.*, 1392-1394.

Okahata, Y., Ijiro, K. (1992), Preparation of a lipid-coated lipase and catalysis of glyceride ester syntheses in homogeneous organic solvents, *Bull. Chem. Soc. Jpn.* **65**, 2411-2420.

Okahata, Y., Fujimoto, Y., Ijiro, K. (1995a), A lipid-coated lipase as an enantioselective ester synthesis catalyst in homogeneous organic solvents, *J. Org. Chem.* **60**, 2244-2250.

Okahata, Y., Hatano, A., Ijiro, K. (1995b), Enhancing enantioselectivity of a lipid-coated lipase via imprinting methods for esterification in organic solvents, *Tetrahedron: Asymmetry* **6**, 1311-1322.

Okumura, S., Iwai, M., Tominaga, Y. (1984), Synthesis of ester oligomer by *Aspergillus niger* lipase, *Agric. Biol. Chem.* **48**, 2805-2808.

Oladepo, D. K., Halling, P. J., Larsen, V. F. (1994), Reaction rates in organic media show similar dependence on water activity with lipase catalyst immobilized on different supports, *Biocatalysis* **8**, 283-287.

Oladepo, D. K., Halling, P. J., Larsen, V. F. (1995), Effect of different supports on the reaction rate of *Rhizomucor miehei* lipase in organic media, *Biocatal. Biotransform.* **12**, 47-54.

Olivieri, R., Fascetti, E., Angelini, L., Degen, L. (1979), Enzymatic conversion of *N*-carbamoyl-D-amino acids to D-amino acids, *Enzyme Microb. Technol.* **1**, 201-204.

Olivieri, R., Fascetti, E., Angelini, L., Degen, L. (1981), Microbial transformation of racemic hydantoins to D-amino acids., *Biotechnol. Bioengineer.* **23**, 2173-2183.

Ollis, D. L., Cheah, E., Cygler, M., Dijkstra, B., Frolow, F., Franken, S., Harel, M., Remington, S. J., Silman, I. (1992), The a/b hydrolase fold, *Protein Eng.* **5**, 197-211.

Omar, I. C., Saeki, H., Nishio, N., Nagai, S. (1989), Synthesis of acetone glycerol acyl esters by immobilized lipase of *Mucor miehei*, *Biotechnol. Lett.* **11**, 161-166.

Onakunle, O. A., Knowles, C. J., Bunch, A. W. (1997), The formation and substrate specificity of bacterial lactonases capable of enantioselective resolution of racemic lactones, *Enzyme Microb. Technol.* **21**, 245-251.

Onumonu, A. N., Colocoussi, A., Mathews, C., Woodland, M. P., Leak, D. J. (1994), Microbial alkene epoxidation - merits and limitations, *Biocatalysis* **10**, 211-218.

Ooi, Y., Mitsuo, N., Satoh, T. (1985), Enzymatic synthesis of glycosides of racemic alcohols using β-galactosidase and separation of the diasteromers by high-performance liquid chromatograhy using a conventional column, *Chem. Pharm. Bull.* **33**, 5547-5550.

Orrenius, C., Mattson, A., Norin, T. (1994), Preparation of 1-pyridinylethanols of high enantiomeric purity by lipase catalyzed transesterifications, *Tetrahedron: Asymmetry* **5**, 1363-1366.

Orrenius, C., Norin, T., Hult, K., Carrea, G. (1995a), The *Candida antarctica* lipase B catalyzed kinetic resolution of seudenol in non-aqueous media of controlled water activity, *Tetrahedron: Asymmetry* **6**, 3023-3030.

Orrenius, C., Oehrner, N., Rotticci, D., Mattson, A., Hult, K., Norin, T. (1995b), *Candida antarctica* lipase B catalyzed kinetic resolutions: substrate structure requirements for the preparation of enantiomerically enriched secondary alcohols, *Tetrahedron: Asymmetry* **6**, 1217-1220.

Orru, R. V. A., Kroutil, W., Faber, K. (1997), Deracemization of (±)-2,2-disubstituted epoxides *via* enantioconvergent chemoenzymic hydrolysis using *Nocardia* EH1 epoxide hydrolase and sulfuric acid, *Tetrahedron Lett.* **38**, 1753-1754.

Orru, R. V. A., Archelas, A., Furstoss, R., Faber, K. (1998a), Epoxide hydrolases and their synthetic applications, *Adv. Biochem. Eng./Biotechnol.* **63**, 145-167.

Orru, R. V. A., Mayer, S. F., Kroutil, W., Faber, K. (1998b), Chemoenzymatic deracemization of (±)-2,2-disubstituted oxiranes, *Tetrahedron* **54**, 859-874.

Orru, R. V. A., Osprian, I., Kroutil, W., Faber, K. (1998c), An efficient large-scale synthesis of (*R*)-(−)-mevalonolactone using simple biological and chemical catalysts, *Synthesis*, 1259-1263.

Orsat, B., Alper, P. B., Moree, W., Mak, C.-P., Wong, C.-H. (1996), Homocarbonates as substrates for the enantioselective enzymatic protection of amines, *J. Am. Chem. Soc.* **118**, 712-713.

Osprian, I., Kroutil, W., Mischitz, M., Faber, K. (1997), Biocatalytic resolution of 2-methyl-2-(aryl)alkyloxiranes using novel bacterial epoxide hydrolases, *Tetrahedron: Asymmetry* **8**, 65-71.

Otto, R., Bornscheuer, U. T., Syldatk, C., Schmid, R. D. (1998a), Lipase-catalyzed synthesis of arylaliphatic esters of β-D(+)-glucose, alkyl- and arylglucosides and characterization of their surfactant properties, *J. Biotechnol.* **64**, 231-237.

Otto, R. T., Bornscheuer, U. T., Syldatk, C., Schmid, R. D. (1998b), Synthesis of aromatic *n*-alkyl-glucoside esters in a coupled ß-glucosidase and lipase reaction, *Biotechnol. Lett.* **20**, 437-440.

Ottolina, G., Bovara, R., Riva, S., Carrea, G. (1994), Activity and selectivity of some hydrolases in enantiomeric solvents, *Biotechnol. Lett.* **16**, 923-928.

Oyama, K. (1992), Industrial production of aspartame, in.: *Chirality in Industry* (Collins, A. N., Sheldrake, G. N. , Crosby, J.; Eds.), pp. 237-247. Chichester: Wiley.

Ozaki, E., Sakimae, A. (1997), Purification and characterization of recombinant esterase from *Pseudomonas putida* MR-2068 and its application to the optical resolution of dimethyl methyl-succinate, *J. Ferment. Bioeng.* **83**, 535-539.

Ozegowski, R., Kunath, A., Schick, H. (1994a), Enzymes in organic synthesis. 19. The enzyme-catalyzed sequential esterification of (±)-*anti*-2,4-dimethylglutaric anhydride - an efficient route to enantiomerically enriched mono- and diesters of *anti*-2,4-dimethylglutaric acid, *Liebigs Ann.*, 215-217.

Ozegowski, R., Kunath, A., Schick, H. (1994b), Lipase-catalyzed conversion of (±)-2-methylglu-taric anhydride into (*S*)-2-methyl- and (*R*)-4-methyl-δ-valerolactone via a regio- and enantiose-lective sequential esterification, *Liebigs Ann. Chem.*, 1019-1023.

Ozegowski, R., Kunath, A., Schick, H. (1995a), The different behaviour of *syn*- and *anti*-2,3-dimethylbutanedioic anhydride in the lipase-catalyzed enantioselective alcoholysis, *Tetrahe-dron: Asymmetry* **6**, 1191-1194.

Ozegowski, R., Kunath, A., Schick, H. (1995b), Lipase-catalyzed sequential esterification of (±)-2-methylbutanedioic anhydride - a biocatalytical access to an enantiomerically pure 1-monoester of (*S*)-2-methylbutanedioic acid, *Liebigs Ann.*, 1699-1702.

Ozegowski, R., Kunath, A., Schick, H. (1996), Synthesis of enantiomerically enriched 2,3- and 3,4-dimethylpentan-5-olides by lipase-catalyzed regio- and enantioselective alcoholysis of *cis*- and *trans*-2,3-dimethylpentanedioic anhydrides, *Liebigs Ann. Chem.*, 1443-1448.

Padt, A. van der, Keurentjes, J. T. F., Sewalt, J. J. W., van Dam, E. M., van Dorp, L. J., van't Riet, K. (1992), Enzymatic synthesis of monoglycerides in a membrane bioreactor with an in-line ad-sorption column, *J. Am. Oil Chem. Soc.* **69**, 748-754.

Pai, Y. C., Fang, J. M., Wu, S. H. (1994), Resolution of homoallylic alcohols containing dithio-ketene acetal functionalities - synthesis of optically active γ-lactones by a combination of chemical and enzymatic methods, *J. Org. Chem.* **59**, 6018-6025.

Pallavicini, M., Valoti, E., Villa, L., Piccolo, O. (1994), Lipase-catalyzed resolution of glycerol-2,3-carbonate, *J. Org. Chem.* **59**, 1751-1754.

Pamperin, D., Schulz, C., Hopf, H., Syldatk, C., Pietzsch, M. (1998), Chemoenzymatic synthesis of optically pure planar chiral (*S*)-(−)-5-formyl-4-hydroxyl[2.2]paracyclophane, *Eur. J. Org. Chem.* 1441-1445.

Panunzio, M., Camerini, R., Mazzoni, A., Donati, D., Marchioro, C., Pachera, R. (1997), Lipase-catalysed resolution of *cis*-1-ethoxycarbonyl-2-hydroxy-cyclohexane: enantioselective total synthesis of 10-ethyl-trinem, *Tetrahedron: Asymmetry* **8**, 15-17.

Panza, L., Brasca, S., Riva, S., Russo, G. (1993a), Selective lipase-catalyzed acylation of 4,6-*O*-benzylidene-D-glucopyranosides to synthetically useful esters, *Tetrahedron: Asymmetry* **4**, 931-932.

Panza, L., Luisetti, M., Crociati, E., Riva, S. (1993b), Selective acylation of 4,6-*O*-benzylidene glycopyranosides by enzymic catalysis, *J. Carbohydr. Chem.* **12**, 125-130.

Parida, S., Dordick, J. S. (1991), Substrate structure and solvent hydrophobicity control lipase catalysis and enantioselectivity in organic media, *J. Am. Chem. Soc.* **113**, 2253-2259.

Park, H. G., Chang, H. N., Dordick, J. S. (1994), Enzymatic synthesis of various aromatic polyes-ters in anhydrous organic solvents, *Biocatalysis* **11**, 263-271.

Parker, M.-C., Besson, T., Lamare, S., Legoy, M.-D. (1996), Microwave radiation can increase the rate of enzyme-catalyzed reactions in organic media, *Tetrahedron Lett.* **37**, 8383-8386.

Parmar, V. S., Prasad, A. K., Sharma, N. K., Singh, S. K., Pati, H. N., Gupta, S. (1992), Regiose-lective deacylation of polyacetoxy aryl-methyl ketones by lipases in organic solvents, *Tetrahe-dron* **48**, 6495-6498.

Parmar, V. S., Prasad, A. K., Sharma, N. K., Varan, A., Pati, H. N., Sharma, S. K., Bisht, K. S. (1993a), Lipase-catalyzed selective deacylation of peracetylated benzopyranones, *J. Chem. Soc., Chem. Commun.*, 27-29.

Parmar, V. S., Sinha, R., Bisht, K. S., Gupta, S., Prasad, A. K., Taneja, P. (1993b), Regioselective esterification of diols and triols with lipases in organic solvents, *Tetrahedron* **49**, 4107-4116.

Parmar, V. S., Pati, H. N., Sharma, S. K., Singh, A., Malhotra, S., et al. (1996), Hydrolytic reactions on polyphenolic perpropanoates by porcine pancreatic lipase immobilized in microemulsion-based gels, *Bioorg. Med. Chem. Lett.* **6**, 2269-2274.

Parmar, V. S., Kumar, A., Bisht, K., Mukherjee, S., Prasad, A. K., Sharma, S. K., Wengel, J., Olsen, C. E. (1997), Novel chemoselective de-esterification of esters of polyacetoxy aromatic acids by lipases, *Tetrahedron* **53**, 2163-2176.

Partali, V., Waagen, V., Alvik, T., Anthonsen, T. (1993), Enzymic resolution of butanoic esters of 1-phenylmethyl and 1-(2-phenylethyl) ethers of 3-chloro-1,2-propanediol, *Tetrahedron: Asymmetry* **4**, 961-968.

Pastor, E., Otero, C., Ballesteros, A. (1995a), Enzymatic preparation of mono- and distearin by glycerolysis of ethyl stearate and direct esterification of glycerol in the presence of a lipase from *Candida antarctica* (Novozym 435), *Biocatal. Biotrans.* **12**, 147-157.

Pastor, E., Otero, C., Ballesteros, A. (1995b), Synthesis of mono- and dioleylglycerols using an immobilized lipase, *Appl. Biochem. Biotechnol.* **50**, 251-263.

Patel, R. N., Howell, J. M., Banerjee, A., Fortney, K. F., Szarka, L. J. (1991), Stereoselective enzymatic esterification of 3-benzoylthio-2-methylpropanoic acid, *Appl. Microbiol. Biotechnol.* **36**, 29-34.

Patel, R. N., Howell, J. M., McNamee, C. G., Fortney, K. F., Szarka, L. J. (1992a), Stereoselective enzymatic hydrolysis of α-[(acetylthio)methyl]benzenepropanoic acid and 3-acetylthio-2-methylpropanoic acid, *Biotechnol. Appl. Biochem.* **16**, 34-47.

Patel, R. N., Liu, M., Banerjee, A., Szarka, L. J. (1992b), Stereoselective enzymatic hydrolysis of (*exo,exo*)-7-oxabicyclo[2.2.1]heptane-2,3-dimethanol diacetate ester in a biphasic system, *Appl. Microbiol. Biotechnol.* **37**, 180-183.

Patel, R. N., Banerjee, A., Ko, R. Y., Howell, J. M., Li, W.-S., Comezoglu, F. T., Partyka, R. A., Szarka, L. (1994), Enzymic preparation of (3R-cis)-3-(acetyloxy)-4-phenyl-2-azetidinone - a taxol side-chain synthon, *Biotechnol. Appl. Biochem.* **20**, 23-33.

Patel, R. N., Banerjee, A., Szarka, L. J. (1997), Stereoselective acetylation of racemic 7-[N,N'-bis(benzyloxycarbonyl)-N-(guanidinoheptanoyl)]-α-hydroxyglycine, *Tetrahedron: Asymmetry* **8**, 1767-1771.

Patti, A., Sanfilippo, C., Piattelli, M., Nicolosi, G. (1996), Enzymatic desymmetrization of conduritol D. Preparation of homochiral intermediates for the synthesis of cyclitols and aminocyclitols, *Tetrahedron: Asymmetry* **7**, 2665-2670.

Payne, M. S., Wu, S., Fallon, R. D., Tudor, G., Stieglitz, B., Jr., I. M. T., Nelson, M. J. (1997), A stereoselective cobalt-containing nitrile hydratase, *Biochemistry* **36**, 5447-5454.

Pearson, A. J., Bansal, H. S., Lai, Y. S. (1987), Enzymatic hydrolysis of 3,7-diacetoxycycloheptene derivatives, *J. Chem. Soc., Chem. Commun.*, 519-520.

Pearson, A. J., Lai, Y. S. (1988), A synthesis of the (+)-Prelog-Djerassi lactone, *J. Chem. Soc., Chem. Commun.*, 442-443.

Pearson, A. J., Srinivasan, K. (1992), Approaches to the synthesis of heptitol derivatives via iron-mediated stereocontrolled functionalization of cycloheptatrienone, *J. Org. Chem.* **57**, 3965-3973.

Pecnik, S., Knez, Z. (1992), Enzymatic fatty ester synthesis, *J. Am. Oil Chem. Soc.* **69**, 261-265.

Pedersen, S. B., Holmer, G. (1995), Studies of the fatty acid specificity of the lipase from *Rhizo-mucor miehei* toward 20:1n-9, 20:5n-3, 22:1n-9 and 22:6n-3, *J. Am. Oil Chem. Soc.* **72**, 239-243.

Pedragosa-Moreau, S., Archelas, A., Furstoss, R. (1993), Microbiological transformations. 28. Enantiocomplementary epoxide hydrolyses as a preparative access to both enantiomers of styrene oxide, *J. Org. Chem.* **58**, 5533-5536.

Pedragosa-Moreau, S., Archelas, A., Furstoss, R. (1995), Epoxydes énantiopurs: obtention par voie chimique ou par voie enzymatique, *Bull. Soc. Chim. Fr.* **132**, 769-800.

Pedragosa-Moreau, S., Archelas, A., Furstoss, R. (1996a), Microbiological transformations. 31: Synthesis of enantiopure epoxides and vicinal diols using fungal epoxide hydrolase mediated hydrolysis, *Tetrahedron Lett.* **37**, 3319-3322.

Pedragosa-Moreau, S., Archelas, A., Furstoss, R. (1996b), Microbiological transformations. 32. Use of epoxide hydrolase mediated biohydrolysis as a way to enantiopure epoxides and vicinal diols: Application to substituted styrene oxides, *Tetrahedron* **52**, 4593-4606.

Pedragosa-Moreau, S., Morisseau, C., Zylber, J., Archelas, A., Baratti, J., Furstoss, R. (1996c), Microbiological transformations. 33. Fungal epoxide hydrolases applied to the synthesis of enantiopure *para*-substituted styrene oxides. A mechanistic approach, *J. Org. Chem.* **61**, 7402-7407.

Pelenc, V. P., Paul, F. M. B., Monsan, P. F. (1993), Method for the enyzmatic production of α-glucosides and esters of α-glucosides, and utilization of products thus obtained, World Patent WO 93/04185 (Chem. Abstr. 119: 7324).

Perona, J. J., Craik, C. S. (1995), Structural basis of substrate specificity in the serine proteases, *Protein Sci.* **4**, 337-360.

Persichetti, R. A., Lalonde, J. J., Govardhan, C. P., Khalaf, N. K., Margolin, A. L. (1996), *Candida rugosa* lipase - enantioselectivity enhancements in organic solvents, *Tetrahedron Lett.* **37**, 6507-6510.

Petschen, I., Malo, E. A., Bosch, M. P., Guerrero, A. (1996), Highly enantioselective synthesis of long chain alkyl trifluoromethyl carbinols and β-thiotrifluoromethyl carbinols through lipases, *Tetrahedron: Asymmetry* **7**, 2135-2143.

Phytian, S. J. (1998), Esterases, in.: *Biotechnology-Series* (Rehm, H. J., Reed, G., Pühler, A., Stadler, P. J. W. , Kelly, D. R.; Eds.), **Vol. 8a**, pp. 193-241. Weinheim: Wiley-VCH.

Pietzsch, M., Vielhauer, O., Pamperin, D., Ohse, B., Hopf, H. (1999), On the kinetics of the enzyme-catalyzed hydrolysis of axial chiral alkyl allenecarboxylates: preparation of optically active *R*-(−)-2-ethyl-4-phenyl-2,3-hexadiene-carboxylic acid and its optically pure *S*-(+)-methyl-ester, *J. Mol. Catal. B: Enzymatic* **6**, 51-57.

Pinot, F., Grant, D. F., Beetham, J. K., Parker, A. G., Borhan, B., Landt, S., Jones, A. D., Hammock, B. D. (1995), Molecular and biochemical evidence for the involvement of the Asp-333-His-523 pair in the catalytic mechanism of soluble epoxide hydrolase, *J. Biol. Chem.* **270**, 7968-7974.

Pisch, S., Bornscheuer, U., Meyer, H. H., Schmid, R. D. (1997), Properties of unusual phospholipids IV: Chemoenzymatic synthesis of phospholipids bearing acetylenic fatty acids, *Tetrahedron* **53**, 14627-14634.

Pleiss, J., Fischer, M., Schmid, R. D. (1998), Anatomy of lipase binding sites: the scissile fatty acid binding site, *Chem. Phys. Lipids* **93**, 67-80.

Plou, F. J., Barandiaran, M., Calvo, M. V., Ballesteros, A., Pastor, E. (1996), High-yield production of mono- and di-oleylglycerol by lipase-catalyzed hydrolysis of triolein, *Enzyme Microb. Technol.* **18**, 66-71.

Pohl, T., Waldmann, H. (1995), Enhancement of the enantioselectivity of penicillin G acylase from *E. coli* by "substrate tuning", *Tetrahedron Lett.* **36**, 2963-2966.

Pohl, T., Waldmann, H. (1996), Enzymatic synthesis of a characteristic phosphorylated and glycosylated peptide fragment of the large subunit of mammalian RNA polymerase II, *Angew. Chem. Intl. Ed. Engl.* **35**, 1720-1723.

Polla, M., Frejd, T. (1991), Synthesis of optically active cyclohexenol derivatives via enzyme-catalyzed ester hydrolysis of 4-acetoxy-3-methyl-2-cyclohexenone, *Tetrahedron* **41**, 5883-5894.

Poppe, L., Novak, L. (1992), *Selective biocatalysis*, Weinheim: Wiley-VCH.

Poppe, L., Novák, L., Kajtár-Peredy, M., Szántay, C. (1993), Lipase catalyzed enantiomer selective hydrolysis of 1,2-diol acetates, *Tetrahedron: Asymmetry* **4**, 2211-2217.

Pottie, M., van der Eycken, J., Vandewalle, M. (1991), Synthesis of optically active derivatives of erythritol, *Tetrahedron: Asymmetry* **2**, 329-330.

Pozo, M., Gotor, V. (1993a), Chiral carbamates through an enzymic alkoxycarbonylation reaction, *Tetrahedron* **49**, 4321-4326.

Pozo, M., Gotor, V. (1993b), Kinetic resolution of vinyl carbonates through a lipase-mediated synthesis of their carbonate and carbamate derivatives, *Tetrahedron* **49**, 10725-10732.

Pozo, M., Pulido, R., Gotor, V. (1992), Vinyl carbonates as novel alkoxycarbonylation reagents in enzymatic synthesis of carbonates, *Tetrahedron* **48**, 6477-6484.

Prasad, A. K., Sorensen, M. D., Parmar, V. S., Wengel, J. (1995), Tri-*O*-acyl 2-deoxy-D-ribofuranose: an effective enzyme-assisted one-pot synthesis from 2-deoxy-D-ribose and transformation into 2'-deoxynucleosides, *Tetrahedron Lett.* **36**, 6163-6166.

Prazeres, D. M. F., Cabral, J. M. S. (1994), Enzymatic membrane bioreactors and their applications, *Enzyme Microb. Technol.* **16**, 738-750.

Provencher, L., Wynn, H., Jones, J. B., Krawczyk, A. R. (1993), Enzymes in organic synthesis 51. Probing the dimensions of the large hydrophobic pocket of the active site of pig liver esterase, *Tetrahedron: Asymmetry* **4**, 2025-2040.

Puertas, S., Brieva, R., Rebolledo, F., Gotor, V. (1993), Lipase catalyzed aminolysis of ethyl propiolate and acrylic esters. Synthesis of chiral acrylamides, *Tetrahedron* **49**, 4007-4014.

Pulido, R., Gotor, V. (1993), Enzymatic regioselective alkoxycarbonylation of hexoses and pentoses with carbonate oxime esters, *J. Chem Soc. Perkin Trans. 1*, 589-592.

Pulido, R., López, F. O., Gotor, V. (1992), Enzymatic regioselective acylation of hexoses and pentoses using oxime esters, *J. Chem Soc. Perkin Trans. 1*, 2981-2988.

Quartey, E. G. K., Hustad, J. A., Faber, K., Anthonsen, T. (1996), Selectivity enhancement of PPL-catalyzed resolution by enzyme fractionation and medium engineering - syntheses of both enantiomers of tetrahydropyran-2-methanol, *Enzyme Microb. Technol.* **19**, 361-366.

Quax, W. J., Broekhuizen, C. P. (1994), Development of a new *Bacillus* carboxyl esterase for use in the resolution of chiral drugs, *Appl. Microbiol. Biotechnol.* **41**, 425-431.

Quinlan, P., Moore, S. (1993), Modification of triglycerides by lipases: process technology and its application to the production of nutritionally improved fats, *Inform* **4**, 579-583.

Quirós, M., Sanchez, V. M., Brieva, R., Rebbolledo, F., Gotor, V. (1993), Lipase-catalyzed synthesis of optically active amides in organic media, *Tetrahedron: Asymmetry* **4**, 1105-1112.

Quyen, D. T., Schmidt-Dannert, C., Schmid, R. D. (1999), High-level formation of active *Pseudomonas cepacia* lipase after heterologous expression of the encoding gene and its modified chaperone in *Escherichia coli* and rapid in vitro refolding, *Appl. Environm. Microbiol.* **65**, 787-794.

Rabiller, C. G., Koenigsberger, K., Faber, K., Griengl, H. (1990), Enzymic recognition of diastereomeric esters, *Tetrahedron* **46**, 4231-4240.

Rakels, J. L. L., Caillat, P., Straathof, A. J. J., Heijnen, J. J. (1994a), Modification of the enzyme enantioselectivity by product inhibition, *Biotechnol. Prog.* **10**, 403-409.

Rakels, J. L. L., Wolff, A., Straathof, A. J. J., Heijnen, J. J. (1994b), Sequential kinetic resolution by two enantioselective enzymes, *Biocatalysis* **9**, 31-47.

Ramaswamy, S., Hui, R. A. H. F., Jones, J. B. (1986), Enantiomerically selective pig liver esterase-catalysed hydrolyses of racemic allenic esters, *J. Chem. Soc., Chem. Commun.*, 1545-1546.

Ramaswamy, S., Oehlschlager, A. C. (1991), Chemico-enzymatic syntheses of racemic and chiral isomers of 7-methyl-1,6-dioxaspiro[4.5]decane, *Tetrahedron* **47**, 1157-1162.

Ramos-Tombo, G. M., Schär, H. P., Busquets, H. P., Fernandez, X., Ghisalba, O. (1986), Synthesis of both enantiomeric forms of 2-substituted 1,3-propanediol monoacetates starting from a common prochiral precursor, using enzymatic transformations in aqueous and in organic media, *Tetrahedron Lett.* **27**, 5707-5710.

Rangheard, M. S., Langrand, G., Triantaphylides, C., Baratti, J. (1989), Multi-competitive enzymatic reactions in organic media: a simple test for the determination of lipase fatty acid selectivity, *Biochim. Biophys. Acta* **1004**, 20-28.

Ransac, S., Carriére, F., Rogalska, E., Verger, R., Marguet, F., Buono, G., Melo., E. P., Cabral, J. M. S., Egloff, M. P. E., van Tilbeurgh, H., Cambillau, C. (1996), The kinetics specificities and structural features of lipases, in.: *Molecular Dynamics of Biomembranes* (op den Kamp, J. A. F., Ed.), **Vol. H 96**, pp. 254-304. Heidelberg: Springer.

Rantakylae, M., Aaltonen, O. (1994), Enantioselective esterification of ibuprofen in supercritical carbon dioxide by immobilized lipase, *Biotechnol. Lett.* **16**, 825-830.

Rathbone, D. A., Holt, P. J., Lowe, C. R., Bruce, N. C. (1997), Molecular analysis of the *Rhodococcus* sp. strain H1 her gene and characterization of its product, a heroin esterase, expressed in *Escherichia coli*, *Appl. Environm. Microbiol.* **63**, 2062-2066.

Rees, G. D., de Nascimento, M. G., Jenta, T. R. J., Robinson, B. H. (1991), Reverse enzyme synthesis in microemulsion-based organo-gels, *Biochim. Biophys. Acta* **1073**, 493-501.

Rees, G. D., Jenta, T. R. J., Nascimento, M. G., Catauro, M., Robinson, B. H., Stephenson, G. R., Olphert, R. D. G. (1993), Use of water-in-oil microemulsions and gelatin-containing microemulsion based gels for lipase -catalysed ester synthesis in organic solvents, *Indian J. Chem., Sect. B* **32B**, 30-34.

Rees, G. D., Robinson, B. H., Stephenson, G. R. (1995), Preparative-scale kinetic resolutions catalyzed by microbial lipases immobilized in AOT-stabilized microemulsion-based organogels: cryoenzymology as a tool for improving enantioselectivity, *Biochim. Biophys. Acta* **1259**, 73-81.

Reetz, M. T. (1997), Entrapment of biocatalysts in hydrophobic sol-gel materials for use in organic chemistry, *Adv. Mater.* **9**, 943-954.

Reetz, M. T., Dreisbach, C. (1994), Highly efficient lipase-catalyzed kinetic resolution of chiral amines, *Chimia* **48**, 570.

Reetz, M. T., Jaeger, K.-E. (1999), Superior biocatalysts by directed evolution, *Topic Curr. Chem.* **200**, 31-57.

Reetz, M. T., Schimossek, K. (1996), Lipase-catalyzed dynamic kinetic resolution of chiral amines: use of palladium as the racemization catalyst, *Chimia* **90**, 668-669.

Reetz, M., Zonta, A., Simpelkamp, J. (1995), Efficient heterogeneous biocatalysts by entrapment of lipases in hydrophobic sol-gel materials, *Angew. Chem. Int. Ed. Engl.* **34**, 301-303.

Reetz, M. T., Zonta, A., Simpelkamp, J. (1996a), Efficient immobilization of lipases by entrapment in hydrophobic sol-gel materials, *Biotechnol. Bioeng.* **49**, 527-534.

Reetz, M. T., Zonta, A., Simpelkamp, J., Konen, W. (1996b), *In situ* fixation of lipase-containing hydrophobic sol-gel materials on sintered glass - highly efficient heterogeneous biocatalysts, *Chem. Commun.*, 1397-1398.

Reetz, M. T., Zonta, A., Schimossek, K., Liebeton, K., Jaeger, K.-E. (1997), Creation of enantioselective biocatalysts for organic chemistry by in vitro evolution, *Angew. Chem. Int. Ed. Engl.* **36**, 2830-2832.

Reetz, M. T., Becker, M. H., Kühling, K. M., Holzwarth, A. (1998), Time-resolved IR-thermographic detection and screening of enantioselectivity in catalytic reactions, *Angew. Chem. Int. Ed. Engl.* **37**, 2647-2650.

Renold, P., Tamm, C. (1995), Comparison of the hydrolysis of cyclic *meso*-diesters with pig liver esterease (PLE) and rabbit liver esterase (RLE), *Biocatal. Biotransform.* **12**, 37-46.

Renouf, P., Poirier, J.-M., Duhamel, P. (1997), Asymmetric hydrolysis of pro-chiral 3,3-disubstituted 2,4-diacetoxycyclohexa-1,4-dienes, *J. Chem. Soc., Perkin Trans. 1*, 1739-1745.

Ricca, J.-M., Crout, D. H. G. (1993), Selectivity and specificity in substrate binding to proteases: novel hydrolytic reactions catalysed by α-chymotrypsin suspended in organic solvents with low water content and mediated by ammonium hydrogen carbonate, *J. Chem. Soc., Perkin Trans. 1*, 1225-1233.

Ricks, E. E., Estrada-Valdes, M. C., McLean, T. L., Iacobucci, G. A. (1992), Highly enantioselective hydrolysis of (*R,S*)-phenylalanine isopropyl ester by subtilisin Carlsberg. Continuous synthesis of (*S*)-phenylalanine in a hollow fiber/liquid membrane reactor, *Biotechnol. Prog.* **8**, 197-203.

Rigby, J. H., Sugathapala, P. (1996), Lipase-mediated resolution and higher-order cycloaddition of substituted tricarbonyl (η^6-cycloheptatriene)chromium(0) complexes, *Tetrahedron Lett.* **37**, 5293-5296.

Rink, R., Fennema, M., Smids, M., Dehmel, U., Janssen, D. B. (1997), Primary structure and catalytic mechanism of the epoxide hydrolase from *Agrobacterium radiobacter* AD1, *J. Biol. Chem.* **272**, 14650-14657.

Riva, S. (1996), Regioselectivity of hydrolases in organic media, in.: *Enzymatic Reactions in Organic Media* (Koskinen, A. M. P. , Klibanov, A. M.; Eds.), pp. 140-169. Glasgow: Chapman & Hall.

Riva, S., Chopineau, J., Kieboom, A. P. G., Klibanov, A. M. (1988), Protease-catalyzed regioselective esterification of sugars and related compounds in anhydrous dimethylformamide, *J. Am. Chem. Soc.* **110**, 584-589.

Riva, S., Danieli, B., Luisetti, M. (1996), A two-step efficient chemoenzymatic synthesis of flavonoid glycoside malonates, *J. Nat. Prod.* **59**, 618-621.

Roberts, S. M. (1989), Use of enzymes as catalysts to promote key transformations in organic synthesis, *Phil. Trans. R. Soc. London B*, **324**, 577-587.

Roberts, S. M. (Ed.) (1992-1996), *Preparative Biotransformations*, New York: Wiley.

Roberts, S. M., Shoberu, K. A. (1991), Enzymic resolution of *cis*- and *trans*-4-hydroxycyclopent-2-enylmethanol derivatives and a novel preparation of carbocyclic 2',3'-dideoxydidehydronucleosides and aristeromycin, *J. Chem. Soc., Perkin Trans. 1*, 2605-2607.

Roberts, S. M., Wan, P. W. H. (1998), Enzyme-catalysed Baeyer-Villiger oxidations, *J. Mol. Catal. B: Enzymatic* **4**, 111-136.

Robinson, G. K., Alston, M. J., Knowles, C. J., Cheetham, P. S. J., Motion, K. R. (1994), An investigation into the factors influencing lipase-catalyzed intramolecular lactonization in microaqueous systems, *Enzyme Microb. Technol.* **16**, 855-863.

Rocco, V. P., Danishefsky, S. J., Schulte, G. K. (1991), Substrate specificity in enzymatically mediated transacetylation reactions of calicheamicinone intermediates, *Tetrahedron Lett.* **32**, 6671-6674.

Rogalska, E., Cudrey, C., Ferrato, F. S., Verger, R. (1993), Stereoselective hydrolysis of triglycerides by animal and microbial lipases, *Chirality* **5**, 24-30.

Rosell, C. M., Vaidya, A. M., Halling, P. J. (1996), Continuous *in situ* water activity control for organic phase biocatalysis in a packed bed hollow fiber reactor, *Biotechnol. Bioeng.* **49**, 284-289.

Rosenquist, Å., Kvarnström, I., Classon, B., Samuelsson, B. (1996), Synthesis of enantiomerically pure bis(hydroxymethyl)-branched cyclohexenyl and cyclohexyl purines as potential inhibitors of HIV, *J. Org. Chem.* **61**,

Rotticci, D., Orrenius, C., Hult, K., Norin, T. (1997), Enantiomerically enriched bifunctional sec-alcohols prepared by *Candida antarctica* lipase B catalysis. Evidence of non-steric interactions, *Tetrahedron: Asymmetry* **8**, 359-362.

Rúa, M. L., Diaz-Maurino, T., Fernandez, V. M., Otero, C., Ballesteros, A. (1993), Purification and characterization of two distinct lipases from *Candida cylindracea*, *Biochim. Biophys. Acta* **1156**, 181-189.

Rubio, E., Fernandez-Mayorales, A., Klibanov, A. M. (1991), Effect of the solvent on enzyme regioselectivity, *J. Am. Chem. Soc.* **113**, 695-696.

Rudolf, M. T., Schultz, C. (1996), Lipase-catalyzed regio- and enantioselective esterification of *rac*-1,2-*O*-cyclohexylidene-*myo*-inositol, *Liebigs Ann. Chem.*, 533-537.

Rupley, J. A., Gratton, E., Careri, G. (1983), Water and globular proteins, *Trends Biochem. Sci.* **8**, 18-22.

Ruppert, S., Gais, H. J. (1997), Activity enhancement of pig liver esterase in organic solvents by colyophilization with methoxypolyethylene glycol: kinetic resolution of alcohols, *Tetrahedron: Asymmetry* **8**, 3657-3664.

Rüsch gen. Klaas, M., Warwel, S. (1996), Chemoenzymatic epoxidation of unsaturated fatty acid esters and plant oils, *J. Am. Oil Chem. Soc.* **73**, 1453-1457.

Rüsch gen. Klaas, M., Warwel, S. (1998), A three-step-one-pot chemo-enzymatic synthesis of epoxyalkanolacylates, *Synth. Commun.*, **28**, 251-260.

Sahai, P., Vishwakarma, R. A. (1997), Phospholipase-A_2-mediated stereoselective synthesis of (*R*)-1-*O*-alkylglycero-3-phosphate and alkyl-acyl analogues: application for synthesis of radio-labelled biosynthetic precursors of cell surface glycoconjugates of *Leishmania donovani*, *J. Chem. Soc., Perkin Trans. 1*, 1845-1849.

Sakagami, H., Samizu, K., Kamikubo, T., Ogasawara, K. (1996a), Enantiocontrolled synthesis and absolute configuration of (+)-crooksidine, an indole alkaloid from *Haplophyton crooksii* using chiral piperideinol block, *Synlett*, 163-164.

Sakagami, H. K., Kamikubo, T., Ogasawara, K. (1996b), Novel reduction of 3-hydroxypyridine and its use in the enantioselective synthesis of (+)-pseudoconhydrine and (+)-*N*-methylpseudo-conhydrine, *Chem. Commun.*, 1433-1434.

Sakagami, H., Kamikubo, T., Ogasawara, K. (1997), A facile synthesis of *R*-4-amino-3-hydroxy-butanoic acid (gabob) from 3-hydroxypyridine, *Synlett*, 221-222.

Sakai, T., Kawabata, I., Kishimoto, T., Ema, T., Utaka, M. (1997), Enhancement of the enantiose-lectivity in lipase-catalyzed kinetic resolutions of 3-phenyl-2H-azirine-2-methanol by lowering the temperature to -40 degree, *J. Org. Chem.* **62**, 4906-4907.

Saksena, A. K., Girijavallabhan, V. M., Lovey, R. G., Pike, R. E., Wang, H., Ganguly, A. K., Morgan, B., Zaks, A., Puar, M. S. (1995), Highly stereoselective access to novel 2,2,4-trisub-stituted tetrahydrofurans by halocyclization: practical chemoenzymic synthesis of SCH 51048, a broad-spectrum orally active antifungal agent, *Tetrahedron Lett.* **36**, 1787-1790.

Sakurai, T., Margolin, A. L., Russell, A. J., Klibanov, A. M. (1988), Control of enzyme enantio-selectivity by the reaction medium, *J. Am. Chem. Soc.* **110**, 7236-7237.

Salazar, L., Sih, C. J. (1995), Optically active dihydropyridines via lipase-catalyzed enantioselec-tive hydrolysis, *Tetrahedron: Asymmetry* **6**, 2917-2920.

Sánchez, V. M., Rebolledo, F., Gotor, V. (1997), *Candida antarctica* lipase catalyzed resolution of ethyl (±)-3-aminobutyrate, *Tetrahedron: Asymmetry* **8**, 37-40.

Sanchez-Montero, J. M., Hamon, V., Thomas, D., Legoy, M. D. (1991), Modulation of lipase hydrolysis and synthesis reactions using carbohydrates, *Biochim. Biophys. Acta* **1078**, 345-350.

Sanfillipo, C., Patti, A., Piattelli, M., Nicolosi, G. (1997), An efficient enzymatic preparation of (+)- and (−)-conduritol E, a cyclitol with C_2 symmetry, *Tetrahedron: Asymmetry* **8**, 1569-1573.

Santaniello, E., Ferraboschi, P., Grisenti, P., Manzocchi, A. (1992), The biocatalytic approach to the preparation of enantiomerically-pure chiral building blocks, *Chem. Rev.* **92**, 1071-1140.

Saraiva-Conçalves, J. C., Razouk, C., Poupaert, J. H., Dumont, P. (1989), High-performance liquid chromatography of chlorambucil prodrugs structurally related to lipids in rat plasma, *J. Chromatogr.* **494**, 389-396.

Sarney, D. B., Kapeller, H., Fregapane, G., Vulfson, E. N. (1994), Chemo-enzymatic synthesis of disaccharide fatty acid esters, *J. Am. Oil Chem. Soc.* **71**, 711-714.

Sarney, D. B., Vulfson, E. N. (1995), Application of enzymes to the synthesis of surfactants, *Trends Biotechnol.* **13**, 164-172.

Sato, M., Ohuchi, H., Abe, Y., Kaneko, C. (1992), Regio-, stereo-, and enantioselective synthesis of cyclobutanols by means of the photoaddition of 1,3-dioxin-4-ones and lipase catalyzed acy-lation, *Tetrahedron: Asymmetry* **3**, 313-328.

Sato, S., Murakata, T., Ochifuji, M., Fukushima, M., Suzuki, T. (1994), Development of immobi-lized enzyme entrapped within inorganic matrix and its catalytic activity in organic medium, *J. Chem. Eng. Jpn.* **27**, 732-736.

Satoshi, O. (Ed.) (1992), *The search for bioactive compounds from microorganisms*, Berlin: Springer.

Sattler, A., Haufe, G. (1995), Synthesis of (9R)- and (9S)-10-fluorodecan-9-olide (fluoro-phora-cantholide I) first lipase-catalyzed enantioselective esterification of β-fluoroalcohols, *Tetrahe-dron: Asymmetry* **6**, 2841-2848.

Savjalov, W. W. (1901), Zur Theorie der Eiweissverdauung, *Pflügers Arch. Ges. Physiol.* **85**, 171.

Sayle, R. A., Milner-White, E. J. (1995), RASMOL: biomolecular graphics for all, *Trends Bio-chem. Sci.* **20**, 374-376.

Schechter, I., Berger, A. (1967), On the active site of proteases. I. Papain., *Biochem. Biophys. Res. Commun.*, **27**, 157-162.

Scheckermann, C., Schlotterbeck, A., Schmidt, M., Wray, V., Lang, S. (1995), Enzymatic monoa-cylation of fructose by two procedures, *Enzyme Microb. Technol.* **17**, 157-162.

Scheib, H., Pleiss, J., Stadler, P., Kovac, A., Potthoff, A. P., Haalck, L., Spener, F., Paltauf, F., Schmid, R. D. (1998), Rational design of *Rhizopus oryzae* lipase with modified stereoselectivity toward triradylglycerols, *Protein Eng.* **11**, 675-682.

Scheib, H., Pleiss, J., Kovac, A., Paltauf, F., Schmid, R. D. (1999), Stereoselectivity of *Mucorales* lipases toward triradylglycerols - a simple solution to a complex problem, *Protein Sci.* **8**, 215-221.

Schellenberger, V., Jakubke, H. D. (1991), Protease-catalyzed kinetically controlled peptide synthesis, *Angew. Chem. Int. Ed. Engl.* **30**, 1437-1449.

Schlotterbeck, A., Lang, S., Wray, V., Wagner, F. (1993), Lipase-catalyzed monoacylation of fructose, *Biotechnol. Lett.* **15**, 61-64.

Schmid, R. D., Verger, R. (1998), Lipases - interfacial enzymes with attractive applications, *Angew. Chem. Int. Ed. Engl.* **37**, 1608-1633.

Schmid, U., Bornscheuer, U. T., Soumanou, M. M., McNeill, G. P., Schmid, R. D. (1998), Optimization of the reaction conditions in the lipase-catalyzed synthesis of structured triglycerides, *J. Am. Oil Chem. Soc.* **75**, 1527-1531.

Schmitke, J. L., Wescott, C. R., Klibanov, A. M. (1996), The mechanistic dissection of the plunge in enzymatic activity upon transition from water to anhydrous solvents, *J. Am. Chem. Soc.* **118**, 3360-3365.

Schneider, M. P., Goergens, U. (1992), An efficient route to enantiomerically pure antidepressants: Tomoxetine, Nisoxetine and Fluoxetine, *Tetrahedron: Asymmetry* **3**, 525-528.

Schneider, M., Engel, N., Hönicke, P., Heinemann, G., Görisch, H. (1984), Hydrolytic enzymes in organic synthesis. 3. Enzymatic syntheses of chiral building-blocks from prochiral *meso*-substrates - preparation of methyl(hydrogen)-1,3-cycloalkanedicarboxylates, *Angew. Chem. Int. Ed. Engl.* **23**, 67-68.

Schoffers, E., Golebiowski, A., Johnson, C. R. (1996), Enantioselective synthesis through enzymatic asymmetrization, *Tetrahedron* **52**, 3769-3826.

Schrag, J. D., Li, Y., Wu, S., Cygler, M. (1991), Ser-His-Glu triad forms the catalytic site of the lipase from *Geotrichum candidum*, *Nature* **351**, 761-764.

Schrag, J. D., Li, Y., Cygler, M., Lang, D., Burgdorf, T., Hecht, H. J., Schmid, R., Schomburg, D., Rydel, T. J., Oliver, J. D., Strickland, L. C., Dunaway, C. M., Larson, S. B., Day, J., McPherson, A. (1997), The open conformation of a *Pseudomonas* lipase, *Structure* **5**, 187-202.

Schricker, B., Thirring, K., Berner, H. (1992), α-Chymotrypsin catalyzed enantioselective hydrolysis of alkenyl α-amino acid esters, *Bioorg. Med. Chem. Lett.* **2**, 387-390.

Schudok, M., Kretzschmar, G. (1997), Enzyme catalyzed resolution of alcohols using ethoxyvinyl acetate, *Tetrahedron Lett.* **38**, 387-388.

Schueller, C. M., Manning, D. D., Kiessling, L. L. (1996), Preparation of (*R*)-(+)-7-oxabicyclo[2.2.1]hept-5-ene-exo-2-carboxylic acid, a precursor to substrates for the ring opening metathesis polymerization, *Tetrahedron Lett.* **37**, 8853-8856.

Schultz, M., Hermann, P., Kunz, H. (1992), Enzymatic cleavage of tert-butyl esters: the mitase catalyzed deprotection of peptides and *O*-glycopeptides, *Synlett*, 37-38.

Schutt, H., Schmidt-Kastner, G., Arens, A., Preiss, M. (1985), Preparation of optically active D-arylglycines for use as side chains for semisynthetic penicillins and cephalosporins using immobilized subtilisins in two-phase systems, *Biotechnol. Bioeng.* **27**, 420-433.

Schwienhorst, A. (1998), Evolutive Methoden im Enzymdesign, *Biospektrum* **4**, 44-47.

Scilimati, A., Ngooi, T. K., Sih, C. J. (1988), Biocatalytic resolution of (±)-hydroxyalkanoic esters. A strategy for enhancing the enantiomeric specificity of lipase-catalyzed ester hydrolysis, *Tetrahedron Lett.* **29**, 4927-4930.

Scott, D. L., White, S. P., Otwinowski, Z., Yuan, W., Gelb, M. H., Sigler, P. B. (1990), Interfacial catalysis: the mechanism of phospholipase A2, *Science* **250**, 1541-1546.

Secundo, F., Riva, S., Carrea, G. (1992), Effects of medium and of reaction conditions on the enantioselectivity of lipases in organic solvents and possible rationales, *Tetrahedron: Asymmetry* **3**, 267-280.

Seebach, D., Eberle, M. (1986), Enantioselective cleavage of *meso*-nitrodiol diacetates by an esterase concentrate from fresh pig liver: preparation of useful nitroaliphatic building blocks for EPC syntheses, *Chimia* **40**, 315-318.

Seemayer, R., Bar, N., Schneider, M. P. (1992), Enzymic preparation of isomerically pure 1,4:3,6-dianhydro-D- glucitol monoacetates - precursors for isoglucitol 2- and 5-mononitrates, *Tetrahedron: Asymmetry* **3**, 1123-1126.

Seidegard, J., de Pierre, J. W. (1983), Microsomal epoxide hydrolase. Properties, regulation and function, *Biochim. Biophys. Acta* **695**, 251-270.

Seki, M., Furutani, T., Miyake, T., Yamanaka, T., Ohmizu, H. (1996), A novel synthesis of a key intermediate for penems and carbapenems utilizing lipase-catalyzed kinetic resolution, *Tetrahedron: Asymmetry* **7**, 1241-1244.

Serdarevich, B. (1967), Glyceride isomerizations in lipid chemistry, *J. Am. Oil Chem. Soc.* **44**, 381-393.

Serebryakov, E. P., Gamalevich, G. D., Strakhov, A. V., Vasil'ev, A. A. (1995), Enhancement and reversal of enantioselectivity of the enzymic hydrolysis of (*RS*)-3-(4-methoxycarbonyl)phenyl-2-methylprop-1-yl acetate upon its transformation into the η^6- arene(tricarbonyl)chromium complex, *Mendeleev Commun.*, 175-176.

Serreqi, A. N., Kazlauskas, R. J. (1994), Kinetic resolution of phosphines and phosphine oxides with phosphorus stereocenters by hydrolases, *J. Org. Chem.* **59**, 7609-7615.

Serreqi, A. N., Kazlauskas, R. J. (1995), Kinetic resolution of sulfoxides with pendant acetoxy groups using cholesterol esterase: substrate mapping and an empirical rule for chiral phenols, *Can. J. Chem.* **73**, 1357-1367.

Seu, Y. B., Kho, Y. H. (1992), Enzymic preparation of optically active 2-acetoxymethylglycidol, a new chiral building block in natural product synthesis, *Tetrahedron Lett.* **33**, 7015-7016.

Sharma, A., Chattopadhyay, S., Mamdapur, V. R. (1995), PPL catalyzed monoesterification of α,ω-dicarboxylic acids, *Biotechnol. Lett.* **17**, 939-942.

Sheldon, R. A. (1993), *Chirotechnology-industrial synthesis of optically-active compounds*, New York: Marcel Dekker.

Shieh, W. R., Gou, D. M., Chen, C. S. (1991), Computer-aided substrate design for biocatalysis: an enzymatic access to optically active propranolol, *J. Chem. Soc., Chem. Commun.*, 651-653.

Shieh, C. J., Akoh, C. C., Koehler, P. E. (1995), Four-factor response surface optimization of the enzymatic modification of triolein to structured lipids, *J. Am. Oil Chem. Soc.* **72**, 619-623.

Shimada, Y., Sugihara, A., Maruyama, K., Nagao, T., Nakayama, S., Nakano, H., Tominaga, Y. (1995), Enrichment of arachidonic acid: selective hydrolysis of a single-cell oil from *Mortierella* with *Candida cylindracea* lipase, *J. Am. Oil Chem. Soc.* **72**, 1323-1327.

Shimada, Y., Sugihara, A., Nakano, H., Kuramoto, T., Nagao, T., Gemba, M., Tominaga, Y. (1997), Purification of docosahexaenoic acid by selective esterification of fatty acids from tuna oil with Rhizopus delemar lipase, *J. Am. Oil Chem. Soc.* **74**, 97-101.

Shimizu, M., Kawanami, H., Fujisawa, T. (1992a), A lipase-mediated asymmetric hydrolysis of 3-acyloxy-1-octynes and 3-(*E*)-acyloxy-1-octenes, *Chem. Lett.* 107-110.

Shimizu, S., Kataoka, M., Shimizu, K., Hirakata, M., Sakamoto, K., Yamada, H. (1992b), Purification and characterization of a novel lactonohydrolase, catalyzing the hydrolysis of aldonate lactones and aromatic lactones, from *Fusarium oxysporum*, *Eur. J. Biochem.* **209**, 383-390.

Shimizu, N., Akita, H., Kawamata, T. (1996), Enantioselective reaction of (±)-1-hydroxy-5-methyl-3-vinylcyclohex-2-ene and its acyl derivatives using lipid-lipase aggregates in organic solvent, *Chem. Pharm. Bull.* **44**, 665-669.

Shimizu, S., Ogawa, J., Kataoka, M., Kobayashi, M. (1997), Screening of novel microbial enzymes for the production of biologically and chemically useful enzymes, *Adv. Biochem. Eng./Biotechnol.* **58**, 45-87.

Short, J. M. (1997), Recombinant approaches for accessing biodiversity, *Nature Biotechnol.* **15**, 1322-1223.

Shuto, S., Imamura, S., Fukukawa, K., Ueda, T. (1988), Phospholipase D-catalyzed trans-alkylphosphorylation: A facile one-step synthesis of 5'-alkylphosphates, *Chem. Pharm. Bull.* **36**, 5020-5023.

Shuto, S., Ueda, S., Imamura, S., Fukukawa, K., Matsuda, A., Ueda, T. (1987), A facile one-step synthesis of 5'-phosphatidylnucleosides by an enzymatic two-phase reaction, *Tetrahedron Lett.* **28**, 199-202.

Sibi, M. P., Lu, J. L. (1994), Synthesis of both enantiomers of C_2 symmetric *trans*-2,5-bis(hydroxymethyl) pyrrolidine. Lipase mediated sequential kinetic resolution, *Tetrahedron Lett.* **35**, 4915-4918.

Sicsic, S., Ikbal, M., Goffic, F. L. (1987), Chemoenzymatic approach to carbocyclic analogs of ribonucleosides and nicotinamide ribose, *Tetrahedron Lett.* **28**, 1887-1888.

Siddiqi, S. M., Schneller, S. W., Ikeda, S., Snoeck, R., Andrei, G., Balzarini, J., de Clerq, E. (1993), *S*-Adenosyl-L-homocysteine hydrolase inhibitors as anti-viral agents: 5'-deoxyaristeromycin, *Nucleos. Nucleot.* **12**, 185-198.

Sieno, H., Uchibiro, T., Nishitani, T., Inamasu, S. (1984), Enzymatic synthesis of carbohydrate esters of fatty acid (I) esterification of sucrose, glucose, fructose and sorbitol, *J. Am. Oil Chem. Soc.* **61**, 1761-1765.

Siezen, R. J., de Vos, W. M., Leunissen, J. A., Dijkstra, B. W. (1991), Homology modelling and protein engineering strategy of subtilases, the family of subtilisin-like serine proteinases, *Protein Eng.* **4**, 719-737.

Siezen, R. J., Leunissen, J. A. M. (1997), Subtilases: The superfamily of subtilisin-like serine proteases, *Protein Sci.* **6**, 501-523.

Sih, J. C. (1996), Application of immobilized lipase in production of Camptosar (CPT-11), *J. Am. Oil Chem. Soc.* **73**, 1377-1378.

Sih, C. J., Wu, S.-H. (1989), Resolution of enantiomers via biocatalysis, in.: *Topics in Stereochemistry* (Eliel, E. L. , Wilen, S. H.; Eds.), **Vol. 19**, pp. 63-125. New York: John Wiley & Sons.

Sih, C. J., Gu, Q. M., Holdgrün, X. K., Harris, K. (1992), Optically active compounds via biocatalytic methods, *Chirality* **4**, 91-97.

Sin, Y. M., Chung, S. H., Park, J. Y., Lee, T. H. (1996), Synthesis of biodegradable emulsifier using lipase in anhydrous pyridine, *Biotechnol. Lett.* **18**, 689-694.

Singh, C. P., Shah, D. O., Holmberg, K. (1994a), Synthesis of mono- and diglycerides in water-in-oil microemulsions, *J. Am. Oil Chem. Soc.* **71**, 583-587.

Singh, C. P., Skagerlind, P., Holmberg, K., Shah, D. O. (1994b), A comparison between lipase-catalyzed esterification of oleic acid with glycerol in monolayer and microemulsion systems, *J. Am. Oil Chem. Soc.* **71**, 1405-1409.

Sinisterra, J. V., Llama, E. F., Campo, C. D., Cabezas, M. J., Moreno, J. M., Arroyo, M. (1994), Stereoselectivity of chemically-modified α-chymotrypsin and immobilized lipases, *J. Chem. Soc. Perkin Trans. 2*, 1333-1346.

Sjursnes, B. J., Anthonsen, T. (1994), Acyl migration in 1,2-dibutyrin dependence on solvent and water activity, *Biocatalysis* **9**, 285-297.

Sjursnes, B. J., Kvittingen, L., Anthonsen, T. (1995), Regioselective lipase-catalyzed transesterification of tributyrin. Influence of salt hydrates on acyl migration, *J. Am. Oil Chem. Soc.* **72**, 533-537.

Skagerlind, P., Jansson, M., Hult, K. (1992), Surfactant interference on lipase catalysed reactions in microemulsions, *J. Chem. Tech. Biotechnol.* **54**, 277-282.

Smidt, H., Fischer, A., Fischer, P., Schmid, R. D. (1996), Preparation of optically-pure chiral amines by lipase-catalyzed enantioselective hydrolysis of *N*-acyl-amines, *Biotechnol. Techn.* **10**, 335-338.

Smith, G. B., Bhupathy, M., Dezeny, G. C., Douglas, A. W., Lander, R. J. (1992), Kinetics of a heterogeneous enzymatic hydrolysis of a prochiral diester, *J. Org. Chem.* **57**, 4544-5446.

Smith, R. E., Finley, J. W., Leveille, G. A. (1994), Overview of SALATRIM, a family of low calorie fats, *J. Agric. Food Chem.* **42**, 432-434.

Solladie-Cavallo, A., Schwarz, J., Burger, V. (1994), A four-step, highly enantioselective synthesis and enzymic resolution of 3,4-dichlorophenylalanine, *Tetrahedron: Asymmetry* **5**, 1621-1626.

Soloshonok, V. A., Kirilenko, A. G., Fokina, N. A., Kukhar, V. P., Galushko, S. V., Svedas, V. K., Resnati, G. (1994a), Chemo-enzymatic approach to the synthesis of each of the four isomers of α-alkyl-β-fluoroalkyl-substituted β-amino acids, *Tetrahedron: Asymmetry* **5**, 1225-1228.

Soloshonok, V. A., Kirilenko, A. G., Fokina, N. A., Shishkina, I. P., Galushko, S. V., Kukhar, V. P., Svedas, V. K., Kozlova, E. V. (1994b), Biocatalytic resolution of β-fluoroalkyl-β-amino acids, *Tetrahedron: Asymmetry* **5**, 1119-1126.

Soloshonok, V. A., Fokina, N. A., Rybakova, A. V., Shishkina, I. P., Galushko, S. V., Sorochinsky, A. E., Kukhar, V. P., Savchenko, M. V., Svedas, V. K. (1995), Biocatalytic approach to enantiomerically pure β-amino acids, *Tetrahedron: Asymmetry* **6**, 1601-1610.

Sonnet, P. E. (1987), Kinetic resolutions of aliphatic alcohols with a fungal lipase from *Mucor miehei*, *J. Org. Chem.* **52**, 3477-3479.

Sonnet, P. E., Foglia, T. A., Feairheller, S. H. (1993), Fatty acid selectivity of lipases: erucic acid from rapeseed oil, *J. Am. Oil Chem. Soc.* **70**, 387-391.

Sonntag, N. O. V. (1982), Glycerolysis of fats and methyl esters - Status, review and critique, *J. Am. Oil Chem. Soc.* **59**, 795A-802A.

Soumanou, M. M., Bornscheuer, U. T., Menge, U., Schmid, R. D. (1997), Synthesis of structured triglycerides from peanut oil with immobilized lipase, *J. Am. Oil Chem. Soc.* **74**, 427-433.

Soumanou, M. M., Bornscheuer, U. T., Schmid, R. D. (1998a), Two-step enzymatic reaction for the synthesis of pure structured triglycerides, *J. Am. Oil Chem. Soc.* **75**, 703-710.

Soumanou, M. M., Bornscheuer, U. T., Schmid, U., Schmid, R. D. (1998b), Synthesis of structured triglycerides by lipase catalysis, *Fett/Lipid* **100**, 156-160.

Spee, J. H., de Vos, W. M., Kuipers, O. P. (1993), Efficient random mutagenesis method with adjustable mutation frequency by use of PCR and dITP, *Nucl. Acids. Res.* **21**, 777-778.

Spelberg, J. H. L., Rink, R., Kellogg, R. M., Janssen, D. B. (1998), Enantioselectivity of a recombinant epoxide hydrolase from *Agrobacterium radiobacter*, *Tetrahedron: Asymmetry* **9**, 459-466.

Spero, D. M., Kapadia, S. R. (1996), Enantioselective synthesis of α,α-disubstituted amino acid derivatives via enzymatic resolution - preparation of a thiazolyl-substituted α-methyl α-benzyl amine, *J. Org. Chem.* **61**, 7398-7401.

Sperry, W. M., Brand, F. C. (1941), A study of cholesterol esterase in liver and brain, *J. Biol. Chem.* **137**, 377-387.

St.Clair, N. L., Navia, M. A. (1992), Cross-linked enzyme crystals as robust biocatalysts, *J. Am. Chem. Soc.* **114**, 7314-7316.

Stadler, P., Kovac, A., Haalck, L., Spener, F., Paltauf, F. (1995), Stereoselectivity of microbial lipases. The substitution at position sn-2 of triacylglycerol analogs influences the stereoselectivity of different microbial lipases, *Eur. J. Biochem.* **227**, 335-343.

Stamatis, H., Xenakis, A., Dimitriadis, E., Kolisis, F. N. (1995), Catalytic behavior of *Pseudomonas cepacia* lipase in w/o microemulsions, *Biotechnol. Bioeng.* **45**, 33-41.

Stead, P., Marley, H., Mahmoudian, M., Webb, G., Noble, D., Ip, Y. T., Piga, E., Rossi, T., Roberts, S., Dawson, M. J. (1996), Efficient procedures for the large-scale preparation of (1*S*,2*S*)-*trans*-2-methoxycyclohexanol, a key chiral intermediate in the synthesis of tricyclic β-lactam antibiotics, *Tetrahedron: Asymmetry* **7**, 2247-2250.

Stecher, H., Faber, K. (1997), Biocatalytic deracemization techniques: dynamic resolutions and stereoinversions, *Synthesis*, 1-16.

Steffen, B., Ziemann, A., Lang, S., Wagner, F. (1992), Enzymatic monoacylation of trihydroxy compounds, *Biotechnol. Lett.* **14**, 773-778.

Steffen, B., Lang, S., Hamann, D., Schneider, P., Cammenga, H. K., Wagner, F. (1995), Enzymic formation and physico-chemical behavior of uncommon monoglycerides, *Fat Sci. Technol.* **97**, 132-136.

Stemmer, W. P. C. (1994a), DNA shuffling by random fragmentation and reassembly: *In vitro* recombination for molecular evolution, *Proc. Natl. Acad. Sci. USA* **91**, 10747-10751.

Stemmer, W. P. C. (1994b), Rapid evolution of a protein *in vitro* by DNA shuffling, *Nature* **370**, 389-391.

Stemmer, W. P. C. (1995), Searching sequence space. Using recombination to search more efficiently and thoroughly instead of making bigger combinatorial libraries, *Bio/Technology* **13**, 549-553.

Stevenson, D. E., Stanley, R. A., Furneaux, R. H. (1993), Glycerolysis of tallow with immobilized lipase, *Biotechnol. Lett.* **15**, 1043-1048.

Stinson, C. (1997), U.K.'s Chirotech goes it alone, *Chem. Eng. News* **27**, 15-16.

Stokes, T. M., Oehlschlager, A. C. (1987), Enzyme reaction in apolar solvents: the resolution of (±)-sulcatol with porcine pancreatic lipase, *Tetrahedron Lett.* **28**, 2091-2094.

Stolz, A., Trott, S., Binder, M., Bauer, R., Hirrlinger, B., Layh, N., Knackmuss, H.-J. (1998), Enantioselective nitrile hydratases and amidases from different bacterial isolates, *J. Mol. Catal. B: Enzymatic* **5**, 137-141.

Straathof, A. J. J., Rakels, J. L. L., van Tol, J. B. A., Heijnen, J. J. (1995), Improvement of lipase-catalyzed kinetic resolution by tandem transesterification, *Enzyme Microb. Technol.* **17**, 623-628.

Stranix, B. R., Darling, G. D. (1995), Functional polymers from (vinyl)polystyrene. Enzyme immobilization through a cysteinyl-*S*-ethyl spacer, *Biotechnol. Tech.* **9**, 75-80.

Suemune, H., Tanaka, M., Obaishi, H., Sakai, K. (1988), Enzymic procedure for the synthesis of prostaglandin A2, *Chem. Pharm. Bull.* **36**, 15-21.

Sugahara, T., Ogasawara, K. (1996), An expedient route to (-)-*cis*-2-oxabicyclo[3.3.0]oct-6-en-3-one via a *meso*-asymmetrization, *Synlett*, 319-320.

Sugahara, T., Satoh, I., Yamada, O., Takano, S. (1991), Efficient enzymic preparation of (+)- and (−)-Corey lactone derivatives, *Chem. Pharm. Bull.* **39**, 2758-2760.

Sugai, T., Mori, K. (1988), Preparative bioorganic chemistry; 8. Efficient enzymatic preparation of (1R,4S)-(+)-4-hydroxy-2-cyclopentenyl acetate, *Synthesis*, 19-22.

Sugai, T., Ohta, H. (1991), A simple preparation of (R)-2-hydroxy-4-phenyl-butanoic acid, *Agric. Biol. Chem.* **55**, 293-294.

Sugai, T., Kakeya, H., Ohta, H. (1990a), Enzymatic preparation of enantiomerically enriched tertiary α-benzyloxy acid esters. application to the synthesis of (S)-(+)-frontalin, *J. Org. Chem.* **55**, 4643-4647.

Sugai, T., Kakeya, H., Ohta, H. (1990b), A synthesis of (R)-(−)-mevalonolactone by the combination of enzymatic and chemical methods, *Tetrahedron* **46**, 3463-3468.

Sugai, T., Ohsawa, S., Yamada, H., Ohta, H. (1990c), Preparation of enantiomerically enriched compounds using enzymes, VII. A synthesis of japanese beetle pheromone utilizing lipase-catalyzed enantioselective lactonization, *Synthesis*, 1112-1114.

Sugai, T., Katoh, O., Ohta, H. (1995), Chemo-enzymatic synthesis of (R,R)-(-)-pyrenophorin, *Tetrahedron* **51**, 11987-11998.

Sugai, T., Ikeda, H., Ohta, H. (1996), Biocatalytic approaches to both enantiomers of (2R,3S)-2-allyloxy-3,4,5,6-tetrahydro-2H-pyran-3-ol, *Tetrahedron* **52**, 8123-8134.

Sugai, T., Yamazaki, T., Yokoyama, M., Ohta, H. (1997), Biocatalysis in organic synthesis: The use of nitrile- and amide-hydrolyzing microorganisms, *Biosci. Biotech. Biochem.* **61**, 1419-1427.

Suginaka, K., Hayashi, Y., Yamamoto, Y. (1996), Highly selective resolution of secondary alcohols and acetoacetates with lipases and diketenes in organic media, *Tetrahedron: Asymmetry* **7**, 1153-1158.

Sugiura, Y., Kuwahara, J., Nagasawa, T., Yamada, H. (1987), Nitrile hydratase: the first non-heme iron enzyme with a typical low-spin Fe(III)-active center, *J. Am. Chem. Soc.* **109**, 5848-5850.

Sundram, H., Golebiowski, A., Johnson, C. R. (1994), Chemoenzymatic synthesis of (2R,3R)-3-hydroxyproline from cyclopentadiene, *Tetrahedron Lett.* **35**, 6975-6976.

Sussman, J. L., .Harel, M., Frolow, F., Oefner, C., Goldman, A., Toker, L., Silman, I. (1991), Atomic structure of acetylcholinesterase from *Torpedo californica*: a prototypic acetylcholine-binding protein., *Science* **253**, 872-879.

Sutherland, A., Willis, C. L. (1997), Enantioselective syntheses of α-amino-β-hydroxy acids, [^{15}N]-L-allothreonine and [^{15}N]-L-threonine, *Tetrahedron Lett.* **38**, 1837-1840.

Suzuki, Y., Marumo, S. (1972), Fungal metabolism of (±)-epoxyfarnesol and its absolute stereochemistry, *Tetrahedron Lett.* **19**, 1887-1890.

Svendsen, A. (1994), Sequence comparisons within the lipase family, in.: *Lipases* (Wooley., P., Peterson, S. B.; Eds.), pp. 1-21. Cambridge: Cambridge University Press.

Svendsen, A., Borch, K., Barfoed, M., Nielsen, T. B., Gormsen, E., Patkar, S. A. (1995), Biochemical properties of cloned lipases from the *Pseudomonas* family, *Biochim. Biophys. Acta* **1259**, 9-17.

Svensson, I., Wehtje, E., Adlercreutz, P., Mattiasson, B. (1994), Effects of water activity on reaction rates and equilibrium positions in enzymatic esterifications, *Biotechnol. Bioeng.* **44**, 549-556.

Svirkin, Y. Y., Xu, J., Gross, R. A., Kaplan, D. L., Swift, G. (1996), Enzyme-catalyzed stereoselective ring-opening polymerization of α-methyl-β-propiolactone, *Macromolecules* **29**, 4591-4597.

Swaving, J., de Bont, J. A. N. (1998), Microbial transformation of epoxides, *Enzyme Microb. Technol.* **22**, 19-26.

Sweers, H. M., Wong, C.-H. (1986), Enzyme-catalyzed regioselective deacylation of protected sugars in carbohydrate synthesis, *J. Am. Chem. Soc.* **108**, 6421-6422.

Syldatk, C., Müller, R., Pietzsch, M., Wagner, F. (1992a), Microbial and enzymatic production of L-amino acids from DL-5-substituted hydantoins, in.: *Biocatalytic production of amino acids and derivatives* (Rozell, J. D. , Wagner, F.; Eds.), pp. 131-176. New York: Hanser.

Syldatk, C., Müller, R., Siemann, M., Wagner, F. (1992b), Microbial and enzymatic production of D-amino acids from DL-5-substituted hydantoins, in.: *Biocatalytic production of amino acids and derivatives* (Rozell, J. D. , Wagner, F.; Eds.), pp. 75-127. New York: Hanser.

Syldatk, C., May, O., Altenbuchner, J., Mattes, R., Siemann, M. (1999), Microbial hydantoinases - industrial enzymes from the origin of life?, *Appl. Microbiol. Biotechnol.*, in press.

Sym, E. A. (1936), Action of esterase in the presence of organic solvents, *Biochem. J.* **30**, 609-617.

Taipa, M. A., Liebeton, K., Costa, J. V., Cabral, J. M. S., Jaeger, K. E. (1995), Lipase from *Chromobacterium viscosum*: biochemical characterization indicating homology to the lipase from *Pseudomonas glumae*, *Biochim. Biophys. Acta* **1256**, 396-402.

Takagi, Y., Teramoto, J., Kihara, H., Itoh, T., Tsukube, H. (1996), Thiacrown ether as regulator of lipase-catalyzed trans-esterification in organic media - practical optical resolution of allyl alcohols, *Tetrahedron Lett.* **37**, 4991-4992.

Takahashi, M., Ogasawara, K. (1996), Lipase-mediated resolution of *trans*-1-azidoindan-2-ol: a new route to optically pure *cis*-1-aminoindan-2-ol, *Synthesis*, 636-638.

Takahashi, S., Ohashi, T., Kii, Y., Kumagai, H., Yamada, H. (1979), Microbial transformation of hydantoins to amino acids. III. Microbial transformation of hydantoins to *N*-carbamyl-D-amino acids, *J. Ferment. Technol.* **57**, 328-332.

Takahashi, T., Ikai, A., Takahashi, K. (1989), Purification and characterization of proline-β-naphthylamidase, a novel enzyme from pig intestinal mucosa, *J. Biol. Chem.* **264**, 11565-11571.

Takahashi, T., Nishigai, M., Ikai, A., Takahashi, K. (1991), Electron microscopy and biochemical evidence that proline-β-naphthylamidase is composed of three identical subunits, *FEBS* **280**, 297-300.

Takahashi, S., Ueda, M., Atomi, H., Beer, H. D., Bornscheuer, U. T., Schmid, R. D., Tanaka, A. (1998), Extracellular production of active *Rhizopus oryzae* lipase by *Saccharomyces cerevisiae*, *J. Ferment. Bioeng.* **86**, 164-168.

Takami, M., Hidaka, N., Suzuki, Y. (1994), Phospholipase D-catalyzed synthesis of phosphatidyl aromatic compounds, *Biosci. Biotech. Biochem.* **58**, 2140-2144.

Takano, S., Inomata, K., Takahashi, M., Ogasawara, K. (1991), Expedient preparation and enantiomerization of optically pure dicyclopentadienone (tricyclo[$5.2.1.0^{2,6}$]deca-4,8-dien-3-one), *Synlett*, 636-638.

Takano, S., Setoh, M., Ogasawara, K. (1992a), Resolution of racemic *O*-(4-methoxyphenyl)glycidol, *Heterocycles* **34**, 173-180.

Takano, S., Yamane, T., Takahashi, M., Ogasawara, K. (1992b), Efficient chiral route to a key building block of 1,25-dihydroxyvitamin D3 via lipase -mediated resolution, *Synlett*, 410-412.

Takano, S., Yamane, T., Takahashi, M., Ogasawara, K. (1992c), Enantiocomplementary synthesis of functionalized cycloalkenol building blocks using lipase, *Tetrahedron: Asymmetry* **3**, 837-840.

Takano, S., Higashi, Y., Kamikubo, T., Moriya, M., Ogasawara, K. (1993a), Enantiodivergent preparation of chiral 2,5-cyclohexadienone synthons, *Synthesis*, 948-950.

Takano, S., Moriya, M., Higashi, Y., Ogasawara, K. (1993b), Enantiodivergent route to conduritol C via lipase -mediated asymmetrization, *J. Chem. Soc., Chem. Commun.*, 177-178.

Takano, S., Setoh, M., Ogasawara, K. (1993c), Enantiocomplementary resolution of 4-hydroxy-5-(4-methoxyphenoxy)-1-pentyne using the same lipase, *Tetrahedron: Asymmetry* **4**, 157-160.

Takano, S., Setoh, M., Yamada, O., Ogasawara, K. (1993d), Synthesis of optically active 4-benzyloxymethyl- and 4-(4-methoxyphenoxy)methyl-buten-2-olides via lipase mediated resolution, *Synthesis*, 1253-1256.

Takano, S., Suzuki, M., Ogasawara, K. (1993e), Enantiocomplementary preparation of optically pure 2-trimethylsilylethynyl-2-cyclopentenol by homochiralization of racemic precursors: a new route to the key intermediate of 1,25-dihydroxycholecalciferol and vincamine, *Tetrahedron: Asymmetry* **4**, 1043-1046.

Takaoka, Y., Kajimoto, T., Wong, C.-H. (1993), Inhibition of *N*-acetylglucosaminyltransfer enzymes: chemical-enzymatic synthesis of new five-membered acetamido azasugars, *J. Org. Chem.* **58**, 4809-4812.

Takayama, S., Moree, W. J., Wong, C. H. (1996), Enzymatic resolution of amines and amino alcohols using pent-4-enoyl derivatives, *Tetrahedron Lett.* **37**, 6287-6290.

Tamai, S., Miyauchi, S., Morizane, C., Miyagi, K., Shimizu, H., Kume, M., Sano, S., Shiro, M., Nagao, Y. (1994), Enzymic hydrolyses of the σ-symmetric dicarboxylic diesters bearing a sulfinyl group as the prochiral center, *Chem. Lett.* 2381-2384.

Tamm, C. (1992), Pig liver esterase catalyzed hydrolysis: substrate specificity and stereoselectivity, *Pure Appl. Chem.* **64**, 1187-1191.

Tan, D. S., Guenter, M. M., Drueckhammer, D. G. (1995), Enzymatic resolution coupled with substrate racemization using a thioester substrate, *J. Am. Chem. Soc.* **117**, 9093-9094.

Tanaka, M., Yoshioka, M., Sakai, K. (1992), Practical enzymic procedure for the synthesis of (−)-aristeromycin, *J. Chem. Soc., Chem. Commun.*, 1454-1455.

Tanaka, K., Shogase, Y., Osuga, H., Suzuki, H., Nakamura, K. (1995), Enantioselective synthesis of helical molecules: lipase-catalyzed resolution of bis(hydroxymethyl)[7]thiaheterohelicene, *Tetrahedron Lett.* **36**, 1675-1678.

Tanaka, M., Norimine, Y., Fujita, T., Suemune, H., Sakai, K. (1996), Chemoenzymatic synthesis of antiviral carbocyclic nucleosides - asymmetric hydrolysis of *meso*-3,5-bis(acetoxymethyl)cyclopentenes using *Rhizopus delemar* lipase, *J. Org. Chem.* **61**, 6952-6957.

Taniguchi, T., Ogasawara, K. (1997), Lipase–triethylamine-mediated dynamic transesterification of a tricyclic acyloin having a latent *meso*-structure: a new route to optically pure oxodicyclopentadiene, *Chem. Commun.*, 1399-1400.

Taniguchi, T., Kanada, R. M., Ogasawara, K. (1997), Lipase-mediated kinetic resolution of tricyclic acyloins, endo-3-hydroxytricyclo[4.2.1.0^{2,5}]non-7-en-4-one and endo-3-hydroxytricyclo[4.2.2.0^{2,5}]dec-7-en-4-one, *Tetrahedron: Asymmetry* **8**, 2773-2780.

Tanimoto, H., Oritani, T. (1996), Practical synthesis of Ambrox from farnesyl acetate involving lipase-catalyzed resolution, *Tetrahedron: Asymmetry* **7**, 1695-1704.

Tanyeli, C., Demir, A. S., Dikici, E. (1996), New chiral synthon from the PLE catalyzed enantiomeric separation of 6-acetoxy-3-methylcyclohex-2-en-1-one, *Tetrahedron: Asymmetry* **7**, 2399-2402.

Taschner, M. J., Black, D. J. (1988), The enzymatic Baeyer-Villiger Oxidation: Enantioselective synthesis of lactones from mesomeric cyclohexanones, *J. Am. Chem. Soc.* **110**, 6892-6893.

Taylor, S. J. C., McCague, R., Wisdom, R., Lee, C., Dickson, K., Ruecroft, G., O'Brien, F., Littlechild, J., Bevan, J., Roberts, S. M., Evans, C. T. (1993), Development of the biocatalytic resolution of 2-azabicyclo[2.2.1]hept-5-en-3-one as an entry to single-enantiomer carbocyclic nucleosides, *Tetrahedron: Asymmetry* **4**, 1117-1128.

Taylor, S. K., Atkinson, R. F., Almli, E. P., Carr, M. D., van Huis, T. J., Whittaker, M. R. (1995), The synthesis of three important lactones via an enzymatic resolution strategy that improves ee's and yields, *Tetrahedron: Asymmetry* **6**, 157-164.

Terao, Y., Murata, M., Achiwa, K. (1988), Highly-efficient lipase-catalyzed asymmetric synthesis of chiral glycerol derivatives leading to practical synthesis of *(S)*-propranolol, *Tetrahedron Lett.* **29**, 5173-5176.

Terao, Y., Tsuji, K., Murata, M., Achiwa, K., Nishio, T., Watanabe, N., Seto, K. (1989), Facile process for enzymic resolution of racemic alcohols, *Chem. Pharm. Bull.* **37**, 1653-1655.

Terradas, F., Teston-Henry, M., Fitzpatrick, P. A., Klibanov, A. M. (1993), Marked dependence of enzyme prochiral selectivity on the solvent, *J. Am. Chem. Soc.* **115**, 390-396.

Testet-Lamant, V., Archaimbault, B., Durand, J., Rigaud, M. (1992), Enzymatic synthesis of structural analogs of PAF-acether by phospholipase D-catalysed transphosphatidylation, *Biochim. Biophys. Acta* **1123**, 347-350.

Theil, F. (1995), Lipase-supported synthesis of biologically active compounds, *Chem. Rev.* **95**, 2203-2227.

Theil, F. (1997), *Enzyme in der Organischen Synthese*, Heidelberg: Spektrum Akademischer Verlag.

Theil, F., Schick, H. (1991), An improved procedure for the regioselective acetylation of monosaccharide derivatives by pancreatin-catalyzed transesterification in organic solvents, *Synthesis*, 533-535.

Theil, F., Schick, H., Nedkov, P., Boehme, M., Haefner, B., Schwarz, S. (1988), Synthesis of enantiomerically pure prostaglandin intermediates by enzymic hydrolysis of (1*SR*,5*RS*,6*RS*,7*RS*)-7-Acetoxy-6-acetoxymethyl-2-oxabicyclo [3.3.0]octan-3-one, *J. Prakt. Chem.* **330**, 893-899.

Theil, F., Schick, H., Winter, G., Reck, G. (1991), Lipase-catalyzed transesterification of *meso*-cyclopentane diols, *Tetrahedron* **47**, 7569-7582.

Theil, F., Kunath, A., Schick, H. (1992), "Double enantioselection" by a lipase-catalyzed transesterification of a meso-diol with a racemic carboxylic ester, *Tetrahedron Lett.* **33**, 3457-3460.

Theil, F., Weidner, J., Ballschuh, S., Kunath, A., Schick, H. (1994), Kinetic resolution of acyclic 1,2-diols using a sequential lipase-catalyzed transesterification in organic solvents, *J. Org. Chem.* **59**, 388-393.

Theil, F., Lemke, K., Ballschuh, S., Kunath, A., Schick, H. (1995), Lipase-catalyzed resolution of 3-(aryloxy)-1,2-propanediol derivatives - towards an improved active site model of *Pseudomonas cepacia* lipase (Amano PS), *Tetrahedron: Asymmetry* **6**, 1323-1344.

Therisod, M., Klibanov, A. M. (1986), Facile enzymatic preparation of monoacylated sugars in pyridine, *J. Am. Chem. Soc.* **108**, 5638-5640.

Therisod, M., Klibanov, A. M. (1987), Regioselective acylation of secondary hydroxyl groups in sugars catalyzed by lipases in organic solvents, *J. Am. Chem. Soc.* **109**, 3977-3981.

Thiem, J. (1995), Applications of enzymes in synthetic carbohydrate chemistry, *FEMS Microbiol. Rev.* **16**, 193-211.

Thuring, J. W. J. F., Klunder, A. J. H., Nefkens, G. H. L., Wegman, M. A., Zwanenburg, B. (1996a), Lipase catalyzed dynamic kinetic resolution of some 5-hydroxy-2(5H)-furanones, *Tetrahedron Lett.* **37**, 4759-4760.

Thuring, J. W. J. F., Nefkens, G. H. L., Wegman, M. A., Klunder, A. J. H., Zwanenburg, B. (1996b), Enzymatic kinetic resolution of 5-hydroxy-4-oxa-*endo*-tricyclo[5.2.1.0^{2,6}]dec-8-en-3-ones - a useful approach to D-ring synthons for strigol analogues with remarkable stereoselectivity, *J. Org. Chem.* **61**, 6931-6935.

Tilbeurgh, H. van, Egloff, M.-P., Martinez, C., Rugani, N., Verger, R., Cambillau, C. (1993), Interfacial activation of the lipase-procolipase complex by mixed micelles revealed by X-ray crystallography, *Nature* **362**,

Tokiwa, Y., Suzuki, T., Ando, T. (1979), Synthesis of copolyamide-esters and some aspects involved in their hydrolysis by lipase, *J. Appl. Polym. Sci.* **24**, 1701-1711.

Tong, J. H., Petitclerc, C., D'Iorio, A., Benoiton, N. L. (1971), Resolution of ring-substituted phenylalanines by the action of α-chymotrypsin on their ethyl esters, *Can. J. Biochem.* **49**, 877-881.

Toone, E. J., Jones, J. B. (1991), Enzymes in organic synthesis. 49. Resolutions of racemic monocyclic esters with pig liver esterase, *Tetrahedron: Asymmetry* **2**, 207-222.

Toone, E. J., Werth, M. J., Jones, J. B. (1990), Enzymes in organic synthesis. 47. Active-site model for interpreting and predicting the specificity of pig liver esterase, *J. Am. Chem. Soc.* **112**, 4946-4952.

Toyooka, N., Nishino, A., Momose, T. (1993), Ring differentiation of the *trans*-decahydronaphthalene system via chemoenzymic dissymmetrization of its s-symmetric glycol: synthesis of a highly functionalized chiral building block for the terpene synthesis, *Tetrahedron Lett.* **34**, 4539-4540.

Trani, M., Lortie, R., Ergan, F. (1992), Synthesis of trierucin from HEAR oil, *Inform* **3**, 482.

Trani, M., Ducret, A., Pepin, P., Lortie, R. (1995), Scale-up of the enantioselective reaction for the enzymic resolution of (*R,S*)-ibuprofen, *Biotechnol. Lett.* **17**, 1095-1098.

Treilhou, M., Fauve, A., Pougny, J. R., Prome, J. C., Veschambre, H. (1992), Use of biological catalysts for the preparation of chiral molecules. 8. Preparation of propargylic alcohols. Application in the total synthesis of leukotriene B4, *J. Org. Chem.* **57**, 3203-3208.

Trollsås, M., Orrenius, C., Sahlén, F., Gedde, U. W., Norin, T., Hult, A., Hermann, D., Rudquist, P., Komitov, L., Lagerwall, S. T., Lindström, J. (1996), Preparation of a novel cross-linked polymer for second-order nonlinear optics, *J. Am. Chem. Soc.* **118**, 8542-8548.

Tsai, S. W., Dordick, J. S. (1996), Extraordinary enantiospecificity of lipase catalysis in organic media induced by purification and catalyst engineering, *Biotechnol. Bioeng.* **52**, 296-300.

Tsai, S. W., Wei, H. J. (1994a), Effect of solvent on enantioselective esterification of naproxen by lipase with trimethylsilyl methanol, *Biotechnol. Bioeng.* **43**, 64-68.

Tsai, S. W., Wei, H. J. (1994b), Enantioselective esterification of racemic naproxen by lipases in organic solvents, *Enzyme Microb. Technol.* **16**, 328-333.

Tsai, S. W., Wei, H. J. (1994c), Kinetics of enantioselective esterification of naproxen by lipase in organic solvents, *Biocatalysis* **11**, 33-45.

Tsuboi, S., Yamafuji, N., Utaka, M. (1997), Lipase-catalyzed kinetic resolution of 3-chloro-2-hydroxyalkanoates. Its application for the synthesis of (−)-disparlure, *Tetrahedron: Asymmetry* **8**, 375-379.

Tsugawa, R., Okumura, S., Ito, T., Katsuga, N. (1966), Production of L-glutamic acid from D,L-5-hydantoin propionic acid by microorganisms, *Agric. Biol. Chem.* **30**, 27-34.

Tsuji, K., Terao, Y., Achiwa, K. (1989), Lipase-catalyzed asymmetric synthesis of chiral 1,3-propanediols and its application to the preparation of optically-pure building block for renin inhibitors, *Tetrahedron Lett.* **30**, 6189-6192.

Tsuzuki, W., Okahata, Y., Katayama, O., Suzuki, T. (1991), Preparation of organic-solvent-soluble enzyme (lipase B) and characterization by gel permeation chromatography, *J. Chem. Soc., Perkin Trans. 1*, 1245-1247.

Tsuzuki, W., Akasaka, K., Kobayashi, S., Suzuki, T. (1995), Kinetics of organic solvent-soluble and native lipase, *J. Am. Oil Chem. Soc.* **72**, 1333-1337.

Tulinsky, A., Blevins, R. A. (1987), Structure of a tetrahedral transition state complex of α-chymotrypsin at 1.8-Å resolution, *J. Biol. Chem.* **262**, 7737-7743.

Tuomi, W. V., Kazlauskas, R. J. (1999), Molecular basis for enantioselectivity of lipase from *Pseudomonas cepacia* toward primary alcohols. Modeling, kinetics and chemical modification of Tyr29 to increase or decrease enantioselectivity, *J. Org. Chem.*, in press.

Turner, N. J., Winterman, J. R., McCague, R., Parratt, J. S., Taylor, S. J. C. (1995), Synthesis of homochiral L-(S)-*tert*-leucine via a lipase catalyzed dynamic resolution process, *Tetrahedron Lett.* **36**, 1113-1116.

Udding, J. H., Fraanje, J., Goubitz, K., Hiemstra, H., Speckamp, W. N., Kaptein, B., Schoemaker, H. E., Kamphuis, J. (1993), Resolution of methyl *cis*-3-chloromethyl-2-tetrahydrofurancarboxylate via enzymic hydrolysis, *Tetrahedron: Asymmetry* **4**, 425-432.

Ueji, S., Fujino, R., Okubo, N., Miyazawa, T., Kurita, S., Kitadani, M., Muromatsu, A. (1992), Solvent-induced inversion of enantioselectivity in lipase-catalyzed esterification of 2-phenoxy-propionic acids, *Biotechnol. Lett.* **14**, 163-168.

Uejima, A., Fukui, T., Fukusaki, E., Omata, T., Kawamoto, T., Sonomoto, K., Tanaka, A. (1993), Efficient kinetic resolution of organosilicon compounds by stereoselective esterification with hydrolases in organic solvent, *Appl. Microbiol. Biotechnol.* **38**, 482-486.

Uemasu, I., Hinze, W. L. (1994), Enantioselective esterification of 2-methylbutyric acid catalyzed via lipase immobilized in microemulsion-based organogels, *Chirality* **6**, 649-653.

Uemura, A., Nozaki, K., Yamashita, J., Yasumoto, M. (1989a), Lipase-catalyzed regioselective acylation of sugar moieties of nucleosides, *Tetrahedron Lett.* **30**, 3817-3818.

Uemura, A., Nozaki, K., Yamashita, J., Yasumoto, M. (1989b), Regioselective deprotection of 3,5-*O*-acylated pyrimidine nucleosides by lipase and esterase, *Tetrahedron Lett.* **30**, 3819-3820.

Uemura, M., Nishimura, H., Yamada, S., Nakamura, K., Hayashi, Y. (1993), Kinetic resolution of hydroxymethyl substituted iron (diene)Fe(CO)$_3$ complexes by lipase, *Tetrahedron Lett.* **34**, 6581-6582.

Uemura, M., Nishimura, H., Yamada, S., Hayashi, Y. (1994), Kinetic resolution of hydroxymethyl-substituted (arene)Cr(CO)$_3$ and (diene)Fe(CO)$_3$ by lipase, *Tetrahedron: Asymmetry* **5**, 1673-1682.

Uemura, T., Furukawa, M., Kodera, Y., Hiroto, M., Matsushima, A., Kuno, H., Matsushita, H., Sakurai, K., Inada, Y. (1995), Polyethylene glycol-modified lipase catalyses asymmetric alcoholysis of δ-decalactone in *n*-decanol, *Biotechnol. Lett.* **17**, 61-66.

Uenishi, J., Nishiwaki, K., Hata, S., Nakamura, K. (1994), An optical resolution of pyridyl and bipyridylethanols and a facile preparation of optically pure oligopyridines, *Tetrahedron Lett.* **35**, 7973-7976.

Uenishi, J., Hiraoka, T., Hata, S., Nishiwaki, K., Yonemitsu, O., Nakamura, K., Tsukube, H. (1998), Chiral pyridines: optical resolution of 1-(2-pyridyl)- and 1-[6-(2,2'-bipyridyl)]ethanols by lipase-catalyzed enantioselective acetylation, *J. Org. Chem.* **63**, 2481-2487.

Ulbrich-Hofmann, R., Haftendorn, R., Dittrich, N., Hirche, F., Aurich, I. (1998), Phospholipid analogs - chemoenzymatic syntheses and properties as enzyme effectors, *Fett/Lipid* **100**, 114-120.

Um, P. J., Drueckhammer, D. G. (1998), Dynamic enzymatic resolution of thioesters, *J. Am. Chem. Soc.* **120**, 5605-5610.

Uppenberg, J., Hansen, M. T., Patkar, S., Jones, T. A. (1994), The sequence, crystal structure determination and refinement of two crystal forms of lipase B from *Candida antarctica*, *Structure* **2**, 293-308.

Uppenberg, J., Öhrner, N., Norin, M., Hult, K., Patkar, S., Waagen, V., Anthonsen, T., Jones, T. A. (1995), Crystallographic and molecular modelling studies of lipase B from *Candida antarctica* reveal a stereospecificity pocket for secondary alcohols, *Biochemistry* **34**, 16838-16851.

Urban, F. J., Breitenbach, R., Vincent, L. A. (1990), Synthesis of optically active 3(*R*)-[(alkylsulfonyl)oxy] thiolanes from 2(*R*)-hydroxy-4-(methylthio)butanoic acid or D-methionine, *J. Org. Chem.* **55**, 3670-3672.

Ushio, K., Nakagawa, K., Nakagawa, K., Watanabe, K. (1992), An easy access to optically-pure (*R*)-malic acid via enantioselective hydrolysis of diethyl malate by *Rhizopus* lipase, *Biotechnol. Lett.* **14**, 795-800.

Uyama, H., Kobayashi, S. (1993), Enzymatic ring-opening polymerization of lactones catalyzed by lipases, *Chem. Lett.* 1149-1150.

Uyama, H., Kobayashi, S. (1994), Lipase-catalyzed polymerization of divinyl adipate with glycols to polyesters, *Chem. Lett.* 1687-1690.

Uyama, H., Takeya, K., Kobayashi, S. (1993), Synthesis of polyesters by enzymic ring-opening copolymerization using lipase catalyst, *Proc. Jpn. Acad.* **69B**, 203-207.

Uyama, H., Takeya, K., Kobayashi, S. (1995), Enzymic ring-opening polymerization lactones to polyesters by lipase catalyst: unusually high reactivity of macrolides, *Bull. Chem. Soc. Jpn.* **68**, 56-61.

Uyttenbröck, W., Hendriks, D., Vriend, G., de Baere, I., Moens, L., Scharpé, S. (1993), Molecular characterization of an extracellular acid-resistant lipase produced by *Rhizopus javanicus*, *Biol. Chem. Hoppe Seyler* **374**, 245-254.

Valis, T. P., Xenakis, A., Kolisis, F. N. (1992), Comparative studies of lipase from *R. delemar* in various microemulsion systems, *Biocatalysis* **6**, 267-279.

Valivety, R. H., Halling, P. J., Macrae, A. R. (1992a), Reaction rate with suspended lipase catalyst shows similar dependence on water activity in different organic solvents, *Biochim. Biophys. Acta* **1118**, 218-222.

Valivety, R. H., Halling, P. J., Peilow, A. D., Macrae, A. R. (1992b), Lipases from different sources vary widely in dependece of catalytic activity on water activity, *Biochim. Biophys. Acta* **1122**, 143-146.

Van Almsick, A., Buddrus, J., Hönicke-Schmidt, P., Laumen, K., Schneider, M. P. (1989), Enzymatic preparation of optically active cyanohydrin acetates, *J. Chem. Soc., Chem. Commun.*, 1391-1393.

Van Deenen, L. L. M., de Haas, G. H. (1963), The substrate selectivity of phospholipase A, *Biochim. Biophys. Acta* **70**, 538-553.

Van den Heuvel, N., Cuiper, A. D., van der Deen, H., Kellogg, R. M., Feringa, B. L. (1997), Optically active 6-acetyloxy-2H-pyran-3(6H)-one obtained by lipase catalyzed transesterification and esterification, *Tetrahedron Lett.* **38**, 1655-1658.

Van der Deen, H., Cuiper, A. D., Hof, R. P., Vanoeveren, A., Feringa, B. L., Kellogg, R. M. (1996), Lipase-catalyzed second-order asymmetric transformations as resolution and synthesis strategies for chiral 5-(acyloxy)-2(5H)-furanone and pyrrolinone synthons, *J. Am. Chem. Soc.* **118**, 3801-3803.

Van der Eycken, J., Vandewalle, M., Heinemann, G., Laumen, K., Schneider, M. P., Kredel, J., Sauer, J. (1989), Enzymic preparation of optically active bicyclo[2.2.1]heptene derivatives, building blocks for terpenoid natural products. An attractive alternative to enantioselective Diels-Alder syntheses, *J. Chem. Soc., Chem. Commun.*, 306-308.

Van Tol, J. B. A., Jongejan, J. A., Duine, J. A. (1995a), Description of hydrolase-enantioselectivity must be based on the actual kinetic mechanism: analysis of the kinetic resolution of glycidyl (2,3-epoxy-1-propyl) butyrate by pig pancreas lipase, *Biocatal. Biotransform.* **12**, 99-117.

Van Tol, J. B. A., Kraayveld, D. E., Jongejan, J. A., Duine, J. A. (1995b), The catalytic performance of pig pancreas lipase in enantioselective transesterification in organic solvents, *Biocatal. Biotransform.* **12**, 119-136.

Vänttinen, E., Kanerva, L. T. (1994), Lipase-catalysed transesterification in the preparation of optically active solketal, *J. Chem. Soc., Perkin Trans 1*, 3459-3463.

Vänttinen, E., Kanerva, L. T. (1995), Combination of the lipase-catalysed resolution with the Mitsunobu esterification in one pot, *Tetrahedron: Asymmetry* **6**, 1779-1786.

Vänttinen, E., Kanerva, L. T. (1997), Optimized double kinetic resolution for the preparation of (*S*)-solketal, *Tetrahedron: Asymmetry* **8**, 923-933.

Vartanian, J.-P., Henry, M., Wain-Hobson, S. (1996), Hypermutagenic PCR involving all four transitions and a sizeable proportion of transversions, *Nucl. Acids Res.* **24**, 2627-2631.

Verger, R. (1997), 'Interfacial activation' of lipases: facts and artefacts, *Trends Biotechnol.* **15**, 32-38.

Vermuë, M. H., Tramper, J., de Jong, J. P. J., Oostrom, W. H. M. (1992), Enzyme transesterification in near-critical carbon dioxide: Effect of pressure, Hildebrand solubility parameter and water content, *Enzyme Microb.Technol.* **14**, 649-655.

Vic, G., Scigelova, M., Hastings, J. J., Howarth, O. W., Crout, D. H. G. (1996), Glycosidase-catalyzed synthesis of oligosaccharides: trisaccharides with the α-D-Gal-(1->3)-D-Gal terminus responsible for the hyperacute rejection response in cross-species transplant rejection from pigs to man, *Chem. Commun.*, 1473-1474.

Villeneuve, P., Foglia, T. A. (1997), Lipase specificities: Potential applications in lipid bioconversions, *Inform* **8**, 640-650.

Vogel, K., Cook, J., Chmielewski, J. (1996), Subtilisin-catalyzed religation of proteolyzed hen egg-white lysozyme: investigation of the role of disulfides, *Chem. Biol.* **3**, 295-299.

Von der Osten, C. H., Sinskey, A. J., III, C. F. B., Pederson, R. L., Wang, Y.-F., Wong, C.-H. (1989), Use of a recombinant bacterial fructose-1,6-diphosphate aldolase in aldol reactions: preparative synthesis of 1-deoxynojirimycin, 1-deoxymannojirimycin, 1,4-dideoxy-1,4-imino-D-arabinitol, and fagomine, *J. Am. Chem. Soc.* **111**, 3924-3927.

Vörde, C., Högberg, H. E., Hedenström, E. (1996), Resolution of 2-methylalkanoic esters - enantioselective aminolysis by (*R*)-1-phenylethylamine of ethyl 2-methyloctanoate catalysed by lipase B from *Candida antarctica*, *Tetrahedron: Asymmetry* **7**, 1507-1513.

Vorderwülbecke, T., Kieslich, K., Erdmann, H. (1992), Comparison of lipases by different assays, *Enzyme Microb. Technol.* **14**, 631-639.

Vriesema, B. K., ten Hoeve, W., Wynberg, H., Kellogg, R. M., Boesten, W. H. J., Meijer, E. M., Schoemaker, H. E. (1986), Resolution of 2-amino-5-(thiomethyl)pentanoic acid (homomethionine) with aminopeptidase from *Pseudomonas putida* or chiral phosphoric acids, *Tetrahedron Lett.* **27**, 2045-2048.

Vulfson, E. N. (1994), Industrial applications of lipases, in.: *Lipases their structure biochemistry and application* (Wooley., P. , Peterson, S. B.; Eds.), pp. 271-288. Cambridge: Cambridge University Press.

Waagen, V., Hollingstaeter, I., Partali, V., Thorstad, O., Anthonsen, T. (1993), Enzymatic resolution of butanoic esters of 1-phenyl, 1-phenylmethyl, 1-[2-phenylethyl] and 1-[2-phenoxyethyl]ethers of 3-methoxy-1,2-propanediol, *Tetrahedron: Asymmetry* **4**, 2265-2274.

Wagegg, T., Enzelberger, M. M., Bornscheuer, U. T., Schmid, R. D. (1998), The use of methoxy acetoxy esters significantly enhances reaction rates in the lipase-catalyzed preparation of optical pure 1-(4-chloro-phenyl) ethyl amines, *J. Biotechnol.* **61**, 75-78.

Walde, P., Han, D., Luisi, P. L. (1993), Spectroscopic and kinetic studies of lipases solubilized in reverse micelles, *Biochemistry* **32**, 4029-4034.

Waldinger, C., Schneider, M., Botta, M., Corelli, F., Summa, V. (1996), Aryl propargylic alcohols of high enantiomeric purity via lipase catalyzed resolutions, *Tetrahedron: Asymmetry* **7**, 1485-1488.

Waldmann, H. (1989), A new access to chiral 2-furylcarbinols by enantioselective hydrolysis with penicillin acylase, *Tetrahedon Lett.* **30**, 3057-3058.

Waldmann, H., Naegele, E. (1995), Synthesis of the palmitoylated and farnesylated C-terminal lipohexapeptide of the human N-ras protein by employing an enzymically removable urethane protecting group, *Angew. Chem. Intl. Ed. Engl.* **34**, 2259-2262.

Waldmann, H., Sebastian, D. (1994), Enzymic protecting group techniques, *Chem. Rev.* **94**, 911-937.

Waldmann, H., Braun, P., Kunz, H. (1991), New enzymic protecting group techniques for the construction of peptides and glycopeptides, *Biomed. Biochim. Acta* **50**, S243-S248.

Wallace, J. S., Morrow, C. J. (1989a), Biocatalytic synthesis of polymers. Synthesis of an optically active, epoxy-substituted polyester by lipase-catalyzed polymerization, *J. Polym. Sci., Part A: Polym. Chem.* **27**, 2553-2567.

Wallace, J. S., Morrow, C. J. (1989b), Biocatalytic synthesis of polymers. II. Preparation of [AA-BB]x polyesters by porcine pancreatic lipase catalyzed polymerization, *J. Polym. Sci. A: Polymer Chem.* **27**, 3271-3284.

Wallace, J. S., Reda, K. B., Williams, M. E., Morrow, C. J. (1990), Resolution of a chiral ester by lipase-catalyzed transesterification with polyethylene glycol in organic media, *J. Org. Chem.* **55**, 3544-3546.

Wallace, J. S., Baldwin, B. W., Morrow, C. J. (1992), Separation of remote diol and triol stereoisomers by enzyme-catalyzed esterification in organic media or hydrolysis in aqueous media, *J. Org. Chem.* **57**, 5231-5239.

Walts, A. E., Fox, E. M. (1990), A lipase fraction for resolution of glycidyl esters to high enantiomeric excess, *US Patent* US 4 923 810 (Genzyme) (Chem. Abstr. 113: 113 879).

Wandel, U., Mischitz, M., Kroutil, W., Faber, K. (1995), Highly selective asymmetric hydrolysis of 2,2-disubstituted epoxides using lyophilized cells of *Rhodococcus* sp. NCIMB11216, *J. Chem. Soc., Perkin Trans. 1*, 735-736.

Wang, Q., Withers, S. G. (1995), Substrate-assisted catalysis in glycosidases, *J. Am. Chem. Soc.* **117**, 10137-10138.

Wang, Y. F., Wong, C. H. (1988), Lipase -catalyzed irreversible transesterification for preparative synthesis of chiral glycerol derivatives, *J. Org. Chem.* **53**, 3127-3129.

Wang, Y.-F., Chen, C.-S., Girdaukas, G., Sih, C. J. (1984), Bifunctional chiral synthons via biochemical methods. 3. Optical purity enhancement in enzymic asymmetric catalysis, *J. Am. Chem. Soc.* **106**, 3695-3696.

Wang, Y. F., Lalonde, J. J., Momongan, M., Bergbreiter, D. E., Wong, C.-H. (1988), Lipase-catalyzed irreversible transesterifications using enol esters as acylating reagents: preparative enantio- and regioselective syntheses of alcohols glycerol derivatives sugars and organometallics, *J. Am. Chem. Soc.* **110**, 7200-7205.

Wang, Y. F., Chen, S. T., Liu, K. K. C., Wong, C. H. (1989), Lipase-catalyzed irreversible transesterification using enol esters: Resolution of cyanohydrins and syntheses of ethyl (*R*)-2-hydroxy-4-phenylbutyrate and (*S*)-Propranolol, *Tetrahedron Lett.* **30**, 1917-1920.

Wang, P., Schuster, M., Wang, Y.-F., Wong, C.-H. (1993), Synthesis of phospholipid-inhibitor conjugates by enzymatic transphosphatidylation with phospholipase D, *J. Am. Chem. Soc.* **115**, 10487-10491.

Wang, Y. F., Yakovlevsky, K., Margolin, A. L. (1996), An efficient synthesis of chiral amino acid and peptide alkylamides via CLEC-subtilisin catalyzed-coupling and *in situ* resolution, *Tetrahedron Lett.* **37**, 5317-5320.

Wang, X. Q., Wang, C. S., Tang, J., Dyda, F., Zhang, X. J. C. (1997a), The crystal structure of bovine bile salt activated lipase - insights into the bile salt activation mechanism, *Structure* **5**, 1209-1218.

Wang, Y. F., Yakovlevsky, K., Zhang, B. L., Margolin, A. L. (1997b), Cross-linked crystals of subtilisin: versatile catalyst for organic synthesis, *J. Org. Chem.* **62**, 3488-3495.

Ward, R. S. (1995), Dynamic kinetic resolution, *Tetrahedron: Asymmetry* **6**, 1475-1490.

Watanabe, N., Sugai, T., Ohta, H. (1992), Preparation of enantiomerically enriched compounds using enzymes. Part 17. Enzymatic preparation of glycerol-related chiral pool possessing *tert*-alkoxy group, *Chem. Lett.* 657-660.

Weber, H. K., Stecher, H., Faber, K. (1995a), Sensitivity of microbial lipases to acetaldehyde formed by acyl-transfer reactions from vinyl esters, *Biotechnol. Lett.* **17**, 803-808.

Weber, H. K., Stecher, H., Faber, K. (1995b), Some properties of commercially available crude lipase preparations, in.: *Preparative Biotransformations* (Roberts, S. M.; Ed.), pp. 5:2.1. New York: Wiley.

Wehrli, H. P., Pomeranz, Y. (1969), Synthesis of galactosyl glycerides and related lipids, *Chem. Phys. Lipids* **3**, 357-370.

Wehtje, E., Kaur, J., Adlercreutz, P., Chand, S., Mattiasson, B. (1997), Water activity control in enzymatic esterification processes, *Enzyme Microb. Technol.* **21**, 502-510.

Wehtje, E., Svensson, I., Adlercreutz, P., Mattiasson, B. (1993), Continuous control of water activity during biocatalysis in organic media, *Biotechnol. Techn.* **7**, 873-878.

Weijers, C. A. G. M. (1997), Enantioselective hydrolysis of aryl, alicyclic and aliphatic epoxides by *Rhodotorula glutinis*, *Tetrahedron: Asymmetry* **8**, 639-647.

Weijers, C. A. G. M., Botes, A. L., van Dyk, M. S., de Bont, J. A. M. (1998), Enantioselective hydrolysis of unbranched aliphatic 1,2-epoxides by *Rhodotorula glutinis*, *Tetrahedron: Asymmetry* **9**, 467-473.

Weiss, A. (1990), Enzymic preparation of solid fatty acid monoglycerides, *Fat Sci. Technol.* **92**, 392-396.

Weissfloch, A. N. E., Kazlauskas, R. J. (1995), Enantiopreference of lipase from *Pseudomonas cepacia* toward primary alcohols, *J. Org. Chem.* **60**, 6959-6969.

Wells, J. A., Powers, D. B., Bott, R. R., Graycar, T. P., Estell, D. A. (1987), Designing substrate specificity by protein engineering of electrostatic interactions, *Proc. Natl. Acad. Sci. USA* **84**, 1219-1223.

Werschkun, J., König, W. A., Kren, V., Thiem, J. (1995), Substrate structure and incubation-parameter-dependent selectivities in chiral discrimination of galactopyranosides by β-galactosidase hydrolasis, *J. Chem. Soc., Perkin Trans. 1*, 2459-2466.

Westermann, B., Scharmann, H. G., Kartmann, I. (1993), PLE-catalyzed resolution of α-substituted ß-ketoesters, application to the synthesis of (+)-nitramine and (-)-isonitramine, *Tetrahedron: Asymmetry* **4**, 2119-2122.

Wie, Y., Swenson, L., Kneusel, R. E., Matern, U., Derewenda, Z. S. (1996), Crystallization of a novel esterase which inactivates the macrolide toxin brefeldin A, *Acta Cryst.* **D52**, 1194-1195.

Wieser, M., Nagasawa, T. (1999), Stereoselective nitrile-converting enzymes, in.: *Stereoselective Biocatalysis* (Patel, R.; Ed.), in press. New York: Marcel Dekker.

Wieser, M., Takeuchi, K., Wada, Y., Yamada, H., Nagasawa, T. (1999), Low-molecular-weight nitrile hydratase from *Rhodococcus rhodochrous* J1: Purification, substrate specificity and comparison with the analogous high-molecular-weight enzyme, *FEMS Lett.*, in press.

Williams, R. M. (1989), *Synthesis of optically active α-amino acids*, pp. 257-279. Oxford: Pergamon.

Wilson, W. K., Baca, S. B., Barber, Y. J., Scallen, T. J., Morrow, C. J. (1983), Enantioselective hydrolysis of 3-hydroxy-3-methylalkanoic acid esters with pig-liver esterase, *J. Org. Chem.* **48**, 3960-3966.

Wimmer, Z. (1992), A suggestion to the PPL active site model dilemma, *Tetrahedron* **48**, 8431-8436.

Winkler, F. K., D'Arcy, A., Hunziker, W. (1990), Structure of human pancreatic lipase, *Nature* **343**, 771-774.

Wipff, G., Dearing , A., Weiner, P. K., Blaney, J. M., Kollman, P. A. (1983), Molecular mechanics studies of enzyme-substrate interactions: the interaction of L- and D-*N*-acetyltryptophanamide with α-chymotrypsin, *J. Am. Chem. Soc.* **105**, 997-1005.

Wirz, B., Soukup, M. (1997), Enzymatic preparation of homochiral 2-isobutyl succinic acid derivatives, *Tetrahedron: Asymmetry* **8**, 187-189.

Wirz, B., Spurr, P. (1995), Enantio- and regioselective monohydrolysis of diethyl 2-ethoxysuccinate, *Tetrahedron: Asymmetry* **6**, 669-670.

Wirz, B., Walther, W. (1992), Enzymic preparation of chiral 3-(hydroxymethyl)piperidine derivatives, *Tetrahedron: Asymmetry* **3**, 1049-1054.

Wirz, B., Schmid, R., Foricher, J. (1992), Asymmetric enzymatic hydrolysis of prochiral 2-*O*-allylglycerol ester derivatives, *Tetrahedron: Asymmetry* **3**, 137-142.

Wirz, B., Barner, R., Huebscher, J. (1993), Facile chemoenzymic preparation of enantiomerically pure 2-methylglycerol derivatives as versatile trifunctional C4-synthons, *J. Org. Chem.* **58**, 3980-3984.

Witte, K., Sears, P., Martin, R., Wong, C.-H. (1997), Enzymatic glycoprotein synthesis: preparation of ribonuclease glycoforms via enzymatic glycopeptide condensation and glycosylation, *J. Am. Chem. Soc.* **119**, 2114-2118.

Wolff, A., Straathof, A. J. J., Heijnen, J. J. (1994), Enzymatic resolution of racemates contaminated by racemic product, *Biocatalysis* **11**, 249-261.

Wong, C.-H. (1995), Enzymatic and chemo-enzymatic synthesis of carbohydrates, *Pure Appl. Chem.* **67**, 1609-1616.

Wong, C.-H., Whitesides, G. M. (1994), *Enzymes in synthetic organic chemistry*, Pergamon Press: Oxford.

Woolley, P., Petersen, S. B. (Ed.) (1994), *Lipases: Their structure, biochemistry, and application*, Cambridge University Press: Cambridge.

Wright, C. S., Alden, R. A., Kraut, J. (1969), Structure of subtilisin BPN' at 2.5 A resolution, *Nature* **221**, 235-242.

Wu, S.-H., Zhang, L.-Q., Chen, C. S., Girdaukas, G., Sih, C. J. (1985), Bifunctional chiral synthons via biochemical methods. VII. Optically active 2,2'-dihydroxy-1,1'-binaphthyl, *Tetrahedron Lett.* **26**, 4323-4326.

Wu, S. H., Guo, Z. W., Sih, C. J. (1990), Enhancing the enantioselectivity of *Candida* lipase-catalyzed ester hydrolysis via noncovalent enzyme modification, *J. Am. Chem. Soc.* **112**, 1990-1995.

Wu, S.-H., Chu, F.-Y., Chang, C.-H., Wang, K.-T. (1991), The synthesis of D-isoglutamine by a chemoenzymatic method, *Tetrahedron Lett.* **32**, 3529.

Wu, D. R., Cramer, S. M., Belfort, G. (1993), Kinetic resolution of racemic glycidyl butyrate using a multiphase membrane enzyme reactor: experiments and model verification, *Biotechnol. Bioeng.* **41**, 979-990.

Wünsche, K., Schwaneberg, U., Bornscheuer, U. T., Meyer, H. H. (1996), Chemoenzymatic route to β-blockers via 3-hydroxy esters, *Tetrahedron: Asymmetry* **7**, 2017-2022.

Wyatt, J. M., Linton, E. A. (1988), The industrial potential of microbial nitrile biochemistry, in.: *Ciba Geigy Symp.: Cyanide Compounds in Biology)*, **Vol. 140**, pp. 32-48. New York: John Wiley & Sons.

Xie, Z. F. (1991), *Pseudomonas fluorescens* lipase in asymmetric synthesis, *Tetrahedron: Asymmetry* **2**, 733-750.

Xie, Z. F., Suemune, H., Sakai, K. (1990), Stereochemical observation on the enantioselective hydrolysis using *Pseudomonas fluorescens* lipase, *Tetrahedron: Asymmetry* **1**, 395-402.

Xie, Z. F., Suemune, H., Sakai, K. (1993), Synthesis of chiral building blocks using *Pseudomonas fluorescens* lipase catalyzed asymmetric hydrolysis of *meso* diacetates, *Tetrahedron: Asymmetry* **4**, 973-980.

Xu, J. H., Kawamoto, T., Tanaka, A. (1995a), Efficient kinetic resolution of dl-menthol by lipase catalyzed enantioselective esterification with acid anhydride in fed-batch reactor, *Appl. Microbiol. Biotechnol.* **43**, 402-407.

Xu, J. H., Kawamoto, T., Tanaka, A. (1995b), High-performance continuous operation for enantioselective esterification of menthol by use of acid anhydride and free lipase in organic solvent, *Appl. Microbiol. Biotechnol.* **43**, 639-643.

Xu, J., Gross, R. A., Kaplan, D. L., Swift, G. (1996), Chemoenzymatic synthesis and study of poly(α-methyl-β-propiolactone) stereocopolymers, *Macromolecules* **29**, 4582-4590.

Yadwad, V. B., Ward, O. P., Noronha, L. C. (1991), Application of lipase to concentrate the docosahexaenoic acid (DHA) fraction of fish oil, *Biotechnol. Bioeng.* **38**, 956-959.

Yamada, O., Ogasawara, K. (1995), Lipase-mediated preparation of optically pure four-carbon di- and triols from a *meso*-precursor, *Synthesis*, 1291-1294.

Yamaguchi, S., Mase, T. (1991), High-yield synthesis of monoglyceride by mono- and diacylglycerol lipase from *Penicillium camembertii* U-150, *J. Ferment. Bioeng.* **72**, 162-167.

Yamaguchi, Y., Komatsu, A., Moroe, T. (1976), Optical resolution of menthols and related compounds. Part III. Preliminary fractionation of microbial menthyl ester hydrolases and esterolysis by commercial lipases, *J. Agric. Chem. Soc. Jpn.* **50**, 619-620.

Yamamoto, K., Komatsu, K.-I. (1991), Purification and characterization of nitrilase responsible for the enantioselective hydrolysis from *Acinetobacter* sp. AK 226, *Agric. Biol. Chem.* **55**, 1459-1466.

Yamamoto, K., Nishioka, T., Oda, J., Yamamoto, Y. (1988), Asymmetric ring opening of cyclic acid anhydrides with lipase in organic solvents, *Tetrahedron Lett.* **29**, 1717-1720.

Yamamoto, K., Ueno, Y., Otsubo, K., Kawakami, K., Komatsu, K.-I. (1990a), Production of *S*-(+)-Ibuprofen from a nitrile compound by *Acinetobacter* sp. strain AK226, *Appl. Environm. Microbiol.* **56**, 3125-3129.

Yamamoto, Y., Iwasa, M., Sawada, S., Oda, J. (1990b), Asymmetric synthesis of optically active 3-substituted δ-valerolactones using lipase in organic solvnets, *Agric. Biol. Chem.* **54**, 3269-3274.

Yamamoto, K., Oishi, K., Fujimatsu, I., Komatsu, K.-I. (1991), Production of R-(−)-mandelic acid from mandelonitrile by *Alcaligenes faecalis* ATCC 8750, *Appl. Environm. Microbiol.* **57**, 3028-3032.

Yamamoto, K., Fujimatsu, I., Komatsu, K.-I. (1992), Purification and characterization of the nitrilase from *Alcaligenes faecalis* ATCC 8750 responsible for enantioselective hydrolysis of mandelonitrile, *J. Ferment. Bioeng.* **73**, 425-430.

Yamane, T. (1987), Enzyme technology for the lipids industry: An engineering overview, *J. Am. Oil Chem. Soc.* **64**, 1657-1662.

Yamane, T., Hoq, M. M., Shimizu, S. (1983), Continuous synthesis of glycerides by lipase in a microporous membrane bioreactor, *Ann. N. Y. Acad. Sci.* **434**, 558-568.

Yamane, T., Hoq, M. M., Itoh, S., Shimizu, S. (1986), Glycerolysis of fat by lipase, *J.Jpn.Oil.Chem.Soc.* **35**, 625-631.

Yamane, T., Kang, S. T., Kawahara, K., Koizumi, Y. (1994), High-yield diacylglycerol formation by solid-phase enzymatic glycerolysis of hydrogenated beef tallow, *J. Am. Oil Chem. Soc.* **71**, 339-342.

Yamano, T., Tokuyama, S., Aoki, I., Nishiguchi, Y., Nakahama, K., Takanohashi, K. (1993), Synthesis of *d*-biotin chiral intermediates via a biochemical method, *Bull. Chem. Soc. Jpn.* **66**, 1456-1460.

Yamazaki, Y., Hosono, K. (1990), Facile resolution of planar chiral organometallic alcohols with lipase in organic solvents, *Tetrahedron Lett.* **31**, 3895-3896.

Yamazaki, T., Ohnogi, T., Kitazume, T. (1990), Asymmetric synthesis of both enantiomers of 2-trifluoromethyl-4-aminobutyric acid, *Tetrahedron: Asymmetry* **1**, 215-218.

Yamazaki, Y., Morohashi, N., Hosono, K. (1991), Lipase-mediated homotopic and heterotopic double resolution of a planar chiral organometallic alcohol, *Biotechnol. Lett.* **13**, 81-86.

Yang, F., Hoenke, C., Prinzbach, H. (1995a), Biocatalytic resolutions in total syntheses of purpurosamine and sannamine/sporamine type building blocks of aminoglycoside antibiotics, *Tetrahedron Lett.* **36**, 5151-5154.

Yang, H., Cao, S. G., Han, S. P., Feng, Y., Ding, Z. T., Sun, L. F., Cheng, Y. H. (1995b), Optical resolution of *(R,S)* 2-octanol with lipases in organic solvent, *Ann. N. Y. Acad. Sci.* **750**, 250-254.

Yang, H., Henke, E., Bornscheuer, U. T. (1999), The use of vinyl esters significantly enhanced enantioselectivities and reaction rates in lipase-catalyzed resolutions of arlyaliphatic carboxylic acids, *J. Org. Chem.*, in press.

Yasufuku, Y., Ueji, S. (1995), Effect of temperature on lipase-catalyzed esterification in organic solvent, *Biotechnol. Lett.* **17**, 1311-1316.

Yasufuku, Y., Ueji, S. (1996), Improvement (5-fold) of enantioselectivity for lipase-catalyzed esterification of a bulky substrate at 57°C in organic solvent, *Biotechnol. Tech.* **10**, 625-628.

Yasufuku, Y., Ueji, S. (1997), High temperature-induced high enantioselectivity of lipase for esterifications of 2-phenoxypropionic acids in organic solvent, *Bioorg. Chem.* **25**, 88-99.

Yee, N. K., Nummy, L. J., Byrne, D. P., Smith, L. L., Roth, G. P. (1998), Practical synthesis of enantiomerically pure trans-4,5-disubstituted 2-pyrrolidinone via enzymatic resolution. Preparation of the LTB4 inhibitor BIRZ-227, *J. Org. Chem.* **63**, 326-330.

Yennawar, H. P., Yennawar, N. H., Farber, G. K. (1995), A structural explanation for enzyme memory in nonaqueous solvents, *J. Am. Chem. Soc.* **117**, 577-585.

Yokoyama, M., Sugai, T., Ohta, H. (1993), Asymmetric hydrolysis of a disubstituted malonontrile by the aid of a microorganism, *Tetrahedron: Asymmetry* **4**, 1081-1084.

Yonezawa, T., Sakamoto, Y., Nogawa, K., Yamazaki, T., Kitazume, T. (1996), Highly efficient synthetic metod of optically active 1,1,1-trifluoro-2-alkanols by enzymatic hydrolysis of the corresponding 2-chloroacetates, *Chem. Lett.*

You, L., Arnold, F. H. (1994), Directed evolution of subtilisin E in *Bacillus subtilis* to enhance the total activity in aqueous dimethylformamide, *Protein Eng.* **9**, 77-83.

Zaidi, N. A., O'Hagan, D., Pitchford, N. A., Howard, J. A. K. (1995), The solid state structure of the 34-membered macrocyclic diolide of 16-hydroxyhexadecanoic acid, formed by porcine pancreatic lipase mediated cyclisation in hexane, *J. Chem. Res., Synop.* 427.

Zaks, A., Gross, A. T. (1990a), Enzymatic production of monoglycerides containing omega-3 fatty acids, *World Patent* WO 90/13 656. (Enzytech Inc.) (Chem. Abstr. 114:120 510).

Zaks, A., Gross, A. T. (1990b), Production of monoglycerides by enzymatic transesterification, *World Patent* WO 90/040 333. (Enzytech Inc.) (Chem. Abstr. 113:76 643).

Zaks, A., Klibanov, A. M. (1984), Enzymatic catalysis in organic media at 100°C, *Science* **224**, 1249-1251.

Zaks, A., Klibanov, A. M. (1985), Enzyme-catalyzed processes in organic solvents, *Proc. Natl. Acad. Sci. U. S. A.* **82**, 3192-3196.

Zhang, J., Reddy, J., Roberge, C., Senanayake, C., Greasham, R., Chartrain, M. (1995), Chiral bioresolution of racemic indene oxide by fungal epoxide hydrolase, *J. Ferment. Bioeng.* **80**, 244-246.

Zhang, L. H., Chung, J. C., Costello, T. D., Valvis, I., Ma, P., Kauffman, S., Ward, R. (1997), The enantiospecific synthesis of an isoxazoline. A RGD mimic platelet GPIIb/IIIa antagonist, *J. Org. Chem.* **62**, 2466-2470.

Zhang, X. M., Archelas, A., Furstoss, R. (1991), Microbiological transformations. 19. Asymmetric dihydroxylation of the remote double bond of geraniol: A unique stereochemical control allowing easy access to both enantiomers of geraniol-6,7-diol, *J. Org. Chem.* **56**, 3814-3817.

Zhao, H., Giver, L., Affholter, J. A., Arnold, F. H. (1998), Molecular evolution by staggered extension process (StEP) in vitro recombination, *Nature Biotechnol.* **16**, 258-261.

Zhu, L.-M., Tedford, M. C. (1990), Applications of pig liver esterases (PLE) in asymmetric synthesis, *Tetrahedron* **46**, 6587-6611.

Zmijewski, M. J., Jr., Briggs, B. S., Thompson, A. R., Wright, I. G. (1991), Enantioselective acylation of a β-lactam intermediate in the synthesis of loracarbef using penicillin G amidase, *Tetrahedron Lett.* **32**, 1621-1622.

Index

Page numbers are given in plain (specific) or *italic* (schemes or tables) letters.

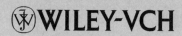